W9-CRB-475

Multimetallic and Macromolecular Inorganic Photochemistry

MOLECULAR AND SUPRAMOLECULAR PHOTOCHEMISTRY

Series Editors

V. RAMAMURTHY

Professor
Department of Chemistry
Tulane University
New Orleans, Louisiana

KIRK S. SCHANZE

Professor
Department of Chemistry
University of Florida
Gainesville, Florida

1. Organic Photochemistry, *edited by V. Ramamurthy and Kirk S. Schanze*
2. Organic and Inorganic Photochemistry, *edited by V. Ramamurthy and Kirk S. Schanze*
3. Organic Molecular Photochemistry, *edited by V. Ramamurthy and Kirk S. Schanze*
4. Multimetallic and Macromolecular Inorganic Photochemistry, edited *by V. Ramamurthy and Kirk S. Schanze*

ADDITIONAL VOLUMES IN PREPARATION

Multimetallic and Macromolecular Inorganic Photochemistry

edited by

V. Ramamurthy
Tulane University
New Orleans, Louisiana

Kirk S. Schanze
University of Florida
Gainesville, Florida

 MARCEL DEKKER, INC.   NEW YORK · BASEL

ISBN: 0-8247-7392-6

This book is printed on acid-free paper.

Headquarters
Marcel Dekker, Inc.
270 Madison Avenue, New York, NY 10016
tel: 212-696-9000; fax: 212-685-4540

Eastern Hemisphere Distribution
Marcel Dekker AG
Hutgasse 4, Postfach 812, CH-4001 Basel, Switzerland
tel: 41-61-261-8482; fax: 41-61-261-8896

World Wide Web
http://www.dekker.com

The publisher offers discounts on this book when ordered in bulk quantities. For more information, write to Special Sales/Professional Marketing at the headquarters address above.

Current printing (last digit):
10 9 8 7 6 5 4 3 2 1

PRINTED IN THE UNITED STATES OF AMERICA

Preface

As with many areas of molecular science, over the past decade inorganic photo-chemistry has entered a period where emphasis has shifted from the study of "simple" atomic and molecular systems to the study of complex supramolecular systems. This shift has occurred partly because the fundamentals of molecular inorganic chemistry are now reasonably well understood, and partly because synthetic methods and tools for spectroscopic characterization of complex materials have improved drastically over the past decade. The fourth volume of the Molecular and Supramolecular Photochemistry series, which can be read as a reflection of this shift, focuses on new developments in the field of supramolecular inorganic chemistry.

Chapter 1 (W. E. Jones et al) is concerned with the emerging area of π-conjugated polymers that contain transition metal ions. The authors discuss the synthesis and physical characterization of metal-organic π-conjugated polymers. These materials, while fundamentally interesting, are also finding application in the rapidly emerging area of luminescence (optical) sensors.

Yam and Lo (Chap. 2) focus on the photophysics of a broad family of supramolecular metal complex clusters derived from the group Ib transition metal ions (Cu, Ag, and Au). This chapter discusses basic types of excited states found in clusters that contain two, three, and four metal ions held together by ligands such as halide, phosphine, and acetylide. Continuing on this theme are S. W. Jones and coworkers (Chap. 4), who outline the photophysical and electrochemical properties of ruthenium and osmium metal complex clusters that are coordinated with polypyridine ligands. The clusters described by the authors have primarily been used in fundamental studies of electron and energy transfer processes; however, there has recently been interest in the application of some of these systems in optical sensing.

Ogawa (Chap. 3) reviews the use of peptide-bridged metal complexes in the study of electron transfer reactions. This area, to which Ogawa has been a major contributor, has seen significant activity over the past decade. This chapter gives an outstanding overview of the factors that control long-range electron transfer from the perspective of biological model systems. The chapter not only reviews earlier work in the field, but also contains highlights of recent findings from ongoing work in the author's laboratory.

Another chapter examines recent work in the area of metal-organic complexes derived from palladium(II) and platinum(II) diimine-dithiolate complexes. Pilato (Chap. 5), who is a leader in the development of new luminescent complexes of this family, has pioneered their application to sensing organophosphates and oxygen. His chapter describes the synthesis of these complexes, provides a comprehensive overview of their photophysical properties, and gives some examples of their application to the development of solid-state sensors.

Finally, porphyrins and metalloporphyrins have been the focus of countless studies in supramolecular photochemistry over the past several decades. In keeping with the importance of this area to the field of supramolecular photochemistry, Chapter 6 is dedicated to the properties of supramolecular porphyrin and metalloporphyrin complexes. Takagi and Inoue have put together an excellent chapter that exhaustively explores the fundamental properties of metalloporphyrins. In addition, they describe the supramolecular porphyrin complexes that are constructed from porphyrin units by using covalent and noncovalent approaches.

Multimetallic and Macromolecular Inorganic Photochemistry provides a snapshot view of significant topics in the emerging field of supramolecular inorganic photochemistry. Although the volume is not comprehensive in its coverage, it provides a survey of contemporary research being carried out by leaders in the field. A common theme that emerges from this volume is the synergism between fundamental problems, such as molecular energy transduction and charge transfer, and applications of basic science to real-world problems, such as chemical sensing. We believe that the next great advances in supramolecular inorganic photochemistry will emerge from this synergism between fundamentals and applications.

Kirk S. Schanze
V. Ramamurthy

Contents

Contributors

Karen J. Brewer, Ph.D.   Department of Chemistry, Virginia Polytechnic Institute and State University, Blacksburg, Virginia

Leoné Hermans, Ph.D.   Department of Chemistry, State University of New York at Binghamton, Binghamton, New York

Haruo Inoue, Ph.D.   Department of Applied Chemistry, Graduate Course of Engineering, Tokyo Metropolitan University, Tokyo, Japan

Biwang Jiang, Ph.D.   Department of Chemistry and Institute for Materials Research, State University of New York at Binghamton, Binghamton, New York

Sumner W. Jones, Ph.D.   Department of Chemistry, Virginia Polytechnic Institute and State University, Blacksburg, Virginia

Wayne E. Jones, Jr., Ph.D.   Department of Chemistry and Institute for Materials Research, State University of New York at Binghamton, Binghamton, New York

Michael R. Jordan, Ph.D.   Department of Chemistry, Virginia Polytechnic Institute and State University, Blacksburg, Virginia

Kenneth K.-W. Lo, Ph.D.   Department of Chemistry, University of Hong Kong, Hong Kong, China

Michael Y. Ogawa, Ph.D.   Department of Chemistry and Center for Photochemical Sciences, Bowling Green State University, Bowling Green, Ohio

Robert S. Pilato, Ph.D.   Department of Chemistry and Biochemistry, University of Maryland, College Park, Maryland

Shinsuke Takagi, Ph.D.   Department of Applied Chemistry, Graduate Course of Engineering, Tokyo Metropolitan University, Tokyo, Japan

Kelly A. Van Houten, Ph.D.   Department of Chemistry and Biochemistry, University of Maryland, College Park, Maryland

Vivian W.-W. Yam, Ph.D.   Department of Chemistry, University of Hong Kong, Hong Kong, China

Contents of Previous Volumes

Volume 2.   Organic and Inorganic Photochemistry

1

Metal-Organic Conducting Polymers: Photoactive Switching in Molecular Wires

Wayne E. Jones, Jr., Leoné Hermans, and Biwang Jiang
State University of New York at Binghamton, Binghamton, New York

I. INTRODUCTION

Interest in conjugated polymers as a new class of advanced electronic materials began in the late 1970s with the discovery of metallic, electrical conductivity in oxidatively doped polyacetylene $(CH)_x$ [1,2]. Since then, tremendous effort has been devoted to the design, synthesis, and study of conjugated polymers. Substantial π-electron delocalization along the conjugated polymer backbone gives rise to unique optical, electronic, and nonlinear optical properties in these materials [3–10]. These properties have led to a wide variety of proposed technological applications such as electrical conductors, nonlinear optical devices, polymer light-emitting diodes (LEDs), sensors, electrochromic devices, batteries, antistatic coatings, and transistors [7–14].

Despite the intense research effort, the euphoria of the 1970s has not completely materialized. "Molecular wires" exist, but challenges including air and

1

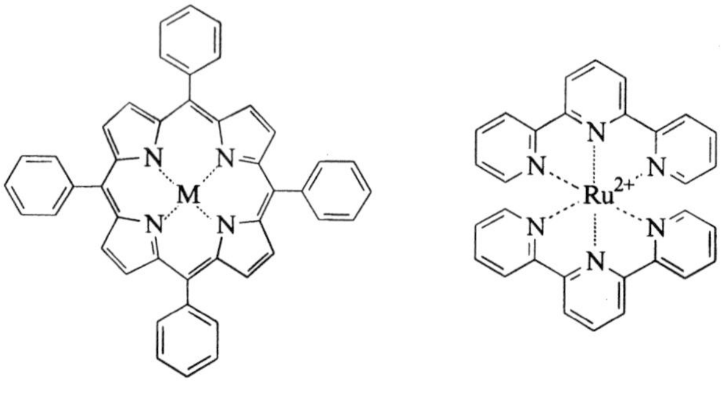

Porphyrin Monomer   Ruthenium Terpyridine Complex

Figure 1 Representative structures of inorganic monomer units incorporated into organic polymer structures.

thermal stability, in addition to the manufacturing processibility, have limited many of the applications originally envisioned [9,11,12,15]. Recently, researchers have turned to inorganic-organic hybrid conjugated materials in order to prepare a new generation of conjugated polymers with improved properties for electronic applications. Initially this work involved pendant inorganic structures for the study of electron and energy transport [16–19]. More recently, structures have been prepared that incorporate transition metal binding sites into the conjugation of organic polymers [20–29]. The results are novel materials that provide an opportunity to combine the molecular wire aspects of conjugated polymer systems with the stability and photochemical tunability of inorganic monomers including porphyrins and metal-to-ligand charge transfer (MLCT) excited states (Fig. 1).

II. CONJUGATED POLYMERS

A great effort has been devoted to the synthesis of a wide variety of conjugated organic polymers in order to control the structure, chemical properties, and function of the resultant materials [15,30–32]. The creative design and development strategies of conjugated polymers have led to enhanced performance and new function for these materials. Some important examples of conjugated polymers include polyacetylene, polyphenylene vinylene (PPV), polyphenylene, polythiophene, polyphenylene ethynylene, and polyaniline derivatives. Their representative molecular structures are shown in Fig. 2. A common feature of the conjugated polymer structure is the alternating double and single bonds that lead to π-electron delocalization along the polymer backbone.

Rigid conjugated polymers, such as poly(p-phenylene), poly(p-phenylene vinylene), and poly(p-phenylene ethynylene), are a class of polymers that have

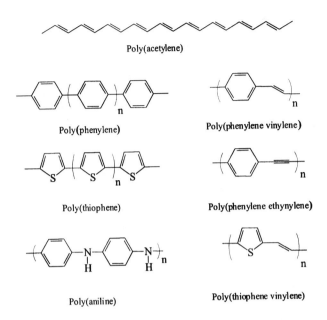

Poly(acetylene)

Poly(phenylene)

Poly(phenylene vinylene)

Poly(thiophene)

Poly(phenylene ethynylene)

Poly(aniline)

Poly(thiophene vinylene)

Figure 2   Molecular structures of some representative conjugated polymers.

attracted a great deal of attention both in the scientific community and in the realm of technological applications. These materials possess interesting properties such as electroluminescence [6,13,14,33–38], liquid crystallinity [39–43], high third-order optical nonlinearities [44–46], and impressive mechanical properties [47,48]. However, the rigid, conjugated backbone that is generally responsible for these attributes also renders these polymers insoluble, infusible, and therefore nonprocessible. Furthermore, classical syntheses generally resulted in oligomeric materials since the polymers precipitate from solution before high molecular weights can be obtained [49]. As a result, great effort has been devoted to circumventing this insolubility/intractability problem, largely through the development of two basic strategies.

     The first strategy involved the incorporation of long, flexible side chains that disrupted the crystallinity and exploited the entropy of solvent–side chain mixing to enhance solubility [50–58]. The second strategy has been to synthesize a soluble precursor polymer. A high-quality thin film obtained by spin-coating of the precursor polymer solution can then be treated with heat or light to give the desired insoluble conjugated polymer. These two approaches have been applied extensively in the syntheses of poly(p-phenylene) [59,60], poly(p-phenylene ethynylene), poly(p-phenylene vinylene) [61–64], polyacetylene [65–71], and their derivatives.

     Conjugated organic polymers can now be prepared at high molecular weights and relatively low polydispersities [15,30]. Most conjugated polymers are semiconductors or insulators in their neutral form. Their electronic structures are diverse with strong absorptions in the visible region of the spectrum. Conjugated polymers can be prepared in either an n-doped or p-doped form using electron donors (alkali metals, etc.) or acceptors (iodine, H_2SO_4, AsF_5, etc.), respectively [72–74].

$$(C)_x^+ A^- + e^- \rightleftharpoons (C)_x + A^- \qquad (1)$$

In Eq. (1), C is a representative conductive polymer containing cationic radicals stabilized over x units, and A is the counterion that is used to maintain charge neutrality. The formation of radical cations as a result of doping can lead to charge defects giving rise to polaron or bipolaron (bivalent cation for a two-electron loss) energy levels responsible for conduction [75,76].

A number of novel optical, electrical, and magnetic phenomena in these materials have been explained by solid-state physicists using new concepts such as solitons, polarons, and bipolarons [12,33,76]. These elementary excitations can move along the quasi-one-dimensional polymer chains, inducing geometrical changes and, in some instances, carrying the charge associated with electrical conductivity.

III. METAL-ORGANIC CONJUGATED POLYMERS

The preparation of inorganic and organometallic analogs to traditional organic polymers has been an active area of research for more than three decades [77,78]. Examples of inorganic polymers would include the silanes, siloxanes, phosphazenes, coordination polymers, and metallocenes. Recently, new classes of metal-organic polymers have been prepared in which electroactive and photoactive inorganic structures were incorporated into traditional organic polymers. The goal of many of these efforts was to use the structure (both physical and electronic) of the polymer as a molecular wire on which to build molecular level electronic and photonic devices [79–81]. However, in the process, new materials have been prepared with application to sensors, electron transfer theory, and phototherapeutic agents.

Early examples of molecular wires concentrated on the polymer materials as a backbone with a pendant inorganic functionality. For example, Meyer et al. produced a series of polystyrene-based polymers with pendant MLCT complexes of Ru and Os (Fig. 3) [19]. The photophysical analysis of this class of polymers demonstrated that energy and electron transfer between different pendant groups was controlled to a large extent by the distance between the groups. Furthermore, the distance was found to depend more on the stoichiometric loading of the metal dications than the inherent structure of the polymer [81]. Several groups prepared

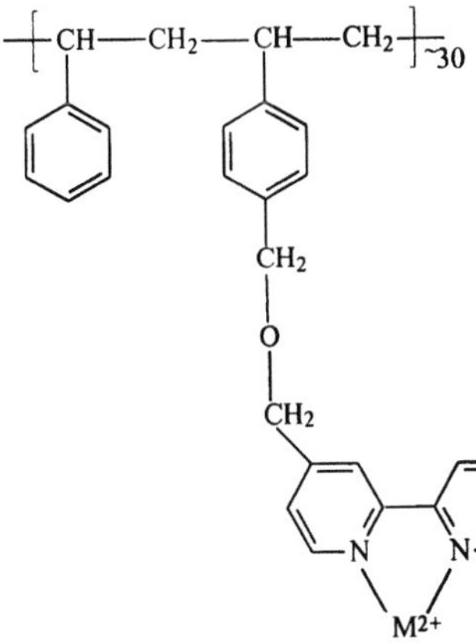

Figure 3 Polystyrene-based polymers with pendant metal complexes (M = Ru and Os).

rigid polymers that used block polymers [82,83] or helical peptides [84,85] to control structure and spacing of the electro- and photoactive groups.

An alternative example of metal-organic copolymer structures involves the preparation of modified electrodes by electropolymerization, as first demonstrated [86] and later reviewed by Abruna [87]. More recently, this work has been reviewed by Deronzier and Moutet [88]. The basic methodology involves electropolymerization of electroactive groups such as pyrrole or thiophene that are covalently attached to metal-coordinated polypyridyl ligands (Fig. 4). Metallo-porphyrin derivatives of pyrrole have also been prepared. Conductive and highly stable films of polypyrrole are prepared by anodic oxidation in organic solvents by cycling to a switching potential of 1.0–1.3 V vs. SCE. A variety of interesting applications of these materials have been developed including rectifying the behavior of bilayer assemblies. More recently, electrochemical redox catalysts of Rh, Ir, and Pd have been prepared for reduction of protons and the hydrogenation of unsaturated organic substrates [89].

Electropolymerization of conjugated polymers with pendant transition metals effectively couples the photoactive and electroactive substituents that are covalently bound within the film. The electronic coupling can lead to rapid electron and energy transfer processes suitable for photoconversion. A number of multilayer assemblies have been created that generate appreciable photocurrents with per-photon quantum efficiencies of greater than 10% [90]. More efficient photoelectrodes have been prepared by selectively designing molecular architectures with vectorial control of the electron transfer process [91]. Vectorial electron transfer can be achieved through the use of molecular assemblies such as the Ru[II] terpyridine structure shown in Fig. 5 [92]. In this case, the electropoly-

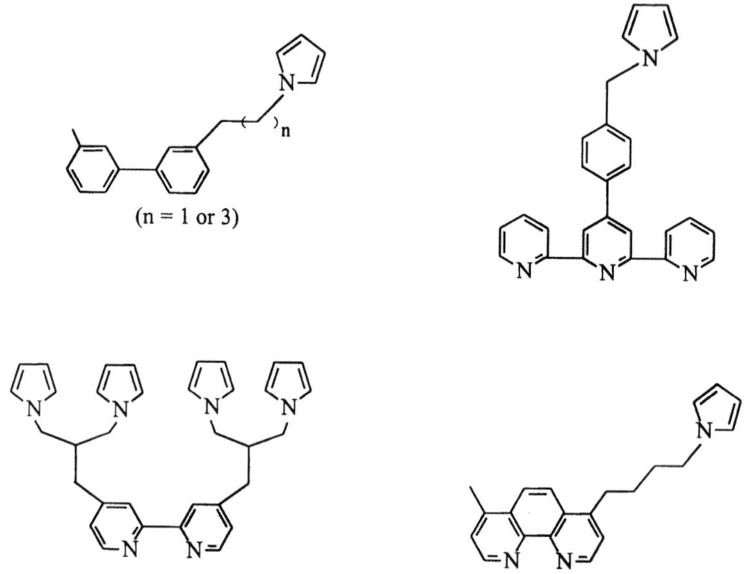

Figure 4 Polypyridyl structures with electropolymerizable substituents.

merization occurred in excess pyrrole, so that the chromophoric site was dilute. Alternatively, an electron transfer cascade mechanism was established by building multiple conjugated polymer layers containing different electro- and photoactive groups [93,94]. Each polymer layer away from the electrode was thermodynamically lower in energy. The result in either case was an increase in the photocurrent efficiency of the system.

A major challenge that exists in the preparation of pyrrole- and thiophene-based pendant metal-organic polymers is the strong absorption in the visible re-

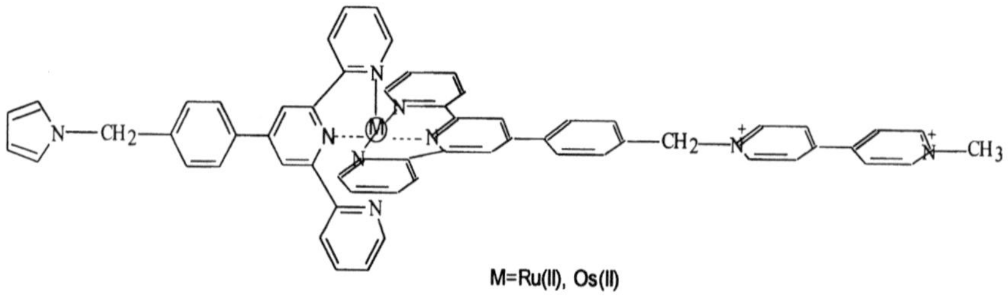

M=Ru(II), Os(II)

Figure 5 Monomer with vectorial electron transport.

gion from the polymer backbone. We have recently prepared copolymer systems containing Ru(ttpy-pyr)$_2^{2+}$ [ttpy-pyr is 4'(*para*-pyrrolylmethylphenyl)-2,2':6',2"-terpyridine] and the transmissive conjugated polymer polyisothianaphthene (PITN) [95–97]. The two monomers can be coelectropolymerized with different molar ratios of Ru(ttpy-pyr)$_2$PF$_6$ under moderate conditions as shown in Fig. 6. The bandgap of PITN in its conducting state is shifted from the visible into the near-IR region of the spectrum, resulting in optically transparent copolymer films.

While systems containing pendant groups demonstrated the successful incorporation of metals into organic polymer systems, the molecular wire concept was not being fully utilized. A system that directly incorporates the chromophores and electrophores into the backbone of a conjugated polymer would appear to

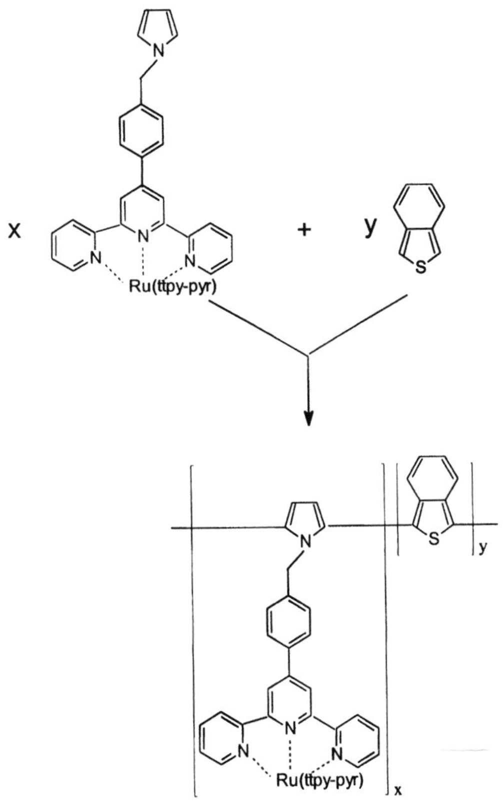

Figure 6 Electropolymerization of pendant ruthenium pyrrole monomer with isothianaphthene.

be more efficient. Conjugated hybrid polymers of this type involve redox sites that are delocalized over large distances. This is distinctly different from pendant redox polymers that have localized redox sites.

As early as 1993, Swager et al. demonstrated that electronic communication was very rapid along a conjugated polymer containing a crown ether substituent [98]. In subsequent studies, the crown ether–derivatized polymer was found to be a model for fluorescent sensors with increased sensitivity (Fig. 7). Significantly enhanced fluorescence quenching was observed in a polymer system with large numbers of crown ether binding sites relative to fluorescence sensors that contain one binding site per fluorophore [99]. This result demonstrated facile exciton mobility along the backbone, which allowed binding of only one site on the polymer to effectively quench the emission from the entire system.

Given recent advances in preparing soluble, organic conjugated polymers, several groups have turned to preparing polymers that incorporate Lewis base ligands directly into the backbone. Several groups recently reported examples of bipyridyl groups in conjugation with organic polymers. Yamamoto et al. demonstrated one of the first examples involving poly(2,2′-bipyridine-5,5′-diyl) and its Ru complex [100]. This polymer system was limited by solubility and processibility difficulties, similar to those of the conjugated polymers outlined previously. Petersen et al. improved on those problems through the preparation of the conjugated polymer poly[1-(2,2′-bipyridine-4-yl)-1,4-diazabutadiene-4,4′-diyl] (polyazabpy), which could be metallated at the bipyridine site as shown in Fig. 8 [101]. The emission lifetime of these materials was decreased relative to the Ru(bpy)$_3^{2+}$ monomer due to energy gap law considerations and an increase in the nonradiative decay rate. However, there was no evidence for significant chromophore interaction along the polymer chain.

Two other structures have recently been published that more clearly demonstrate strong electronic coupling along the polymer chain. Cameron and Pickup prepared a benzimidazole-based conjugated polymer as shown in Fig. 9A [102]. Electrochemical analysis of the metallated form of this material demonstrated

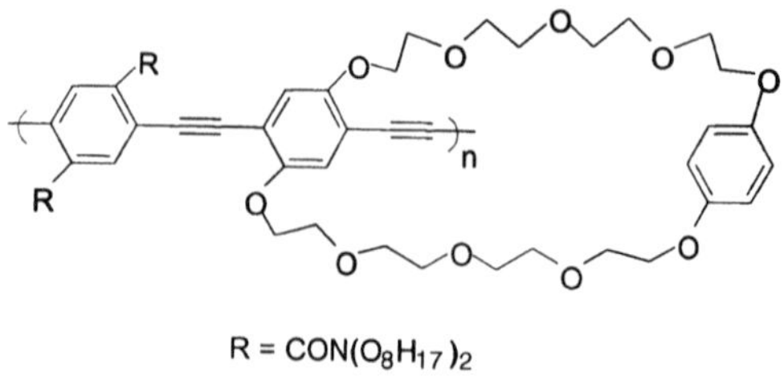

R = CON(O_8H_{17})$_2$

Figure 7 Structure of fully conjugated molecular wire sensor.

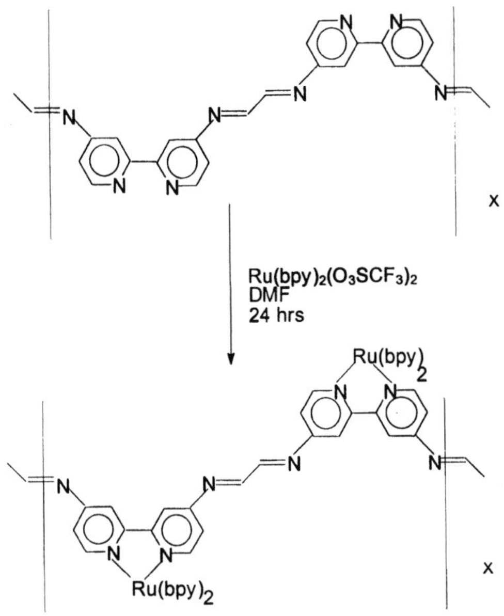

Figure 8 Metallated poly[1-(2,2'-bipyridine-4-yl)-1,4-diazabutadiene-4,4'-diyl](poly-azabpy).

that the conjugated polymer played a role in enhanced electronic communication between the Ru^{2+} metal centers. The increased polymer conjugation resulted in red shifts in the observed emission spectra of up to 120 nm. Wasielewski et al. synthesized a bipyridine containing analog of polyphenylene vinylene as shown in Fig. 9B. This new structure was soluble in common organic solvents due to the alkyl substituents on the phenylene groups. However, the nonmetallated form of this polymer was only partially conjugated [103].

Many structural factors control the level of conjugation along the polymer

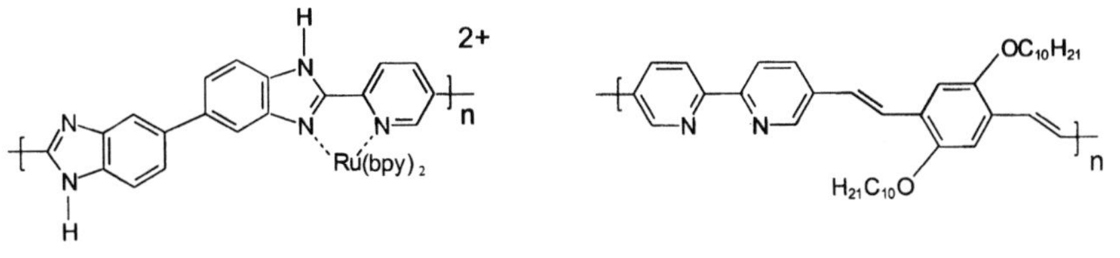

Figure 9 Heterocyclic conjugated polymers A (left) and B (right).

backbone in metal-organic conjugated polymers. For example, the extent of delocalization can be modulated through variations in the coupling chemistry. Schanze et al. have reported several examples of π-conjugated polymers based on arylenethynylene architectures [104–107] with Re^I MLCT chromophores built into the π structure (Scheme 1). Fluorescence studies on this material demonstrate two distinct, noninteracting chromophores, suggesting limited electronic communication along the polymer. By varying the stoichiometric composition of the Re chromophore within the material, some quenching of the higher energy π-π* state of the conjugated polymer is observed. This has been related to enhanced electronic interactions leading to quenching by the MLCT excited state [107].

Recently, a new approach to the preparation of conjugated metal-organic polymers has been reported based on porphyrin copolymer assemblies. The well-known Wittig coupling reaction has been used in constructing porphyrin dimers, trimers, and a star-like pentamer [75,108,109]. Using this chemistry, linear, conjugated copolymers have been prepared with controlled porphyrin separations as depicted in Scheme 2 [27–29,95]. The result is a highly fluorescent PPV-like material with the porphyrin units covalently incorporated within the conjugated backbone at structurally defined positions.

There are several distinct advantages to these new polymer systems for the study of electron and energy transfer. The chromophore separation can be systematically controlled through variations in the PPV bridge length. The reaction is facile even at room temperature, making modification of the polymer and chromophore easily accessible. The solubility of the copolymers is substantially increased due to the long alkoxyl substituents attached to the oligophenylene vinylene (OPV) units and the mesityl substituents on the porphyrin rings [110].

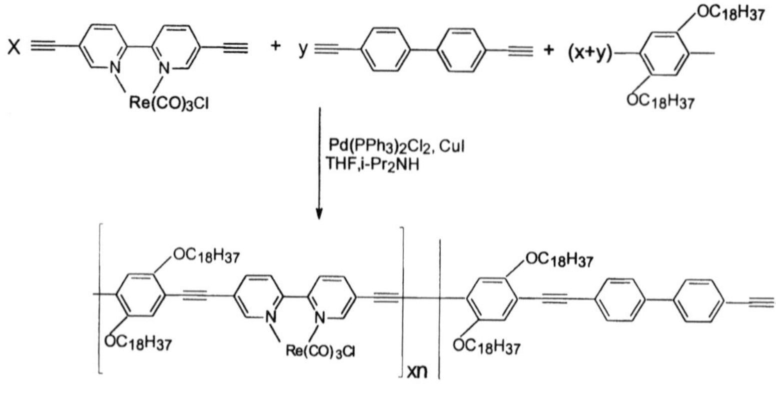

Scheme 1 Schematic for the preparation of the Re^I chromophore–substituted conjugated polymers.

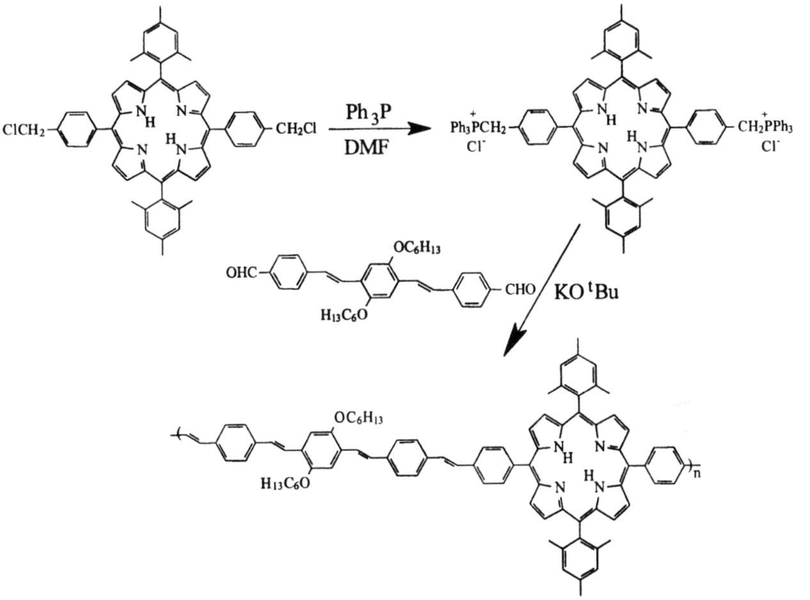

Scheme 2 Example of the Wittig reaction for the synthesis of a porphyrin-PPV polymer.

Finally, the thermal stability of the polymer in the solid state is enhanced over pure PPV polymer systems, based on differential scanning colorimetry (DSC) and thermal gravitational analysis (TGA) measurements [28].

An alternative approach to modulating the electronic communication in metal-organic polymer systems was demonstrated by solution and thin-film UV-Visible (UV-Vis) and fluorescence spectroscopy. The absorption maxima of the porphyrin Soret bands in solution were all slightly red-shifted and broader than that of the corresponding porphyrin monomer due to π-conjugation effects [112]. No splitting of the Soret absorption bands was observed, suggesting that there may be limited ground state electronic interaction between the porphyrin units in solution [113], though a lack of splitting does not preclude electronic communication [114].

The fluorescence spectra of the polymers in tetrahydrofuran (THF) solution shown in Fig. 10 (top) clearly show two emission manifolds. The lower energy bands (640–750 nm) are independent of the length of oligophenylene vinylene bridges. Based on the similarity to the monomer fluorescence properties and comparison of excitation and absorption spectra [27], this can be attributed to the porphyrin subunits within the polymer chain. The higher energy emission bands (450–600 nm) increase in intensity and red-shift with increasing conjugated

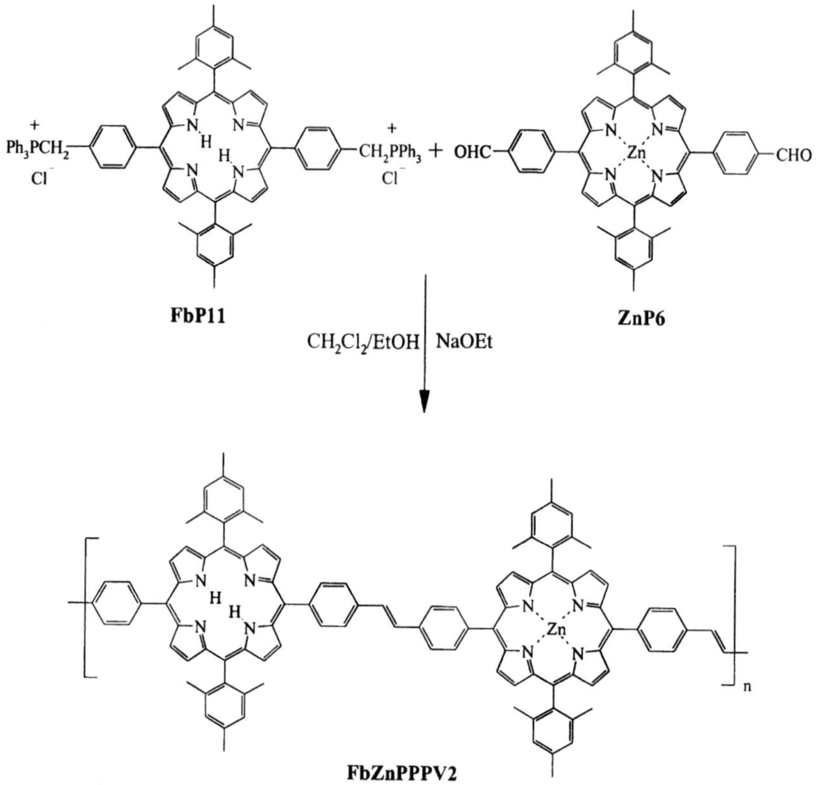

Scheme 3 Synthetic scheme for the heterometallic conjugated porphyrin copolymers.

bridge length. These features can be assigned to the oligophenylene vinylene bridge, based on model studies of the bridge prior to polymerization. The emission lifetime of each band is a single exponential suggesting two noninteracting chromophore sites as observed in the related bipyridine systems reported by Schanze et al. [104,105].

High-quality thin films of the PPV copolymers were spin-cast onto glass slides. The maximum absorption of the undoped film at 430 nm was significantly red-shifted compared to the spectra in THF solution. More interestingly, there was a significant shift in the emission intensities of the polymers when cast as thin films (Fig. 10, bottom). The porphyrin bands at low energy dominated the thin-film fluorescence spectra with a slight red shift from the solution spectra. Noticeably absent in the thin-film spectra are the emission bands at higher energy associated with the oligophenylene vinylene bridge.

It was concluded that energy transfer from the higher energy bridge state

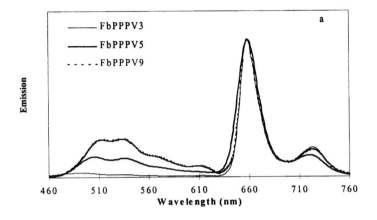

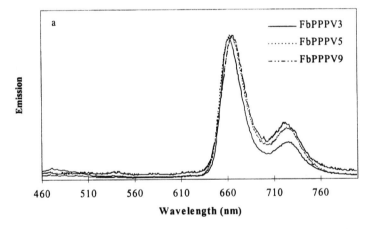

Figure 10 Emission spectra of polymers in THF (top) and as solid films (bottom) (422 nm).

to the lower energy porphyrin units was taking place in the solid state but not in solution. Further support for this conclusion was found by comparing the absorption and excitation spectra [95]. The weak intramolecular communication in solution is explained based on steric hindrance between the β-pyrrole proton of the porphyrin and the ortho proton of the bridge phenyl group. This forces the porphyrin rings out of the plane of conjugation when in solution. It has previously been observed for phenyl-substituted porphyrin dimers and oligomers, [114,115] and was recently supported by a combination of two-dimensional nuclear magnetic resonance (2-D NMR) and molecular modeling [111].

In the solid state there was enhanced electronic communication, leading to

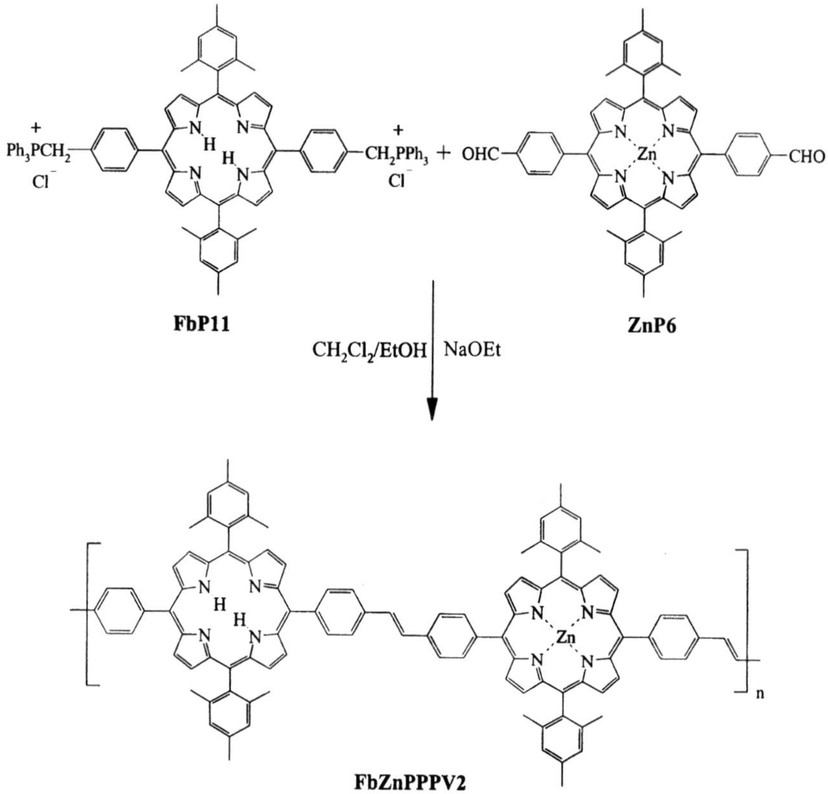

Figure 11 Synthetic scheme for heterometallic conjugated porphyrin copolymers.

energy transfer from the higher energy bridge to the porphyrins. Emission studies on dilute solid samples of the PP9 polymer in polystyrene or polymethyl methacrylate matrices reveals a substantial increase in the bridge-based emission with decreasing concentrations of the polymer. This inverse relationship suggests that the process leading to the observed loss of the dual emission is intermolecular in nature. This intermolecular interaction in the solid state could involve the porphyrin subunits preferentially stacking that would force the porphyrin rings to be in the same plane as the conjugated bridge in the solid state [116,117]. This would lead to enhanced *intrastrand* electronic coupling. An alternative hypothesis would be that in the solid state an increase in *interstrand* energy transfer leads to the observed decrease in emission intensity from the bridge-based state.

A better measure of electronic coupling through these conjugated porphyrin copolymer systems would be to explore interactions between two different chro-

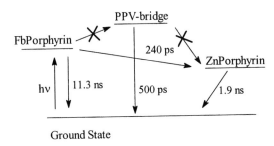

Figure 12 Jablonski diagram of energy transfer in the heterobimetallic porphyrin copolymers.

mophores in the same conjugated chain (Scheme 3). This has recently been achieved by preparing heterometallated porphyrin polymers in which the free-base porphyrin was copolymerized with a metallated porphyrin, where M = Zn, Ni, and Cu (Fig. 11) [95]. The absorption spectra in solution were essentially a superposition of the homometallated porphyrin polymers that were prepared as model structures. However, the fluorescence spectrum of the heterometallic, donor–acceptor copolymer was not a superposition of the two porphyrin emission manifolds. A comparison of the fluorescence spectra of an equimolar mixture of the free-base phenylene vinylene porphoryin polymer, FbPP3, and the derivatized porphyrin with M = Zn, ZnPP3, in addition to the copolymer, ZnFbPP3, clearly showed a substantial decrease in the emission intensity from the higher energy ZnPP3 chromophore relative to the lower energy FbPP3.

Based on the combined fluorescence results cited above, it was concluded that following excitation, energy transfer occurred from the higher lying Zn π-π^* state to the free base as outlined in Fig. 12. The mechanism of this energy transfer process may or may not include the participation of the PPV bridge. In previous investigations on porphyrin dimers, several groups have demonstrated that rapid energy transfer can occur through a Förster-type mechanism [118–123]. This mechanism would not require direct electronic coupling. Given the limited electronic coupling observed between the porphyrins and the conjugated bridges in the homopolymers, it is also likely that a Förster mechanism is involved in the energy transfer.

IV. PHOTOACTIVE SWITCHING

In recent years numerous macromolecular photoinduced electron transfer systems ranging from molecular donor–acceptor assemblies to polymer derivatives have been developed [16,19,91,122]. The extensive electronic coupling and delocalization of the electronic states in conducting polymers provides an opportunity to create photoactive copolymer films. Photoactive conjugated copolymers can provide further insight into the dynamics of long-range ground state and excited state electronic coupling. The goal is to create a macromolecular polymer structure that when irradiated with electromagnetic radiation will undergo electron and energy transfer and trapping that will result in reversible photoconductive switching.

The formation of charged species in conjugated systems can be achieved using charge transfer assemblies similar to those employed in donor–acceptor models. For example, a model has been reported that involved a reversible charge transfer that utilized molecular oxygen as the acceptor and poly(3-alkythiophene) as the donor [124]. To improve the efficiency of long-range electron transport and to create electron traps, MLCT chromophores have now been incorporated into conducting conjugated polymers [125].

By careful selection of excited state oxidation and reduction potentials, a chromophore/conducting polymer system can be prepared whereby initial excitation into the chromophore results in an increase in charge carriers and a change in the conductivity. Heeger et al. first demonstrated a working intramolecular system based on this premise [127–129]. It possessed a conjugated polymer as the chromophoric site and exhibited photoinduced electron transfer to a covalently bound buckminster fullerene, C_{60}, electron trap. The advantage of this first system was that it demonstrated an alternative method of doping conducting polymer through internal derivatization.

A more efficient system would employ a chromophore with a long-lived excited state that was capable of excited state electron transfer to form the oxidized polymer directly. A key example involved excitation into the MLCT band of $[Ru(dmb)_3]^{2+}$ (dmb = 4,4′-dimethyl-2,2′-bipyridine) which results in an excited state reduction potential of +0.77 V vs. SCE in chloroform [126]. The oxidation potential required to switch the emeraldine base form (EB) of either polyaniline [PANi$_{(EB)}$] or poly(2,5-dimethoxyaniline) [PDMA$_{(EB)}$] to the conductive form is approximately +0.40 V vs. SCE [130]. Thus, the photoinduced oxidation of either polymer by $[Ru(dmb)_3]^{2+*}$ is thermodynamically feasible according to the excited state reaction illustrated in Eq. (2).

$$Ru^{2+} + h\nu \rightarrow Ru^{2+*} + PANi_{(EB)} \rightarrow Ru^+ + PANi^+_{(EB)} \qquad (2)$$

Initial investigations of the photoinduced electron transfer events were carried out in solution. The conjugated polymers PANi$_{(EB)}$ and PDMA$_{(EB)}$ were readily synthesized and characterized according to literature methods [131,132]. Stern-Volmer analysis of the emission spectrum of $[Ru(dmb)_3]^{2+*}$ at 612 nm as a function of increasing PDMA$_{(EB)}$ concentration demonstrated that the conjugated polymer system was capable of quenching the MLCT state (Fig. 13). In this experiment, the emission intensities were corrected for the small amount of competitive absorption at 458 nm from the polymer, and pure polymer samples dem-

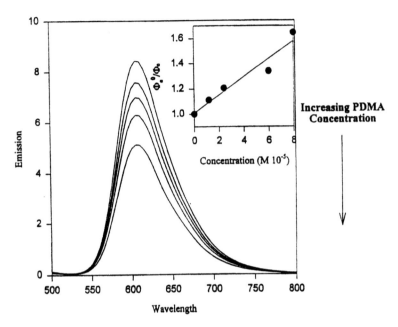

Figure 13 Emission spectra of [Ru(dmb)$_3$]$^{2+}$ and PDMA in CHCl$_3$ solution.

onstrated no measurable luminescence. The quenching rate constant, k_q, was calculated to be $\sim 10^8$ M^{-1} s^{-1} without adjusting for the radius of gyration of the polymer in solution [133,134]. Transient absorption spectroscopy on dilute chloroform solutions of 2.4×10^{-5} M [Ru(dmb)$_3$]$^{2+}$ and 8×10^{-4} M PDMA$_{(EB)}$ excited at 532 nm (5 mJ; 6-ns pulses) showed a decrease in absorption at 600+ nm consistent with loss of absorption from the PDMA in this region. This loss of absorption would be consistent with Eq. (2) and has been assigned to a photoinduced electron transfer event that results in the conducting form of the polymer, PDMA$^+$ [135,136].

While the solution measurements suggested that photoinduced electron transfer would occur, enhancements in the conductivity of the polymer are most easily determined in the solid state. Solid-state samples were prepared by mixing 1–30% of the chromophore, [Ru(dmb)$_3$](PF$_6$)$_2$ [137], by weight with the polymer. The samples were homogenized by cogrinding, and a free-standing pellet was prepared at high pressure in a die press. Resistance measurements of the solid samples under varying illumination conditions were obtained using the in-line four-probe method [135]. Initial resistance measurements were made under dark conditions at ambient temperature and showed conductivities of $>10^{-5}$ S/cm. A decrease in the resistance of the [Ru(dmb)$_3$]$^{2+}$/PANi$_{(EB)}$ mixtures was observed when the samples were excited using visible light (150-W Xe lamp). The decrease in resistance was found to be linear over one order of magnitude change in light intensity (Fig. 14).

Resistance changes of greater than one order of magnitude have been achieved depending on the original resistance of the sample [140]. The time-resolved characteristics of the change in resistance are shown in Fig. 15. Following excitation, a complex rise and decay are observed which could only be fit by a multiexponential analysis. However, the change in resistance was completely reversible, returning to the original, pre-excitation resistance within several seconds. It has been proposed that the kinetics of the process will be highly dependent on the anions and cations generated by the photoinduced electron transfer event. Experiments to explore this hypothesis are ongoing [139,140].

The theory of photoconductivity describes the change in conductivity under the action of light [Eq. (3)] where $\Delta\sigma$ is the change in conductivity, e is the charge of the electron, $\Delta\mu$ is the carrier mobility, and Δn or Δp is the change in the electron or hole concentration, respectively [141].

$$\Delta\sigma = e(\Delta\mu_n \, \Delta n + \Delta\mu_p \, \Delta p) \tag{3}$$

One possible source of the change in resistance observed following excitation involves an increased sample temperature due to light absorption. This would lead to an increase in carrier mobility. Given the relatively small temperature

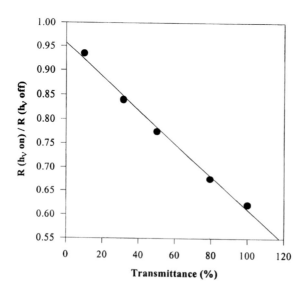

Figure 14 Resistance Percent Transmittance for 10% [Ru(dmb)$_3$]$^{2+}$ in PANi$_{(EB)}$.

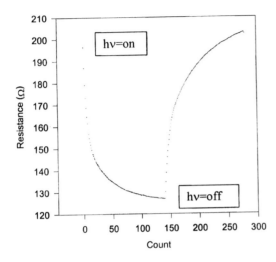

Figure 15 Time-resolved resistance measurements.

changes observed during the illumination period (from 297 to 305 K) and control experiments monitored in the dark over the temperature range of 293 to 318 K, it was concluded that temperature alone was not sufficient to account for the change in resistance [138]. Further evidence that the decrease in resistance was not due to a thermal effect in $PANi_{(EB)}$ has been reported by Misurkin and coworkers who demonstrated that heating to 340 K caused no observable change in the optical absorption spectrum [142].

The decrease in the resistance for the $PANi_{(EB)}$ containing $[Ru(dmb)_3]^{2+}$ could also be attributed to an energy transfer from the predominantly triplet excited state [143] of the Ru chromophore to the polymer. An energy transfer process would be consistent with the observed decrease in the emission intensity of the chromophore. This is possible for the conducting polymer systems as indicated by the presence of a low-energy absorption for $PANi_{(EB)}$ at $\lambda = 620$ nm that leads to the formation of a polaron band at 820 nm [134,136]. The lack of enhanced conductivity observed for direct excitation of this low-energy band in $PANi_{(EB)}$ would seem to preclude the sensitized excitation as a possible mechanism for enhanced conductivity.

The decrease of resistance in the $PANi_{(EB)}$ samples containing the $[Ru(dmb)_3](PF_6)_2$ chromophore was attributed to an excited state intermolecular charge transfer reaction as described in Eq. (2). This reaction mechanism would result in the formation of charge defects that enhance the overall conductivity [126,127]. Both the emission quenching experiment and the transient absorption measurements in solution are consistent with the formation of charge transfer products. Due to the lack of transparency in conducting polymer systems, it is difficult to further characterize the species involved in this reversible change in conductivity. Further, the excitation of $[Ru(dmb)_3]^{2+}$ is restricted to the surface of the pellets and does not include the bulk sample. This results in relatively small changes in the resistance of $PANi_{(EB)}$ under illumination.

While conjugated polymer systems are attractive for long-range electron and energy transfer studies, a major limitation exists in their use for fundamental studies of electron transport. In smaller molecular systems such as diads, triads, and oligomers, the conjugated bridges are not of sufficient length to establish a band-gap transition. As the bridge length increases, band-gap transitions typically occur in the visible region of the spectrum, resulting in polymers that are strongly colored. In order to overcome this limitation, we have begun a series of experiments based on polybenzothiophene derivatives such as PITN, whose resonance forms are shown in Fig. 16 [144,145]. These materials have band-gap transitions in the near-IR region of the electromagnetic spectrum and still retain high conductivities ($>10^{-2}$ S/cm) [146].

In order to covalently incorporate inorganic chromophores within the framework of a conducting polymer matrix, coelectropolymerization can be utilized as described in the paragraph above Fig. 6. The structure of the ruthenium-based chromophore used in this experiment, $[Ru(ttpy-pyr)_2]^{2+}$, is shown in Fig. 17. $[Ru(ttpy-pyr)_2](PF_6)_2$ was prepared by modifying the literature methods developed by Sauvage et al. [147]. This compound has previously been electropolymerized to form a pyrrole-based conducting polymer with pendant $[Ru(ttpy)_2]^{2+}$ (ttpy = $2,2':6',2''$-terpyridine) chromophores. It has also been shown that the electropolymerization can be achieved in the presence of additional monomeric pyrrole, resulting in a copolymer of pyrrole and Ru-substituted pyrrole [147]. When the applied oxidation potential was relatively low (less than $+0.8$ V), the films consisted of stoichiometric quantities of pure pyrrole [148]. This has also been achieved for polypyridyl-based polymer films [147].

The same electropolymerization technique can be applied to the copolymerization of $[Ru(ttpy-pyr)_2](PF_6)_2$ and the isothianaphthene (ITN) monomer. Previously it has been shown that ITN will copolymerize with pyrrole [149]. In that case it was shown that the ITN monomer is covalently bound to the ortho position of the pyrrole. Under conditions of 10:1 molar excess of ITN in anhydrous aceto-

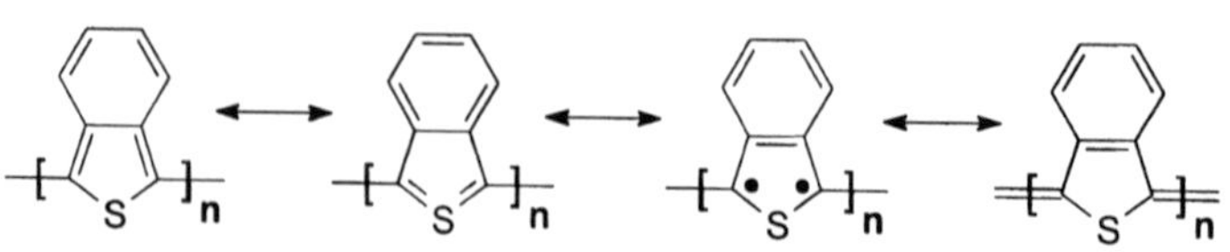

Figure 16   Resonance forms of polyisothianaphthene.

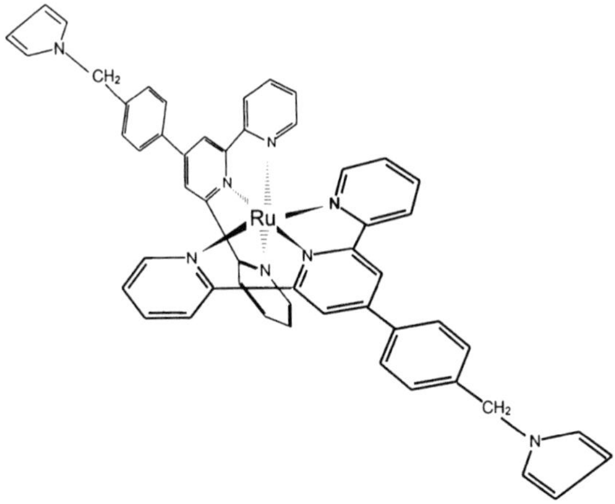

Figure 17 Structure of the electropolymerizable polypyridyl $[\text{Ru(ttpy-pyr)}_2]^{2+}$.

nitrile (with 0.1 M tetraethylammonium toluene-4-sulfonate as supporting electrolyte), copolymer films could be prepared on indium tin oxide (ITO) electrode surfaces. The potential was cycled between -0.3 V and 1.25 V vs. Ag/AgCl to ensure oxidation of the Ru-substituted pyrrole group and copolymerization with the ITN monomer unit [139].

Doping–undoping cycles of PITN are electrochemically reversible and are accompanied by an electrochromic change in the visible spectrum [150]. In the undoped state, thin films of PITN are blue; upon doping, the films become optically transparent.

By performing the copolymerization of the $[\text{Ru(ttpy-pyr)}_2](\text{PF}_6)_2$/ITN mixture on ITO electrodes, subsequent spectroscopic characterization can be facilitated. Shown in Fig. 18 is the UV-Vis spectrum for a thin copolymer film deposited from a $[\text{Ru(ttpy-pyr)}_2](\text{PF}_6)_2$/ITN solution as described above. Clearly evident in the spectrum is the broad adsorption due to PITN ($\lambda_{\text{max}} = 802$ nm) and a sharp absorption at $\lambda_{\text{max}} = 500$ nm due to $[\text{Ru(ttpy-pyr)}_2](\text{PF}_6)_2$.

Preliminary conductivity measurements on the PITN copolymers showed moderate conductivity values of 10^{-2}–10^{-4} S/cm depending on the amount of electrochemical doping. Modulation of the conductivity was also observed in the presence of visible light excitation. In all cases the changes in conductivity were completely reversible [151]. Experiments exploring the spectroscopic changes that occur following excitation are ongoing.

Reynolds and co-workers have observed similar electrochromic changes in their ethylenedioxythiophene (EDOT)–based polymers in which the thiophene rings are substituted with an oxygen-containing bridge between the 3 and 4 posi-

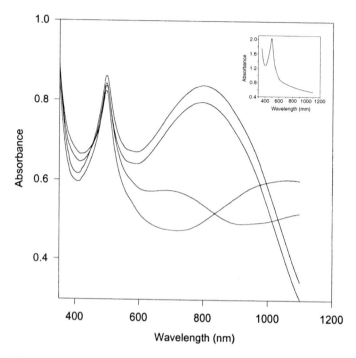

Figure 18 UV-Vis spectrum of the 1:1 PITN/[Ru(ttpy-pyr)$_2$](PF$_6$)$_2$ copolymer. (a) -0.6 V; (b) 0 V; (c) 0.6 V; (d) 1.1 V.

tions [152–157]. This substitution minimizes the steric hindrance between side groups on adjacent thiophene rings, which could lead to a reduction in conjugation along the backbone. These materials exhibit high thermal, chemical, and ambient stability, good ionic and electronic transport properties, in addition to maintaining a high conductivity (200 S/cm) [156]. Similarly, a solution of various ratios of the [Ru(ttpy-pyr)$_2$](PF$_6$)$_2$ chromophore and EDOT can be coelectropolymerized onto ITO electrodes. Shown in Fig. 19 is the spectrum of the copolymer film with the broad absorption due to the polymerized form of EDOT (PEDOT) ($\lambda_{max} = 925$ nm) and the sharp absorption at $\lambda_{max} = 500$ nm due to the [Ru(ttpy-pyr)$_2$](PF$_6$)$_2$. Both the parent PEDOT polymer and the copolymer [Ru(ttpy-pyr)$_2$](PF$_6$)$_2$/PEDOT exhibit optical changes during redox switching. As the potential is increased from -0.8 V to 0.8 V the broad π-to-π^* absorption of the polymer decreases as a lower energy absorption is initiated in the near-IR region [152]. Currently, the kinetics of the photoactive switching in these systems is being explored.

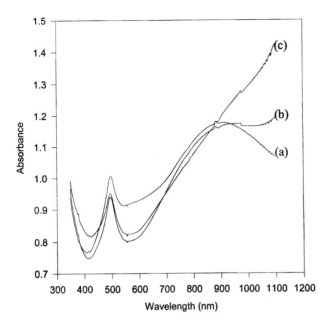

Figure 19 UV-Vis spectrum of a 1:1 PEDOT/[Ru(ttpy-pyr)$_2$](PF$_6$)$_2$ copolymer. (a) −0.8 V; (b) 0 V; (c) 0.8 V.

V. FUTURE EFFORTS AND APPLICATIONS

The development of metal-organic polymers over the past two decades demonstrates the applicability of conjugated polymer systems to a variety of developing technologies. From sensors to molecular electronics, the tunability and enhanced physical properties that result from incorporation of transition metals into polymer systems are likely to play a significant role in the future. Examples of the next generation of metal-organic polymers are already beginning to appear.

Fluorescent sensors are a rapidly expanding technology. Building on the research of Swager [98,99], Wasielewski [103], and others, several systems have recently been prepared that involve metal-organic polymers. For example, a highly fluorescent polymer with transition metal binding sites has been prepared [157–159]. The transition metal-bound, metal-organic polymer is not fluorescent. Early results demonstrate that these materials are highly sensitive to many of the transition metals on the Environmental Protection Agency's list of priority pollutants.

Molecular electronics is another area of research that may see new metal-organic polymer materials. The stabilizing changes observed in the physical properties represent a critical step in preparing polymer materials suitable to withstand the processing conditions of electronic circuit boards. However, there is a trade-off between conductivity and thermal stability that must be further addressed.

A general trend in the future development of metal organic polymers will likely involve multidimensional ordering. Extended π systems in conjugated polymers are typically viewed as polymer ribbons. The problem with quasi-one-dimensional systems, as observed in the porphyrin polymers described above, involves ensuring maximum overlap. Andersen has suggested that a more effective conjugation can be established by constructing three-dimensional lattices. For example, self-assembling, conjugated porphyrin ladders have been prepared covalently using a butadiyne linking structure [160]. Recently, multidimensional ordering of conjugated porphyrin polymers has also been achieved using coordinate covalent bonding of Lewis bases to metalloporphyrins [161]. These strategies and alternative applications of secondary structure could provide new opportunities for the design of highly conjugated metal organic polymers.

ACKNOWLEDGMENTS

Generous support from the Petroleum Research Fund, the Research Foundation of the State University of New York, and the Integrated Electronics and Engineering Center (IEEC) is gratefully acknowledged. The IEEC receives funding from the New York State Science and Technology Foundation, the National Science Foundation, and a consortium of industrial members. WEJ also wishes to thank the students and postdoctoral fellows who have worked on these and related projects as listed in the referenced publications.

REFERENCES

1. Ito, T.; Shirakawa, H., Ikeda, S. *J. Polym. Sci. Chem. Ed.* **1974**, *12*, 11.
2. Chiang, C.K.; Park, Y.W.; Heeger, A.J.; Shirakawa, H.; Louis, E.J., MacDiarmid, A.G. *Phys. Rev. Lett.* **1977**, *39*, 1098.
3. Skotheim, T.J., Ed. *Handbook of Conducting Polymers*, Marcel Dekker: New York, 1986.
4. Salaneck, W.R.; Clark, D.T.; Samuelsen, E.J., Ed. *Science and Applications of Conducting Polymers*, Adam Hilger, ed.; Bristol: Philadelphia, 1990.
5. Etemad, S.; Heeger, A.J.; MacDiarmid, A.G. *Annu. Rev. Phys. Chem.* **1982**, *33*, 443.
6. Greenham, N.C.; Friend, R.H. *Solid State Phys.* **1995**, *49*, 1–149.
7. Hide, F.; Diza-Garcia, M.A.; Schwartz, B.J.; Andersson, M.R.; Pei, Q.; Heeger, A.J. *Science* **1996**, *273*, 1833.
8. Kraft, A.; Grimsdale, A.C.; Holmes, A.B. *Angew. Chem. Int. Ed.* **1998**, *37*, 402–428.

9. Novak, P.; Miller, K.; Santhanam, K.S.V. *Chem. Rev.* **1997**, *97*, 207–281.
10. Gorman, C.B.; Grubbs, R.H. in *Conjugated Polymers: The Novel Science and Technology of Conducting and Nonlinear Optically Active Materials*, Bredas, J. L.; Silbey, R., Ed.; Kluwer Academic, Dordrecht, 1992.
11. Aldissi, M., Ed. *Intrinsically Conducting Polymers: An Emerging Technology*, Kluwer Academic, Dordrecht, 1993.
12. Schopf, G.; Kobmehl, G.; Ed. *Advances In Polymer Science: Polythiophenes— Electrically Conductive Polymers*, Springer-Verlag, Berlin, 1997.
13. Buroughes, J.H.; Bradley, D.D.; Brown, A.R.; Marks, R.N.; MacKay, K.; Friend, R.H.; Burn, P.L. and Holmes, A.B. *Nature* **1990**, *347*, 539.
14. Bradley, D.D.C. *Synth. Met.* **1993**, *54*, 401.
15. Roncali, J. *Chem. Rev.* **1997**, *97*, 173–205.
16. Webber, S.E. *Chem. Rev.* **1990**, *90*, 1469.
17. Hisada, K.; Ito, S.; Yamamoto, M. *J. Phys. Chem. B* **1998**, *102*, 4075.
18. Jones Jr., W.E.; Baxter, S.M.; Strouse, G. F.; Meyer, T.J. *J. Am. Chem. Soc.* **1993**, *115*, 7363.
19. Meyer, T.J. *Coord. Chem. Rev.* **1991**, *111*, 47.
20. Taylor, P.; Wylie, A.; Huuskonen, J.; Andersen, H. *Angew. Chem. Int. Ed.* **1998**, *37*, 986.
21. Andersen, H.L.; Martin, S.J.; Bradley, D.D.C. *Angew. Chem. Int. Ed.* **1994**, *33*, 655.
22. Wang, B.; Wasielewski, M.R. *J. Am. Chem. Soc.* **1997**, *119*, 12.
23. Marsella, J.J.; Newland, R.J.; Carroll, P.J.; Swager, T.M. *J. Am. Chem. Soc.* **1995**, *117*, 9842–9848.
24. Swager, T.M.; Gil, C.J.; Wrighton, M.S. *J. Phys. Chem.* **1995**, *99*, 4886–4893.
25. Marsella, M.J.; Carroll, P.J.; Swager, T.M. *J. Am. Chem. Soc.* **1995**, *117*, 9832.
26. Ley, K.D.; Whittle, C.E.; Bartberger, M.D.; Schanze, K.S. *J. Am. Chem. Soc.* **1997**, *119*, 3423.
27. Jiang, B.; Yang, S.-W.; Jones, Jr., W.E. *Chem. Mater.* **1997**, *9*, 2031.
28. Jiang, B.; Jones, Jr., W.E. *Macromolecules*, **1997**, *30*, 5575.
29. Jiang, B.; Yang, S.-W.; Niver, R.; Jones, Jr., W.E. *Synth. Met.* **1998**, *94(2)*, 205.
30. Feast, W.J.; Tsibouklis, J.; Pouwer, K.L.; Groenendaal, L.; Meijer, E.W. *Polymer* **1996**, *37*, 5017.
31. McCullough, R.D. *Adv. Mater.* **1998**, *10*, 93–116.
32. Higgins, S. *Chem. Soc. Rev.* **1997**, *26*, 247.
33. Pei, Q.; Yu, G.; Zhang, C.; Yang, Y.; Heeger, A.H. *Science* **1995**, *269*, 1086.
34. Wu, C.C.; Chun, J.K.M.; Burrows, P.E.; Sturm, J.C.; Thompson, M.E.; Forrest, S.R.; Register, R.A. *Appl. Phys. Lett.* **1995**, *28*, 4525.
35. Hilberer, A.; Brouwer, H.-J.; van der Scheer, B.-J.; Wildeman, J.; Hadziioannou, G. *Macromolecules* **1995**, *28*, 4525.
36. Grem, G.; Paar, C.; Stamp, J.; Leising, G.; Huber, J.; Scherf, U. *Chem. Mater.* **1995**, *69*, 415.
37. Herold, M.; Gmeiner, J.; Riess, W.; Schwoerer, M. *Syn. Met.* **1996**, *76*, 109.
38. Meghdadi, F.; Leising, G.; Fischer, W.; Stelzer, F. *Syn. Met.* **1996**, *76*, 113.
39. Nehring, J.; Amstutz, H.; Holmes, P.A.; Nevin, A. *Appl. Phys. Lett.* **1987**, *51*, 1283.

40. Noll, A.; Siegfried, N.; Heitz, W. *Macromol. Chem., Rapid Commun.* **1990**, *11*, 485.
41. Witteler, H.; Lieser, G.; Wegner, G.; Schulze, M. *Macromol. Chem. Rapid Commun.* **1993**, *14*, 418.
42. Maddux, T.; Li, W.; Yu, L. *J. Am. Chem. Soc.* **1997**, *119*, 844–845.
43. Grell, M.; Bradley, D.D.C.; Inbasekaran, M.; Woo, E.P. *Adv. Mater.* **1997**, 798.
44. Wautelet, P.; Moroni, M.; Osqald, L.; Le Moigne, J.; Pham, A.; Bigot, J.-Y.; Luzzati, S. *Macromolecules* **1996**, *29*, 446.
45. Hwang, D.-H.; Lee, J.-I.; Shim, H.-K.; Lee, G.J. *Synth. Met.* **1995**, *71* 1721.
46. Yamamoto, T.; Yamada, W.; Takagi, M.; Kizu, K.; Maruyama, T.; Ooba, N.; Tomaru, S.; Kurihara, T.; Kaino, T.; Kubota, K. *Macromolecules* **1994**, *27*, 6620.
47. Gale, D.M. *J. Appl. Polym. Sci.* **1978**, *22*, 1971.
48. Marrocco, M.; Gagne, R.R. U.S. Patent 5,227, 457, 1993.
49. Kovacic, P.; Jones, M.B. *Chem. Rev.* **1987**, *87*, 357.
50. Tour, J.M. *Adv. Mater.* **1994**, *6*, 190.
51. Percec, V.; Hill, D.H. *ACS Symp. Ser.* **1996**, 624, 2.
52. Vanhee, S.; Rulkens, R.; Lehmann, U.; Rosenauer, C.; Schulze, M.; Kohler, W.; Wegner, G. *Macromolecules* **1996**, *29*, 5136.
53. Percec, V.; Zhao, M.; Bae, J.-Y.; Hill, D.H. *Macromolecules* **1996**, *29*, 3727.
54. Moroni, M.; Le Moigne, J.; Luzzati, S. *Macromolecules* **1994**, *27*, 6620.
55. Weiss, K.; Michel, A.; Auth, E.-M.; Bunz, U.H.F.; Mangel, T.; Mullen, K. *Angew. Chem., Int. Ed. Engl.* **1997**, *36*, 506.
56. Tang, B.Z.; Kong, X.; Wan, X.; Feng, X. *Macromolecules* **1997**, *30*, 5620–5628.
57. Lee, H.; Gal, Y.; Lee, W.; Oh, J.; Jin, S.; Choi, S. *Macromolecules* **1995**, *28*, 1208–1213.
58. Kakuchi, T.; Watanabe, T.; Matsunami, S.; and Kamimura, H. *Polymer* **1997**, *38*, 1233.
59. Gin, D.L.; Conticello, V.P.; Grubbs, R.H.J. *Am. Chem. Soc.* **1994**, *116*, 10507.
60. Gin D.L.; Conticello, V.P.; Grubbs, R.H.J. *Am. Chem. Soc.* **1994**, *116*, 10934.
61. Gagnon, D.R.; Capistran, J.D.; Karasz, F.E.; Lenz, R.W.; Antoun, S. *Polymer* **1987**, *28*, 567.
62. Lenz, R.W.; Han, C.-C.; Stenger-smith, J.; Karasz, F.E. *J. Polym. Sci., Polym. Chem. Ed.* **1988**, *26*, 3241.
63. Burn, P.L.; Bradley, D.D.C.; Friend, R.H.; Halliday, D.A.; Holmes, A.B.; Jackson, R.W.; Kraft, A. *J. Chem. Soc., Perkin Trans. 1* **1992**, 3225.
64. Conticello, V.P.; Gin, D.L.; Grubbs, R.H. *J. Am. Chem. Soc.* **1992**, *114*, 9708.
65. Swager, T.M.; Grubbs, R.H. *J. Am. Chem. Soc.* **1989**, *111*, 4413.
66. Kanga, R.; Hogen-Esch, T.; Randrianalimanan, E.; Soum, A.; Fontanille, M. *Macromolecules* **1990**, *23*, 4241.
67. Feast, W. *Makromol. Chem., Macromol. Symp.* **1992**, *53*, 317.
68. Reibel, D.; Nuffer, R.; Mathis, C. *Macromolecules* **1992**, *25*, 7090.
69. Safir, A.L.; Novak, B.M. *Macromolecules* **1993**, *26*, 4072.
70. Bader, A.; Wunsch, J. *Macromolecules* **1995**, *28*, 3794.
71. Safir, A.L.; Novak, B.M. *Macromolecules* **1995**, *28*, 5396.

72. Kiess, H., Ed. *Conjugated Conducting Polymers*, Springer-Verlag, Berlin, 1992.
73. Skotheim, T.J., Ed. *Handbook of Conducting Polymers*, Marcel Dekker, New York, 1986.
74. Sandman, D.J. *Trends. Polym. Sci.* **1994**, *2*, 44.
75. Kanatzidis, M.G. *Chem. Eng. News* **1990**, *68(49)*, 36.
76. Heeger, A.J.; Kivelson, S.; Schrieffer, J.R.; Su, W.-P. *Rev. Mod. Phys.* **1988**, *60*, 781.
77. Sheats, J.E.; Carraher, Jr., C.E.; Pittman, C.U.; Zeldin, M.; Culbertson, B.M., Eds., *Metal Containing Polymeric Materials*, Plenum Press, New York, 1996.
78. Mark, J.E. *J. Inorg. Organometal. Polym.* **1991**, *1*, 431.
79. Balzani, V.; Scandola, F. *Supramolecular Chemistry*, Ellis Harwood, New York, 1991.
80. Astruc, D. *Acc. Chem. Res.* **1997**, *30*, 383.
81. Jones Jr., W.E.; Baxter, S.M.; Strouse, G.F.; Meyer, T.J.; *J. Am. Chem. Soc.* **1993**, *115*, 7363.
82. Schwab, P.; Grubbs, R.H.; Ziller, J.W. *J. Am. Chem. Soc.* **1996**, *118*, 100.
83. Watkins, D.M.; Fox, M. *J. Am. Chem. Soc.* **1996**, *118*, 4344.
84. Isied, S.; Ogawa, M.; Wishart, J. *Chem. Rev.* **1992**, 381.
85. Fitzsimons, M.P.; Barton, J.K. *J. Am. Chem. Soc.* **1997**, *119*, 3379.
86. Abruna, H.D.; Denisevich, P.; Umana, M.; Meyer, T.J.; Murray, R.W. *J. Am. Chem. Soc.* **1981**, *103*, 1.
87. Abruna, H.D. *Coord. Chem. Rev.* **1988**, *86*, 135.
88. Deronzier, A.; Moutet, J.-C. *Coord. Chem. Rev.* **1996**, *147*, 339.
89. Deronzier, A.; Moutet, J.-C. *Platinum Metals Rev.* **1998**, *42(2)*, 60.
90. Downard, A.J.; Surridge, N.A.; Gould, S.; Meyer, T.J.; Deronzier, A.; Moutet, J.-C. *J. Phys. Chem.* **1990**, *94*, 6754.
91. Sauvage, J.-P.; Collin, J.-P.; Chambron, J.-C.; Guillerez, S.; Coudret, C.; Balzini, V.; Barigelletti, F.; De Cola, L.; Flamigni, L. *Chem. Rev.* **1994**, *94*, 993.
92. Collin, J.-P.; Deronzier, A.; Essakalli, A. *J. Phys. Chem.* **1991**, *91*, 5906.
93. Marfurt, J.; Zhao, W.; Walder, L. *J. Chem. Soc., Chem. Commun.* **1994**, 51.
94. Zhao, W.; Marfurt, J.; Walder, L. *Helv. Chem. Acta* **1994**, *77*, 351.
95. Jiang, B.; Yang, S.-W.; Bailey, S.; Hermans, L.; Niver, R.; Bolcar, M.; Jones, Jr., W.E. *Coord. Chem. Rev.* **1998**, *171*, 365.
96. Wudl, F.; Kobayashi, M.; Heeger, A.J. *J. Org. Chem.* **1984**, *49*, 3382.
97. King, G.; Higgins, S.J. *J. Mater. Chem.* **1995**, *5*, 447.
98. Marsella, M.J.; Swager, T.M. *J. Am. Chem. Soc.* **1993**, *115*, 12214.
99. Zhou, Q.; Swager, T.M. *J. Am. Chem. Soc.* **1995**, *117*, 12593.
100. Yamamoto, Y.; Yoneda, Y.; Maruyama, T. *J. Chem. Soc., Chem. Commun.* **1992**, 1652.
101. Rasmussen, S.C.; Thompson, D.W.; Singh, V.; Petersen, J.D. *Inorg. Chem.* **1996**, *35*, 3449.
102. Cameron, C.G.; Pickup, P.G. *J. Chem. Soc., Chem. Commun.* **1997**, 303.
103. Wang, B.; Wasielewski, M.R. *J. Am. Chem. Soc.* **1997**, *119*, 12.
104. Ley, K.D.; Whittle, C.E.; Bartberger, M.D.; Schanze, K.S. *J. Am. Chem. Soc.* **1997**, *119*, 3423.

105. Ley, K.D.; Schanze, K.S. *Coord. Chem. Rev.* **1998**, *171*, 287.
106. Walters, K.A.; Ley, K.D.; Schanze, K.S. *J. Chem. Soc. Chem. Commun.* **1998**, 1115.
107. Ley, K.D.; Walters K.A.; and Schanze K.S. *Synth. Met.*, in press.
108. Bonfantini, E.E.; Officer, D.L. *Tetrahedron Lett.* **1993**, *34*, 8531.
109. Burrell, A.K.; Officer, D.L.; Reid, D.C.W. *Angew. Chem., Int. Ed. Engl.* **1995**, *34*, 900.
110. Wagner R.W.; Johnson, T.E.; Lindsey, J.S. *J. Am. Chem. Soc.* **1996**, *118*, 11166.
111. C. Murphy, unpublished results.
112. Ono, N.; Tomita, H.; Maruyama, K. *J. Chem. Soc., Perkin Trans. 1* **1992**, 2453.
113. Officer, D.L.; Burrell, A.K.; Reid, D.C.W. *J. Chem. Soc., Chem. Commun.* **1996**, 1657.
114. Wagner, R.W.; Lindsey, J.S.; Seth, J.; Palaniappan, V.; Bocian, D.F. *J. Am. Chem. Soc.* **1996**, *118*, 3996.
115. Hsiao, J.; Krueger, B.P.; Wagner, R.W.; Johnson, T.E.; Delaney, J.K.; Mauzerall, D.C.; Fleming, G.R.; Lindsey, J.S.; Bocian, D.F.; Donohoe, R.J. *J. Am. Chem. Soc.* **1996**, *118*, 11181.
116. Gregg, B.; Fox, M.A.; Bard, A.J. *J. Phys. Chem.* **1989**, *93*, 4227.
117. Hunter, C.A.; Sanders, J.K.M. *J. Am. Chem. Soc.* **1990**, *122*, 5525.
118. Foster, T. *Fluorenzenz Organische Verbindungen*, Vandenhoech and Ruprech, Gottingen; 1951.
119. Priyadarshy, S.; Therien, M.J.; Beratan, D.N. *J. Am. Chem. Soc.*, **1996**, *118*, 1504.
120. Anderson, H.L.; Martin, S.J.; Bradley, D.D.C. *Angew. Chem., Int. Ed. Engl.* **1994**, *33*, 655.
121. Kawabata, S.; Yamazaki, I.; Nishimura, Y.; Osuka, A. *J. Chem. Soc., Perkin Trans. 2*, **1997**, 479.
122. Gust, D.; Moore, T.A.; Moore, A.L. *Acc. Chem. Res.* **1993**, *26*, 198.
123. Jones Jr., W.E.; Fox, M.A. *J. Phys. Chem.* **1994**, *98*, 5095.
124. Abdou, M.S.A.; Orfino, F.P.; Xie, Z.W.; Deen, M.J.; Holdcroft, S. *Adv. Mater.* **1994**, *6*, 838.
125. Geoffroy, G.L.; Wrighton M.S., Ed. *Organometallic Photochemistry*, Academic Press, New York, 1979.
126. Sariciftci, N.S.; Smilowitz, L.; Heeger, A.J.; Wudl, F. *Science* **1992**, 258, 1474.
127. Sariciftci, D.; Braun D.; Zhang, C.; Sdanov, V.I.; Heeger, A.J.; Stucky, F.; Wudl F. *Appl. Phys. Lett.* **1993**, *62*, 585.
128. Kraabel, B.; Lee, C.H.; McBranch, D.; Moses, D.; Sariciftci, N.S.; Heeger, A.J. *J. Chem. Phys. Lett.* **1993**, *313*, 389.
129. Lee, K.; Janssen, R.A.J.; Sariciftci, N.S.; Heeger, A.J.; *Phys. Rev. B* **1994**, *49*, 5781.
130. Paul, W.E.; Ricco, A.J.; Wrighton, M.S. *J. Phys. Chem.* **1985**, *89*, 1441.
131. Chiang, J.C.; MacDiarmid, A. *Synth. Met.* **1986**, *13*, 193.
132. Zotti, G.; Comisso, N.; Aprano, G.D.; Leclerc, M. *Adv. Mater.* **1992**, *4*, 749.
133. The quenching rate constant was limited by our knowledge of the molecular weight of the polymer. Based on the literature preparation used, the molecular weight of the polymer was 1.0×10^6 g/mol [37].
134. Kim, Y.H.; Phillips, S.D.; Nowak, M.J.; Spiegel, D.; Foster, C.M.; Yu, G.; Chiang, J.C.; Heeger, A. *Synth. Met.* **1989**, *29*, E291.

135. Stafström, S. in Conjugated Polymers, Brédas, J.L.; Silbey, R., Eds. 113, Kluwer Academic, Dordrecht, 1991.
136. McCall, R.P.; Roe, M.G.; Ginder, J.M.; Kusumato, T.; Epstein, A.J.; Asturias, G.E.; Scherr, E.M.; MacDiarmid, A.G. *Synth. Met.*, **1989**, *29*, E433.
137. Caspar J.V.; Meyer, T.J. *J. Am. Chem. Soc.*, **1983**, *105*, 5583.
138. Lemmon, J.P.; Gross, S.M.; Jones Jr., W.E. *Polym. Prepr.* **1996**, *37*, 113.
139. Hermans L.G.; Jones Jr., W.E. *J. Mater. Chem.*, **1999**, in press.
140. Personal communication with Prof. Allen Macdiarmid, University of Pennsylvania, Philadelphia.
141. Mylnikov, V. *Adv. Polym. Sci.*, **1994**, *115*, 5–12.
142. Misurkin, I.A.; Zhuravleva, T.S.; Geskin, V.M.; Gulbinas, V.; Pakalnis, S.; Butvilos, V. *Phys. Rev. B* **1994**, *49*, 7178.
143. The MLCT excited state of $[Ru(bpy)_3]^{2+*}$ is predominantly triplet in character and is known to sensitize the formation of organic triplet states such as anthracene.
144. Wudl, F.; Kobayashi, M.; Heeger, A.J. *J. Org. Chem.*, **1984**, *49*, 3382.
145. King, G.; Higgins, S.J. *J. Mater. Chem.*, **1995**, *5*, 447.
146. Higgins, S.J.; Jones, C.; King, G.; Slack, K.H.D.; Pétidy, S. *Synth. Met.*, **1996**, *78*, 155.
147. Sauvage, J.; Collin, J.; Guillerez, S.; Barigelletti, F.; DeCola, L.; Flamigni, L.; Balzani, V. *Inorg. Chem.* **1991**, *30*, 4230.
148. Ellis, C.D.; Margerum, L.D.; Murray, R.W.; Meyer, T.J. *Inorg. Chem.*, **1983**, *22*, 1283.
149. Onada, M.; Morita, S.; Nakayama, H.; Yoshino, K. *Jpn. J. Appl. Phys.*, **1993**, *32*, 3534.
150. Tourillon, G.; Garnier. F. *J. Phys. Chem.*, **1983**, *87*, 2289.
151. Hermans L.; Jones Jr., W.E., manuscript in preparation.
152. Sotzing, G.A.; Reddinger, J.L.; Reynolds.; Steel, P.J. *Synth. Met.* **1997**, *84*, 199.
153. Sankaran, B.; Reynolds, J.R. *Macromolecules*, **1997**, *30*, 2582.
154. Sotzing, G.A.; Reynolds, J.R.; Steel, P.J. *Chem. Mater.* **1996**, *8*, 82.
155. Sotzing, G.A.; Reynolds, J.R. *J. Chem. Soc., Chem. Commun.* **1995**, 703.
156. Heywang, G.; Jonas, F. *Adv. Mater.* **1992**, *4, No. 2*, 117.
157. Kimura, M.; Horai, T.; Hanabusa, K.; Shirai, H. *Adv. Mater.* **1998**, *10, No. 6*, 459
158. Jiang, B.; Sahay, S.; Jones Jr., W.E. *Mater. Res. Soc. Symp. Proc.* **1998**, *488*, 671
159. Jiang, B.; Sahay, S.; Chatterjee, S.; Jones Jr., W.E., manuscript in preparation.
160. Andersen, H.L. *Inorg. Chem.* **1994**, *33*, 972.
161. Sarno D.M.; Grosfeld D.; Snyder J.R.; Jiang B.; Jones Jr., W.E. *Polym. Prepr.*, **1998**, 1101.

2

Luminescence Behavior of Polynuclear Metal Complexes of Copper(I), Silver(I), and Gold(I)

Vivian W.-W. Yam and
Kenneth K.-W. Lo
University of Hong Kong, Hong Kong, China

I. INTRODUCTION

The ruthenium(II) tris(bipyridine) complex $[Ru(bpy)_3]^{2+}$ is an important figure in modern inorganic photochemistry. Over the years there have been numerous reports on the photophysics and photochemistry of related complexes [1–6]. The metal-to-ligand charge transfer (MLCT) excited state chemistry has been well studied. This class of complexes has also been widely used in the development of supramolecular photochemistry [1,5,7–9].

In the early 1980s, the dinuclear platinum(II) pyrophosphite complex $[Pt_2(POP)_4]^{4-}$ was found to be highly luminescent [10]. The photochemistry of this complex is unique and completely different from that of ruthenium(II) polypyridine complexes. The emission has been assigned to arise from a metal-centered excited state strongly modified by metal–metal interaction. The lowest energy optical transition involves the promotion of an electron from a filled d_{σ^*} antibonding orbital (formed by the interaction of two d_{z^2} orbitals, z axis taken to be the metal–metal axis) to an empty p_σ bonding orbital (formed by the interaction of two p_z orbitals). Therefore, this transition is associated with an increase in the bond order between the two platinum(II) centers. It can be described as an establishment of a Pt–Pt bond in the excited state. The spectroscopy, photophysics, and photochemical reactivities of this complex and related d^8–d^8 complexes have been investigated extensively [10,11].

The idea of metal–metal bond establishment in the excited state of this complex has also stimulated the interest on the photophysical and photochemical investigations of related d^{10}–d^{10} complexes. In 1970, Dori and co-workers first reported the luminescence properties of phosphine complexes of d^{10} metal centers such as copper(I), silver(I), gold(I), nickel(0), palladium(0), and platinum(0) [12]. Later in 1985, Caspar reported the luminescence properties of polynuclear nickel(0), palladium(0), and platinum(0) complexes of phosphine, phosphite, and arsine [13]. Similar to the related d^8–d^8 systems, the emissive states of [Pd$_2$(dppm)$_3$] and [Pd$_2$(dpam)$_3$] have been suggested to be metal-centered (d–p) in nature modified by metal–metal interaction.

Gray and co-workers reported the spectroscopy, photophysics, and resonance Raman spectroscopy of related d^{10}–d^{10} metal complexes [M$_2$(dppm)$_3$] (M = Pd, Pt) in detail [14]. The relative energies of the highest occupied orbitals are $d_{xz}, d_{yz} < d_{z^2} < d_{xy}, d_{x^2-y^2}$ for mononuclear ML$_3$ complexes under D_{3h} symmetry. The lowest unoccupied orbital is p_z. A molecular orbital diagram for d^{10} ML$_3$ and d^{10}–d^{10} (ML$_3$)$_2$ complexes is shown in Fig. 1. As the splitting of d_σ and $d_{\sigma*}$ is much greater than that of d_δ and $d_{\delta*}$, it is likely that $d_{\sigma*} \rightarrow p_\sigma$ and $d_{\delta*} \rightarrow p_\sigma$ transitions would occur at similar energy. The spin-allowed and spin-forbidden [$d_{\sigma*} \rightarrow p_\sigma$] and [$d_{\delta*} \rightarrow p_\sigma$] transitions have been identified in the low-temperature electronic absorption spectra of these two complexes in 2-methyltetrahydrofuran. For example, the transition $^1[d_{\sigma*} \rightarrow p_\sigma]$ is associated with an absorption band at 440 nm (ε = 33,800 dm^3 mol^{-1} cm^{-1}) and 487 nm (ε = 27,400 dm^3 mol^{-1} cm^{-1}) for [Pd$_2$(dppm)$_3$] and [Pt$_2$(dppm)$_3$], respectively. In addition, the emission of [Pd$_2$(dppm)$_3$] and [Pt$_2$(dppm)$_3$] in 2-methyltetrahydrofuran at 77 K occurs at 694 and 787 nm, respectively. This emission has been assigned to originate from a $^3[(d_{\sigma*})^1(p_\sigma)^1]$ excited state.

In the past 10 years, the number of reports on the spectroscopy, photophysics, and photoreactivities of polynuclear copper(I), silver(I), and gold(I), or in general, d^{10} metal complexes have been increasing significantly. There are also some excellent review articles on this topic to which readers are referred [15–18].

In fact, polynuclear d^{10} metal complexes are also interesting from a molecular structural point of view [19]. A remarkable feature is the observation of a short metal–metal distance present in these complexes. Especially for gold(I), owing to the relativistic effects, this metal center shows a strong tendency to form dimers, oligomers, and polymers. This observation has been coined as *auro-*

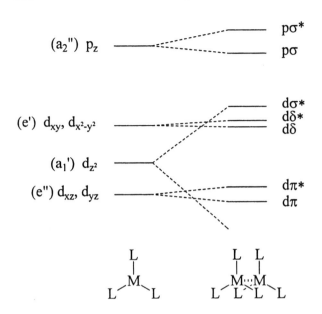

Figure 1   Qualitative MO diagram for d^{10} ML_3 and d^{10}–d^{10} $(ML_3)_2$ complexes. (Adapted from Ref. [14].)

philicity by Schmidbaur [20]. Theoretically, in the absence of $(n + 1)s$ and $(n + 1)p$ functions, interaction between the closed shell d^{10} centers is repulsive in nature. However, configuration mixing of the filled nd orbitals with the empty orbitals derived from higher energy $(n + 1)s$ and $(n + 1)p$ orbitals may establish some weak metal–metal interaction [21–26]. The van der Waals radii of copper, silver, and gold are 1.40, 1.72, and 1.66 Å, respectively [27]. In fact, it is not uncommon to find M–M distances in different compounds shorter than the sum of their van der Waals radii [19]. For example, the shortest Cu–Cu distance (2.35 Å) has been observed for the complex $[Cu_3(\mu_3\text{-}C_6H_4\text{-}CH_3\text{-}4\text{-}NNNNN\text{-}C_6H_4\text{-}CH_3\text{-}4)_3]$ [28]. Besides, the complexes $[M_2(form)_2]$ (M = Cu, Ag) also show short Cu–Cu and Ag–Ag distances of 2.497(2) and 2.705(1) Å, respectively [29]. For the cyclic organometallic complexes $[M\{[C_6H_2(CH_3)_3\text{-}2,4,6]\}_n]^{30}$ (M = Cu, Au, $n = 5$; M = Ag, $n = 4$) short metal–metal separations of 2.437(8)–2.496(9), 2.733(3), and 2.697(1) Å have also been observed for the respective copper(I), gold(I), and silver(I) complexes. Such an observation is common for organometallic complexes of the copper(I) [31,32], silver(I) [32,33], and gold(I) [34,35], in particular those containing electron-deficient bonding systems. In general, short metal–metal contacts do not necessarily indicate the presence of d^{10}–d^{10} bonding interaction, especially under the influence of the bridging ligands [29,36,37].

Nevertheless, the possibility of bonding interaction between closed shell d^{10} metal centers is still attracting a lot of interest [19,38–40].

In this chapter, we will focus on the spectroscopy, photophysics, excited state assignments, and photochemical reactivities of polynuclear copper(I), silver(I), and gold(I) complexes. The sections are generally arranged based on the identity of the coordinating ligands. However, as phosphine ligands have a high affinity for these coinage metal centers and the majority of the examples we discuss contain phosphines as ancillary or bridging ligands, "binary" metal phosphine complexes will be introduced in the first part of each section.

II. COPPER(I) SYSTEMS

A. Phosphine Complexes

Mononuclear copper(I) phosphine complexes such as $[Cu(PPh_3)_2BH_4]$ (1) have been shown to be luminescent [41]. The emission of this complex at room temperature originates from a $[\sigma(Cu-P) \rightarrow \pi^*(PPh_3)]$ state. The luminescence properties of the dinuclear copper(I) phosphine complex $[Cu_2(\mu-dppm)_2(CH_3CN)_4]^{2+}$ (2) were reported in 1992 [42]. The crystal structure of the complex revealed a Cu–Cu separation of 3.757(3) Å [43], indicative of no metal–metal interaction. The emission band of the complex in acetonitrile at room temperature centered at 526 nm ($\tau_0 = 7$ μs, $\Phi = 1.5 \times 10^{-2}$) has been tentatively assigned to be MLCT [Cu → phosphine] in nature. The excited state complex can catalyze the C–C coupling reaction of benzyl chloride, leading to the formation of dibenzyl in 5% yield [42].

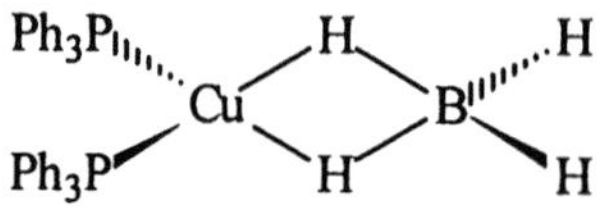

1

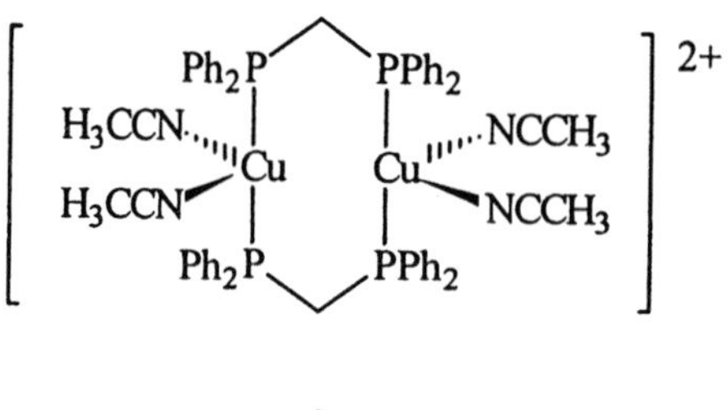

2

Reaction of $[Cu_2(\mu\text{-}dppm)_2(CH_3CN)_4]^{2+}$ with PPh_3 gave $[Cu_2(\mu\text{-}dppm)_2$ $(PPh_3)_2]^{2+}$ (**3**) [44], which in CH_2Cl_2 gave a room temperature UV-Vis absorption spectrum showing a peak at 255 nm ($\varepsilon = 3.00 \times 10^4$ dm^3 mol^{-1} cm^{-1}) and a shoulder at 310 nm ($\varepsilon = 4.76 \times 10^3$ dm^3 mol^{-1} cm^{-1}). The emission of the complex in CH_2Cl_2 solution at room temperature occurs at 550 nm with a very long lifetime of 75 μs.

B. Halide Complexes

The photophysical and photochemical properties of tetranuclear copper(I) halide complexes with the general formula $[Cu_4X_4L_4]$ (X = halide, L = pyridine) have been extensively studied for more than a decade. The molecular geometry of these complexes can be generally described as a distorted cubane structure formed by the four copper(I) centers and four halides, with the four copper centers arranged in the form of a tetrahedron interlocking with a tetrahedral array of halides. Each copper is also coordinated to a pyridine ligand. The luminescence behavior of this class of compounds was first reported by Hardt and co-workers in the 1970s [45]. In 1986, Vogler and co-worker reported the emission spectra of the complexes $[Cu_4I_4(pyridine)_4]$ (**4a**) and $[Cu_4I_4(morpholine)_4]$ (**4b**) [24]. The red luminescence of both complexes has been assigned to a metal-centered excited state $3d^9 4s^1$ of Cu(I) strongly modified by Cu–Cu interaction.

4a: L = pyridine
4b: L = morpholine
4c: L = 4-*tert*-butylpyridine
4d: L = 4-benzylpyridine
4e: L = pyridine-d$_5$
4f: L = 4-phenylpyridine
4g: L = 3-chloropyridine
4h: L = piperidine
4i: L = P^{n}Bu$_3$

In a series of papers, Ford and co-workers described the photophysical and photochemical properties of these luminescent clusters in detail [46–58]. In addition to complexes **4a** and **4b**, time-resolved emission spectra of the tetranuclear copper(I) iodide clusters [Cu$_4$I$_4$(L)$_4$] with a series of substituted pyridines [L = 4-*tert*-butylpyridine (**4c**), 4-benzylpyridine (**4d**), pyridine-d_5 (**4e**), 4-phenylpyridine (**4f**), 3-chloropyridine (**4g**), piperidine (**4h**), P^{n}Bu$_3$ (**4i**)] have also been studied [49]. The photophysical data are summarized in Table 1. In general, in toluene solution at 294 K, the complexes revealed a low-energy emission at 678–698 nm and a weaker, higher energy emission at 473–537 nm. The emission spectrum of **4a** in toluene at 294 K is shown in Fig. 2

The intense low-energy emission band is the dominant feature of the emission spectra for this class of complexes. However, the high-energy emission is only observed for the complexes with aromatic amine ligands. For example, in toluene solution at 294 K, complexes **4b**, **4h**, and **4i** only reveal a low-energy emission band at 671, 680, and 654 nm, respectively. This finding strongly suggests that the high-energy emission is associated with the π* orbitals of the pyridine ligands. In addition, this high-energy emission band is also found to be sensitive to solvent. For example, for **4a**, the high-energy emission occurs at 500 nm (shoulder) in CH$_2$Cl$_2$ and 425 nm in methyl acetate, whereas the low-energy bands occur at the same energy (694 nm). These observations, together with the energy trend observed for the substituted pyridine ligands, suggest that the high-energy emission of these complexes is associated with a metal-to-ligand [Cu → π*(pyridine)] or ligand-to-ligand [I$^-$ → π*(pyridine)] charge transfer excited state.

The low-energy emission of this class of complexes is characteristic of the Cu$_4$I$_4$ core and independent of the identity of the pyridine or amine ligands. It has been suggested that the low-energy emission originates from a cluster-centered excited state where metal–metal bonding is enhanced by depopulation of a d$_{\sigma*}$(Cu–Cu) antibonding orbital and population of a s$_\sigma$(Cu–Cu) bonding orbital. This assignment is also supported by the observation of a very large Stokes shift from the excitation maximum to the low-energy emission maximum (e.g., Stokes shift = 1.64 μm^{-1} for **4a** in toluene solution).

The electronic structures and the origins of the dual luminescence of these tetranuclear copper(I) clusters have been investigated by ab initio molecular orbital studies [51]. The highest occupied molecular orbitals (HOMOs) in **4a** and the model compound [Cu$_4$I$_4$(NH$_3$)$_4$] (**4j**) are largely composed of iodide p orbitals. Therefore, it was suggested by Ford and co-workers that the origin of the high-energy emission is ligand-to-ligand charge transfer [I$^-$ → π* (pyridine)] (XLCT) in character. The results of more thorough calculations which include the factors arising from the electronic reorganization show that the emitting state for the low-energy emission is an admixture of Cu$_4$ [d → s] and XMCT [halide → Cu$_4$] delocalized over the Cu$_4$I$_4$ core.

Table 1 Photophysical Data for **4a–i**, **5a–c**, and **6a–c**

Complex	Medium (T K)	Emission wavelength λ_{max} nm (τ_0 μs)	Ref.
4a	Solid (294)	580 (11.1)	49
	Solid (77)	438 (23.3), 619 (25.5)	
	Toluene (294)	480 (0.45), 690 (10.6)	
	Toluene (77)	436 (32.9), 583 (26.5)	
4b	Toluene (294)	671 (0.51)	49
	Toluene (77)	630	
4c	Solid (294)	623 (18.8)	49
	Solid (77)	437 (29.2), 650 (38.8)	
	Toluene (294)	468 (0.35), 696 (10.3)	
	Toluene (77)	434 (38.7), 595 (43.5)	
4d	Toluene (294)	473 (0.56), 692 (11.0)	49
4e	Solid (294)	580 (11.1)	49
	Solid (77)	433 (20.8), 619 (24.0)	
	Toluene (294)	480 (0.22), 690 (14.1)	
4f	Toluene (294)	520 (0.12), 694 (9.4)	49
	Toluene (77)	505 (63.0)	
4g	Toluene (294)	537 (0.35), 675 (12.7)	49
4h	Solid (294)	590 (12.3)	49
	Solid (77)	590 (13.4)	
	Toluene (294)	680 (0.11)	
4i	Toluene (294)	654 (2.23)	49
	Toluene (77)	620	
5a	Solid (298)	527 (4.0 ± 0.2)	50
	Solid (77)	519 (11.4 ± 0.2)	
	Toluene (77)	577 (21.4 ± 0.1)	
5b	Solid (298)	620 (0.87 ± 0.01)	50
	Solid (77)	613 (3.8 ± 0.1)	
	CH_2Cl_2 (77)	596 (4.6 ± 0.1)	
	Toluene (77)	584 (18.7 ± 0.1)	
5c	Solid (77)	530 (28.2 ± 0.3)	50
	CH_2Cl_2 (77)	560 (9.0 ± 0.5)	
	Toluene (77)	600 (13.6 ± 0.3)	
6a	Solid (298)	500 (2.7)	52
	Solid (77)	505 (10.0)	
	CH_3CN (77)	530 (12.8)	
6b	Solid (298)	480, 626 sh (0.54)	52
	Solid (77)	487 (10.0)	
	CH_3CN (77)	512 (12.5)	
6c	Solid (298)	440, 462 sh, 570 sh (7.5)	52
	Solid (77)	446 sh, 467 (18.1)	
	CH_3CN (77)	500 (15.3)	

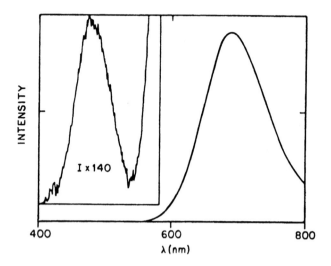

Figure 2   Emission spectrum of **4a** in toluene at 294 K. (Adapted from Ref. [49].)

On the other hand, the luminescence properties of related tetranuclear copper(I) chloride clusters [Cu$_4$Cl$_4$L$_4$] [L = pyridine (**5a**), 4-phenylpyridine (**5b**), diethylnicotinamide (**5c**), morpholine (**5d**), and triethylamine (**5e**)] have also been studied [50]. Emission is only observed for complexes **5a–c** with unsaturated amine ligands. From the energy trend of the single-emission bands of the complexes (Table 1), it is likely that the emission arises from an MLCT or a XLCT excited state. The latter is supported by ab initio studies as the HOMO also possesses a high degree of chloride character [50]. The absence of a low-energy cluster-centered emission band is in accordance with the fact that the Cu–Cu distance of these complexes (>3.0 Å) is longer than the sum of van der Waals radii of Cu(I) (2.80 Å) [27]. This indicates the importance of the Cu–Cu interaction on the nature of the acceptor orbitals in the cluster-centered excited state.

Another series of related copper(I) halide tetramers [Cu$_4$X$_4$(dpmp)$_4$] [X = Cl (**6a**), Br (**6b**), and I (**6c**)] has also been investigated [52]. The Cu–Cu distances of these complexes are very similar (2.9 Å). The photophysical data of the complexes are listed in Table 1. The emission bands of the complexes in the solid state at 77 K are assigned to originate from a XLCT excited state. Although the energy trend is not in agreement with the donating abilities of the halide ions, ab initio studies [54] suggest that the halide ionicity in the clusters decreases from Cl$^-$ to I$^-$, which can compensate for the effects of ionization energy on the clusters' HOMOs. An interesting feature on the luminescence properties of this series of compounds is the observation of a low-energy shoulder at higher temperature (Table 1). These low-energy emission shoulders are similar to the cluster-centered emission observed in the iodide clusters. The interesting emissive behavior of this dpmp series of clusters can be explained by a model outlined in Fig. 3. For **4a**, the lowest energy excited state is the cluster-centered (ds/XMCT) triplet state which leads to the lower energy and more intense emission band. However, for the dpmp complexes, this cluster-centered state is about 1000 cm^{-1} higher in energy than the XLCT triplet, which is associated with the emission at all temperatures and the only emission at low temperature. The poor coupling between the two triplet states for the **4a** cluster is due to the high-energy barrier for the crossing between the two states. However, for the dpmp series, the barrier

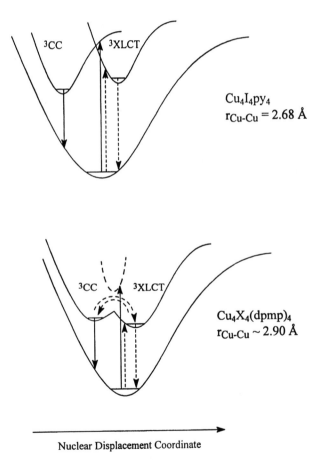

Figure 3 Proposed model for potential energy surfaces for XLCT and CC excited states for **4a** and **6a–c**. (Adapted from Ref. [52].)

is lower in energy and therefore the cluster-centered and XLCT triplets are in thermal equilibrium.

In the cluster-centered ds/XMCT state, the $s_\sigma(Cu-Cu)$ bonding orbital is populated. Therefore, the energy and the shape of this excited state is strongly dependent on the Cu–Cu interaction in the cluster. However, the XLCT triplet state is not sensitive to the Cu–Cu interaction. For **4a**, the packing of large iodide and smaller Cu(I) spheres results in a small Cu_4 tetrahedron with a higher extent of Cu–Cu interaction. However, for the dpmp series, owing to the steric requirements of the bulky pyridine ligands, the Cu–Cu distances for all chloride, bromide, and iodide clusters are held at around 2.9 Å and this leads to a cluster-centered state being higher in energy than the XLCT triplet.

The excited state absorption spectra of these luminescent tetranuclear copper(I) iodide complexes have also been reported [57]. The absorption spectral data of the cluster-centered triplet states of **4a**, **4c**, **4f**, and **4h** are listed in Table 2. The energy of the absorption is in line with the ordering of the π^* orbital energy of the pyridine ligands. For the piperidine analog **4h**, no excited state absorption is observed. It has been suggested that the excited state absorption bands are associated with the transition from a thermally relaxed ^{3}[cluster-centered] state to Franck-Condon states of the ^{3}[XLCT] manifold.

Recently, the effect of pressure-induced luminescence rigidochromism on the emission behavior of **4a** has also been reported [58]. In ambient pressure, the emission of the complex occurs at 695 nm in benzene. However, at a pressure higher than that required to induce the solvent to undergo the room temperature phase transition from fluid to solid, the emission band shifts to 575 nm, which is similar to the solid state emission wavelength (580 nm) of the complex.

Table 2   Excited State Absorption Spectral Data for **4a**, **4c**, **4f**, and **4h**[a]

Complex	Solvent[b]	Absorption wavelength λ_{max} nm	Lifetime τ_0 µs
4a	Toluene	547	7.9
	CH_2Cl_2	550	2.7
	Acetone	557	2.8
4c	Toluene	537	8.0
	CH_2Cl_2	548	4.1
	Acetone	548	4.3
4f	Toluene	565	7.8
	CH_2Cl_2	580	2.7
4h	Toluene	—	—

[a] Excitation wavelength = 337 nm.
[b] Temperature at 294 K.
Source: From Ref. [57].

The photochemistry of this class of complexes has also been studied extensively. For example, the high-energy XLCT emission of **4a** in benzene has been found to be quenched by pyridine [47]. The bimolecular quenching rate constant was estimated to be $5.9 \pm 0.5 \times 10^9$ dm^3 mol^{-1} s^{-1}. The mechanism for the quenching has been ascribed to the exciplex formation between the excited complex and the quencher.

From an estimation of the E^{00} energy of **4a*** of 1.66 μm^{-1} (2.06 eV), and the ground state reduction potential $E_{1/2}$ for **4a**$^+$/**4a** (0.28 V vs. Fc$^+$/Fc), an excited state reduction potential E^0[**4a**$^+$/**4a***] for the tetranuclear copper(I) cluster has been determined to be -1.78 V vs. Fc$^+$/Fc in CH$_2$Cl$_2$ [53]. The quenching of the photoexcited complex by a series of tris(β-dionato)chromium(III) complexes and organic aromatics have been investigated. Energy transfer is essentially a quenching mechanism between the excited complex and the chromium(III) quenchers, and the bimolecular quenching rate constant has been estimated to be 4.6×10^7 dm^3 mol^{-1} s^{-1}. However, for those quenchers with reduction potentials $E_{1/2}(Q/Q^-)$ less negative than -1.4 V vs. Fc$^+$/Fc, electron transfer becomes a competitive quenching pathway. Besides, for 1,2-, 1,3-, 1,4-dinitrobenzene, and 1,4-benzoquinone, the bimolecular quenching rate constants are found to be 7.2×10^6, 4.2×10^6, 2.5×10^8, and 1.3×10^9 dm^3 mol^{-1} s^{-1}, respectively. The π-π* energies for these organic quenchers are too high for any energy transfer to take place. Therefore, the quenching is electron transfer in nature. A comparatively large driving force is required for the electron transfer quenching of the excited copper(I) tetramer to be competitive with radiative and nonradiative deactivation.

A further study using ferrocenium and its methylated analogs as quenchers has also been carried out [55]. The second-order quenching rate constants are in the order of 10^{10} dm^3 mol^{-1} s^{-1}, approaching the estimated diffusion limit. These additional quenching rate data reveal a very large reorganization energy (λ_{total} = 1.89 eV). The outer-sphere contribution (λ_{out}) was estimated to be 0.5 eV and therefore the inner-sphere reorganization energy (λ_{in}) was about 1.4 eV. This large value has been rationalized by the enhanced Cu–Cu bonding in the excited complex associated with the electron transfer, which is also in line with the large Stokes shift of the excitation and the emission bands.

On the other hand, in 1989, Zink and co-workers also reported the photophysics of a series of tetranuclear copper(I) mixed-metal clusters [(DENC)$_4$Cu$^I_4$Cl$_4$] (**7a**), [(DENC)$_3$Cu$^I_3$M(NS)Cl$_4$] [M = Co^{II} (**7b**), Ni^{II} (**7c**), Cu^{II} (**7d**), Zn^{II} (**7e**)], and [(DENC)$_3$Cu$^I_2$Cu^{II}Co^{II}(NS)$_2$Cl$_4$] (**7f**) [59]. The photophysical data are summarized in Table 3. It has been suggested that the emission of all the clusters originates from the 3d^{9}4s^1 copper(I) state strongly modified by Cu–Cu interaction. However, for the complexes containing Cu(II) and Co(II) centers, the emission lifetimes are comparatively shorter than those of the others. This has been attributed to the presence of a low-lying d-d state that provides an additional deactivation pathway. Besides, the same research group also reported the vibronically

Table 3 Photophysical Data for **7a–f** in CH$_2$Cl$_2$ Glass at 77 K[a]

Complex	Emission wavelength λ_{max} nm (τ_0 μs)
7a	662 (7.3 ± 0.4)
7b	741 (5.9 ± 0.4)
7c	714 (7.0 ± 0.4)
7d	760 (5.4 ± 0.4)
7e	667 (8.3 ± 0.4)
7f	752 (5.5 ± 0.4)

[a] Excitation wavelength = 406 nm.
Source: From Ref. [59].

structured emission spectrum of the copper(I) cluster [Cu$_4$I$_4$(dmpp)$_4$] (**8**) at low temperature [60]. The resonance Raman spectrum of the complex has also been studied. From the distortion of the molecule in the excited state, the emission is assigned to originate from a transition terminating on the phosphole ligand. By comparing the emission energies of the chloride and bromide analogs, the excited state has been proposed to be MLCT [Cu$_4$ → phosphole] in nature.

Apart from tetranuclear species, the photophysics of trinuclear copper(I) halide complexes [Cu$_3$(μ$_3$-dppp)$_2$(CH$_3$CN)$_2$(μ-X)$_2$]$^+$ [X = Cl (**9a**), I (**9b**)] has also been investigated [61]. Upon photoexcitation, the chloride and iodide complexes emit at 530 and 560 nm, respectively. Owing to the similar energies of the emission, the lowest energy excited states of these complexes are assigned to be metal-centered 3d^{9}4s^1 in nature and modified by Cu–Cu interaction and/or mixing with the dppp ligand. The photoexcited complexes are strongly reducing. Photoredox reactions with pyridinium acceptors such as *N*-ethylpyridinum and *N*,2,6-trimethylpyridinium ions have been observed.

9a: X = Cl
9b: X = I

C. Chalcogenide and Chalcogenolate Complexes

Harvey and co-workers studied the luminescence properties of the trimeric copper(I) cluster $[Cu_3(\mu\text{-dppm})_3(\mu_3\text{-OH})]^{2+}$ (**10**) [62]. The emission of the complex occurs at 540 nm ($\tau_0 = 89 \pm 9$ μs) and 480 nm ($\tau_0 = 170 \pm 40$ μs) at 298 and 77 K, respectively. The electronic structure of the cluster has also been studied. The HOMO is composed of approximately 51% in-plane Cu (3d) orbitals ($d_{x^2-y^2}$ and d_{xy}), 10% in-plane Cu (4p) orbitals (p_x and p_y), and 29% of in-plane P (p_x and p_y) orbitals. The lowest unoccupied molecular orbital (LUMO) is composed of 27% of Cu (4s) and 62% of Cu (4p) orbitals, together with around 4% of O (p_z) orbital. Cu–Cu bonding and antibonding interactions have been observed in the LUMO and HOMO, respectively. It has been suggested that the Cu–Cu bond lengths decrease significantly in the excited state as compared to those found in the ground state structure [63]. Besides, the emission of the complex has been found to be quenched by acetate and 4-aminobenzoate ions with bimolecular quenching rate constants of 1.65×10^8 and 5.10×10^8 dm^3 mol^{-1} s^{-1}, respectively. The same research group also reported a dinuclear copper(I) complex with a bridging acetate $[Cu_2(\mu\text{-dppm})_2(\mu\text{-O}_2CCH_3)]^+$ (**11**) [64]. Molecular orbital calculations suggest that the lowest lying excited state is MLCT [copper $\rightarrow$ phosphine/acetate] in nature.

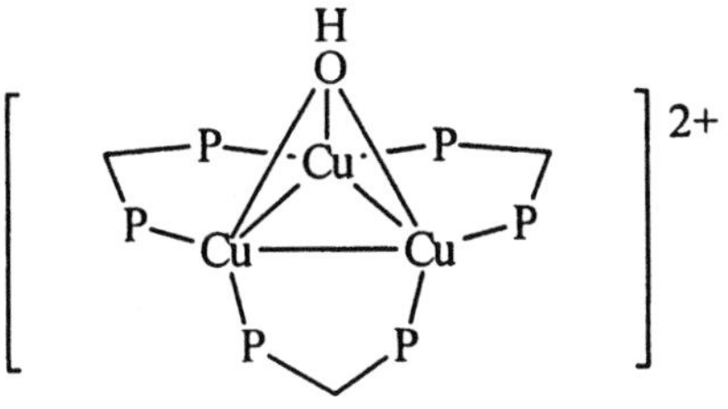

10

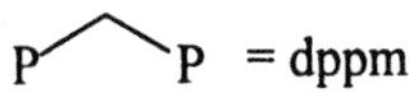

P⌒P = dppm

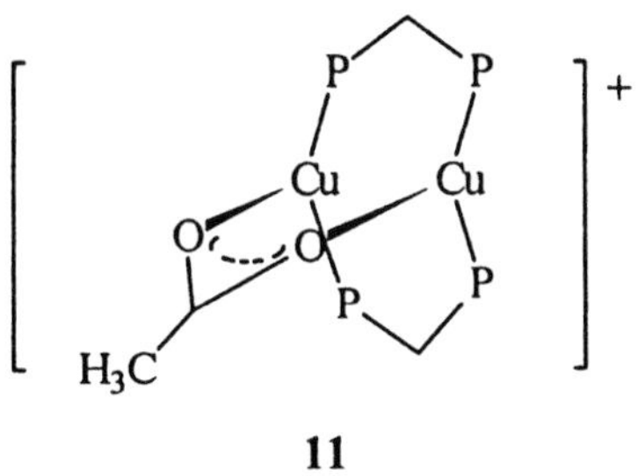

11

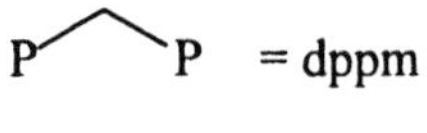

P⌒P = dppm

On the other hand, the first luminescent tetranuclear copper(I) sulfido complex $[Cu_4(\mu\text{-dppm})_4(\mu_4\text{-S})]^{2+}$ (**12a**) was reported by us in 1993 [65]. The X-ray crystal structure of the complex reveals a μ_4-bridging mode of the sulfido ligand. This complex, as well as the structural analogs $[Cu_4(\mu\text{-dppm})_4(\mu_4\text{-Se})]^{2+}$ (**12b**) [66] and $[Cu_4(\mu\text{-dtpm})_4(\mu_4\text{-S})]^{2+}$ (**12c**) [67], displays intense and long-lived orange luminescence in the solid state and in fluid solutions upon photoexcitation. The photophysical data are summarized in Table 4.

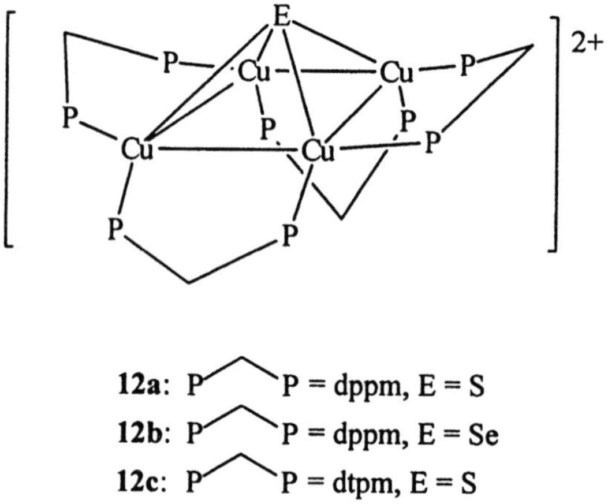

12a: P⌒P = dppm, E = S
12b: P⌒P = dppm, E = Se
12c: P⌒P = dtpm, E = S

Table 4 Photophysical data for **12a–c**

Complex	Medium (T K)	Emission wavelength λ_{max} nm (τ_0 μs)	Quantum yield Φ^a	Ref.
12a	Solid (298)	579 (3.6 ± 0.1)	0.22	65
	Solid (77)	606		
	Acetone (298)	622 (8.1 ± 0.2)		
	CH_3CN (298)	618 (7.8 ± 0.2)		
12b	Solid (298)	595 (3.9 ± 0.2)	0.19	66
	Solid (77)	619		
	Acetone (298)	626 (7.1 ± 0.2)		
	CH_3CN (298)	622 (6.9 ± 0.2)		
12c	Solid (298)	604 (3.5 ± 0.3)	0.26	67
	Solid (77)	658		
	Acetone (298)	622 (8.8 ± 0.4)		
	CH_3CN (298)	620 (7.7 ± 0.4)		

[a] In acetone solution.

The Cu–Cu distances, in the range of 2.869(1)–3.271(4) Å, are comparatively long to justify a pure d-s excited state derived from metal–metal interaction. Excited states of MLCT [$Cu_4 \to$ phosphine] and LLCT [$E^{2-} \to$ phosphine] are not likely owing to the observation that the presence of electron-donating methyl groups on the phenyl rings of the phosphine ligands in **12c** does not shift the emission energy of this complex to the blue compared with the dppm analog **12a**. Based on the strong σ-donating capability of chalcogenides, the excited state of these luminescent copper(I) complexes should bear a high parentage of ligand-to-metal charge transfer LMCT [$E^{2-} \to Cu_4$] triplet character, mixed with a metal-centered (ds/dp) state [65–70]. This assignment is also supported by ab initio molecular orbital calculations [71]. However, one should be aware that the assignments of electronic transitions between metal- and/or ligand-localized orbitals are merely oversimplified descriptions owing to the fact that the states are usually highly mixed in nature as a result of the possible extensive orbital mixing in these complexes.

The lowest energy phosphorescent states of this class of complexes have been found to undergo facile electron transfer reactions with a series of structurally related pyridinium acceptors. Excited state reduction potentials $E^0[Cu_4^{3+}/Cu_4^{2+*}]$ of -1.71 (10), $-1.55(10)$, and $-1.56(10)$ V vs. SSCE for **12a** [65], **12b** [66], and **12c** [72], respectively, have been estimated by a three-parameter, nonlinear, least-squares fit to the equation:

$$(RT/F)\ln k'_q = (RT/F)\ln K\kappa v - (\lambda/4)\,[1 + (\Delta G^{0\prime}/\lambda)]^2$$

where k'_q is the bimolecular quenching rate constant corrected for diffusional effects, $\Delta G^{0\prime} = E^0[Cu_4^{3+}/Cu_4^{2+*}] - E^0[Q^+/Q^0] + \omega_p - \omega_r$, ω_p and ω_r are the respective coulombic work terms to separate the products and to bring the reactants together, $K = k_D/k_{-D}$, κ is the transmission coefficient, v is the nuclear frequency, and λ is the reorganization energy for electron transfer.

The highly negative excited state reduction potentials suggest that the excited complex is strongly reducing. It is interesting to note that the reorganization energies associated with the electron transfers of these complexes are comparatively small compared with that of the tetranuclear copper(I) iodide cluster **4a**. For example, a reorganization energy of 1.12(10) eV has been estimated for **12b** [66].

The photoinduced electron transfer reactions have also been studied with nanosecond transient absorption spectroscopy [67]. The transient absorption difference spectrum for the reaction of **12a*** and 4-(methoxycarbonyl)-N-methylpyridinium is shown in Fig. 4. The difference spectrum is characterized by a sharp intense absorption at approximately 390 nm, a lower intensity band at 484 nm, and an intense broad absorption band at approximately 693 nm. The sharp band at around 390 nm is characteristic of pyridinyl radical absorption. The reaction mechanism is depicted in Scheme 1.

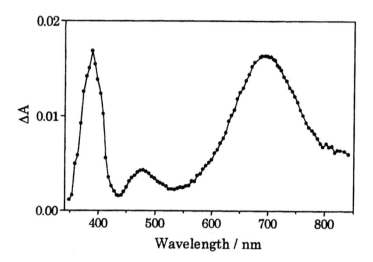

Figure 4 Transient absorption difference spectrum for the reaction of **12a*** and 4-(me-thoxycarbonyl)-*N*-methylpyridinium in degassed acetone (0.1 mol dm^{-3} ^{n}Bu$_4$NPF$_6$). (Adapted from Ref. [67].)

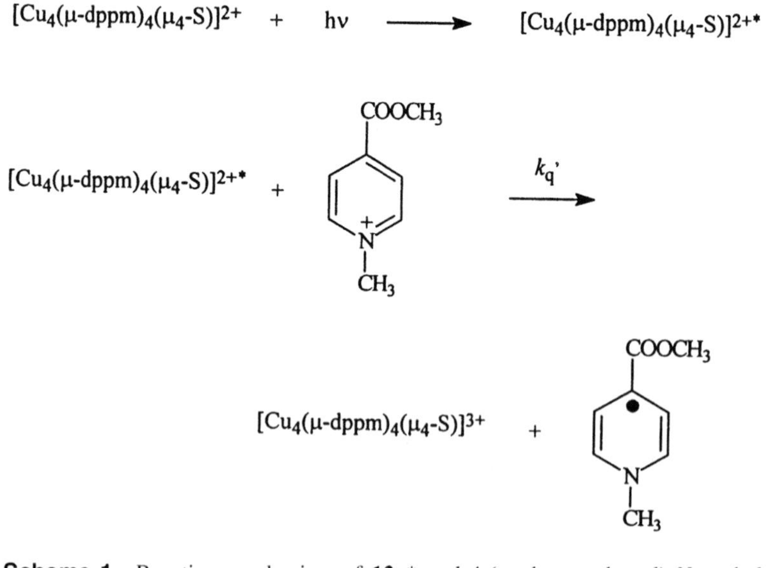

$[Cu_4(\mu\text{-dppm})_4(\mu_4\text{-S})]^{2+}$ $+$ hν $\longrightarrow$ $[Cu_4(\mu\text{-dppm})_4(\mu_4\text{-S})]^{2+*}$

$[Cu_4(\mu\text{-dppm})_4(\mu_4\text{-S})]^{2+*}$ $+$

$[Cu_4(\mu\text{-dppm})_4(\mu_4\text{-S})]^{3+}$ $+$

Scheme 1 Reaction mechanism of **12a*** and 4-(methoxycarbonyl)-*N*-methylpyridin-ium ion.

With a knowledge of the extinction coefficient of 4-(methoxycarbonyl)-*N*-methylpyridinyl radical, the extinction coefficients of the 484-nm and 693-nm bands are estimated to be 1200 and 6700 dm^3 mol^{-1} cm^{-1}, respectively. These absorption bands are characteristic of the one-electron oxidized form of the copper cluster. The possibility of ligand-field transitions of Cu(II) is ruled out owing to the high extinction coefficients of these absorption bands. The 484-nm band is assigned to an LMCT [$S^{2-} \rightarrow$ Cu(II)] transition which is commonly observed for mixed-valence Cu(I,II) thiolate and Cu(II) thioether systems [73]. The low-energy absorption is assigned to an intervalence transfer (IT) transition:

$$Cu(I)Cu(I)Cu(I)Cu(II) + h\nu \rightarrow Cu(I)Cu(I)Cu(II)Cu(I)^*$$

Similar IT transitions have also been observed in a variety of mixed-valence copper(I,II) systems [74–79]. Such a transition has also been observed for **12b**$^+$ and **12c**$^+$ at very similar energies. This indicates that the selenide and phosphine ligands have no significant effects on this metal-centered absorption and further supports the assignment of an IT transition. The occurrence of such absorption bands at fairly high energies has also been reported in dinuclear copper complexes [78,79], in which an assignment of a Cu–Cu centered [$\sigma \rightarrow \sigma^*$] transition which resulted from extensive electron delocalization across the two copper centers has been made. It is interesting to compare the features of the transient absorption difference spectra of these mixed-valence Cu(I,II) species with the electronic absorption spectrum of a class of copper proteins in biological systems, the Cu_A center [80–85]. The purple Cu_A center is composed of a mixed-valence dinuclear Cu_2S_2 structural core. The electronic absorption spectra of Cu_A centers are characterized by a pair of transitions at about 480–485 nm and 530–540 nm, together with a low-energy band at 770–808 nm. The two high-energy absorption bands have been assigned as [S(cysteine) $\rightarrow$ copper] LMCT whereas the low-energy transition is Cu–Cu centered [$\sigma \rightarrow \sigma^*$] [85].

Ford and Vogler reported the photophysical properties of a hexanuclear copper(I) thiolate cluster [$Cu_6(mtc)_6$] (**13**) and the related silver(I) complexes [86]. The copper complex shows an intense emission band at 706 nm ($\tau_0 = 0.99$ μs) and 767 nm ($\tau_0 = 14.0$ μs) in the solid state at 294 and 77 K, respectively. The emissive state has been suggested to derive from an admixture of LMCT [thiolate $\rightarrow$ Cu_6] and metal-centered d-s character.

13

On the other hand, the luminescence properties of a series of trinuclear and hexanuclear copper(I) arenethiolates [CuSC$_6$H$_4${(R)-CH(Me)NMe$_2$}-2]$_3$ (**14a**), [CuSC$_6$H$_4$(CH$_2$NMe$_2$)-2]$_3$ (**14b**), [Cu$_3${SC$_6$H$_4$[CH(Me)NMe$_2$]-2}$_2$(C≡C-^{t}Bu)]$_2$ (**14c**), and [Cu$_3${SC$_6$H$_4$(CH$_2$NMe$_2$)-2}$_2$(C≡C^{t}Bu)]$_2$ (**14d**) have been studied [87]. Complexes **14a** and **14d** show a single emission band at 555 and 530 nm, respectively. However, for **14b**, two emission maxima at 480 and 610 nm are observed. Similarly, two lower energy emission bands at 525 and 650 nm are also observed for **14c**. The optical transitions of these complexes are assigned to be LMCT in nature. The dual luminescence of **14b** and **14c** at low temperature has been suggested as a result of LMCT transitions involving different copper centers of the same complex. This also indicates that the excited state is not delocalized over the three copper(I) centers.

14a

Besides, a tetranuclear copper(I) cluster [Cu$_4${S$_2$P(O^{i}Pr)$_2$}$_4$] (**15**) with dialkyldithiophosphate as ligands has also been reported to be luminescent [88]. The solid-state emission of the complex occurs at 547 nm at 298 K. At 77 K, dual luminescence at 573 and 647 nm has been observed.

The spectroscopic and photophysical properties of a hexanuclear copper(I) complex [(CuPPh$_3$)$_6$L$_2$] (**16**) (H$_3$L = trithiocyanuric acid) have also been reported by Che and co-workers [89]. The UV-Vis absorption spectrum of the complex in CH$_2$Cl$_2$ shows a broad absorption band at 355 nm (ε = 2.03 × 10^4 dm^3 mol^{-1} cm^{-1}) with weaker absorption tails extending to 500 nm. The emission of the complex occurs at 562 nm in the solid state and 580 nm in CH$_2$Cl$_2$ solution. The origin of the luminescence has been suggested to be associated with MLCT [Cu → π*(L)] and/or intraligand [π → π*(L)] transitions of the stacked triazine moieties.

Ford and co-workers reported a series of tetranuclear copper(I) clusters containing tetradentate N$_2$S$_2$ or bidentate N,S Schiff base ligands [90]. The crystal

16

structure of one of these clusters, [Cu$_4$(dpit)$_2$] (**17**), reveals a copper–copper separation in the range of 2.77–2.96 Å. Upon photoexcitation, all of these complexes are luminescent in the solid state and in toluene solution. For example, **17** emits at 599 nm ($\tau_0 = 2.8$ μs) in the solid state and 626 nm ($\tau_0 = 180$ ns) in toluene at 298 K. The emission is suggested to originate from a cluster-centered mixed state of ds/dp and LMCT triplet. The assignment is similar to that for the related tetranuclear copper(I) iodide complexes. The excited state reduction potential for **17**, $E^0[\textbf{17}^+/\textbf{17}^*]$ has been estimated to be -1.84 V vs. Fc$^+$/Fc, indicative of the reducing behavior of the excited complex. A series of chromium(III) and aromatic substrates have been found to quench the emission of the complex. The reorganization energy for the electron transfer reactions has been determined to be 1.37 ± 0.1 eV, which is about 0.5 V smaller than that for **4a**. This is in line with a smaller Stokes shift for the emission of this cluster (1.26 μm^{-1}) compared with that of **4a** (1.64 μm^{-1}). These findings suggest that the excited state of **17** is less distorted from its ground state than that of **4a*** with respect to **4a** and this might be due to the effects of the chelating Schiff base ligands.

The photophysical properties of copper(I) aliphatic thiolate complexes [Cu(SAd)$_2$]$^-$ (**18a**) and [Cu$_5$(SAd)$_6$]$^-$ (**18b**) have also been studied [91]. The electronic absorptions at 253 nm for **18a** and 273 nm and 300 nm (sh) for **18b** have been suggested by the authors to be MLCT transitions. At 140 K, complexes **18a** and **18b** display emission bands at 600 and 618 nm, respectively.

D. Polypyridyl Complexes

The metal-to-ligand charge transfer photochemistry of copper(I) diimine complexes has been studied for 20 years. McMillin and co-workers first reported the photoredox chemistry of [Cu(dmp)$_2$]BF$_4$ in 1977 [92]. The same research group has also reported the photoluminescence properties of different copper(I) diimine complexes. One of the characteristics of this class of complexes is that Lewis bases and coordinating anions and solvents can quench the MLCT excited state of these complexes via an exciplex formation mechanism [93]. Recently, the interaction between copper(I) diimine complexes and DNA has also been investigated extensively [94].

However, compared to mononuclear species, the photophysical and photochemical properties of dinuclear or polynuclear copper(I) diimine complexes are relatively less well documented. The dinuclear copper(I) complex $[\{Cu(PPh_3)_2\}_2(bpm)]^{2+}$ (**19**) has been shown to be strongly luminescent in the solid state [95]. However, in fluid solution the emission behavior disappears. The solid-state emission of the complex has been suggested to originate from the π-stacking of one of the phenyl rings of the phosphines with the 2,2′-bipyrimidine ligand as revealed by X-ray crystallographic studies.

19

The luminescence properties of a 3-catenand (**20**) and related 3-catenates containing a copper(I) center [Cu · **20**]$^+$ were reported in 1991 [96]. The photophysical data are summarized in Table 5. It is interesting to note that while the mononuclear [Cu · **20**]$^+$ shows ligand-centered $^1\pi\pi^*$, $^3\pi\pi^*$, and ^{3}MLCT emission, the dinuclear complex [Cu$_2$ · **20**]$^{2+}$ only reveals a single MLCT phosphorescence band at 700 nm in CH$_2$Cl$_2$. The mixed-metal complex [CuCo^{II} · **20**]$^{3+}$ is not lumi-

Table 5 Photophysical Data for **20**, [Cu·**20**]$^+$ and [Cu$_2$·**20**]$^{2+}$

| Complex | Emission wavelength λ_{max} nm (τ_0) | | |
	LC ($^1\pi\pi^*$)[a]	LC ($^3\pi\pi^*$)[b]	^{3}MLCT[c]
20	401 (2.4 ns)	500 (1.5 s)	—
[Cu·**20**]$^+$	395 (2.4 ns)	501 (1.5 s)	695 (115 ns)
[Cu$_2$·**20**]$^{2+}$	—	—	700 (118 ns)

[a] Air-equilibrated 10^{-5} mol dm^{-3} CH$_2$Cl$_2$ solution at 300 K.
[b] EtOH/MeOH/CH$_2$Cl$_2$ (4:1:1 v/v) at 77 K.
[c] Air-equilibrated 10^{-4} mol dm^{-3} CH$_2$Cl$_2$ solution at 300 K.
Source: From Ref. [96].

nescent due to the presence of low-lying ligand-field (dd) excited states associated with the cobalt center.

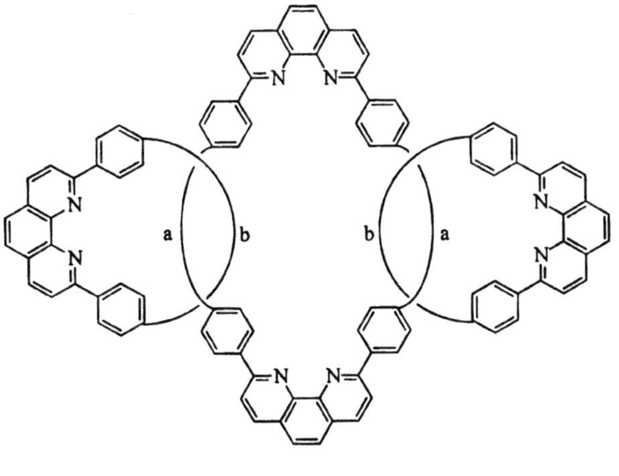

20

$$a = \quad -O-CH_2 \left(C \equiv C \right)_2 CH_2 - O -$$

$$b = \quad -O-CH_2CH_2 \left(OCH_2CH_2 \right)_4 O -$$

A series of dinuclear copper(I) complexes $[\{Cu(PPh_3)_2\}_2(L)]^{2+}$ [L = dpp (**21a**), dmdpq (**21b**), dpq (**21c**), dcdpq (**21d**), dpb (**21e**)] containing a bridging diimine ligand have been synthesized and characterized [97]. The complexes show intense absorption bands, typical of MLCT transition. Upon photoexcitation, the complexes display high-energy ligand-centered emission at around 400–500 nm and low-energy MLCT phosphorescence. The photophysical data are summarized in Table 6. The observed trend in MLCT triplet emission energy is in line with the π^* orbital levels of the bridging diimine ligands.

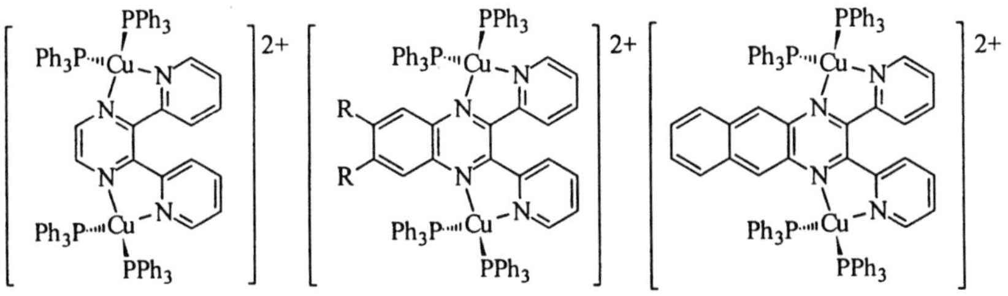

21a

21b: R = CH$_3$
21c: R = H
21d: R = Cl

21e

Table 6 Photophysical Data for **21a–e**

Complex	Medium (T K)	Emission wavelength λ_{max} nm
21a	Solid (298)	635
	Solid (77)	680
	CH$_2$Cl$_2$ (298)	650
21b	Solid (298)	670
	Solid (77)	705
	CH$_2$Cl$_2$ (298)	692
21c	Solid (298)	675
	Solid (77)	708
	CH$_2$Cl$_2$ (298)	708
21d	Solid (298)	732
	Solid (77)	710
	CH$_2$Cl$_2$ (298)	726
21e	Solid (298)	828
	Solid (77)	827
	CH$_2$Cl$_2$ (298)	770

Source: From Ref. [97].

The electronic structure and photoluminescence properties of the dinuclear copper(I) complex [Cu$_2$(μ-Ph-NNN-Ph)$_2$] (**22**) have also been reported [98]. X-ray crystal structure shows a short Cu–Cu separation of 2.45 Å [99]. The complex is fluorescent at 77 K with a structured emission band at around 600 nm. The lowest lying excited state has been assigned to be π-π*/MLCT in nature.

Recently, Zink and co-workers reported the structure and luminescence properties of a mixed-ligand copper(I) polymer [{(Ph$_3$P)$_2$Cu$_2$(μ-Cl)$_2$(μ-pyrazine)}$_n$] (**23**) [100]. The emission band occurs at around 612 nm at 20 K. The complex has also been studied by resonance Raman spectroscopy. It has been found that the largest distortions in the lowest lying triplet excited state occur along totally symmetric modes of the pyrazine ligand, suggesting that the excited state involves a charge transfer to the pyrazine ligand. The origin of the emission has been assigned to arise from an MLCT [Cu → π*(pyrazine)] excited state.

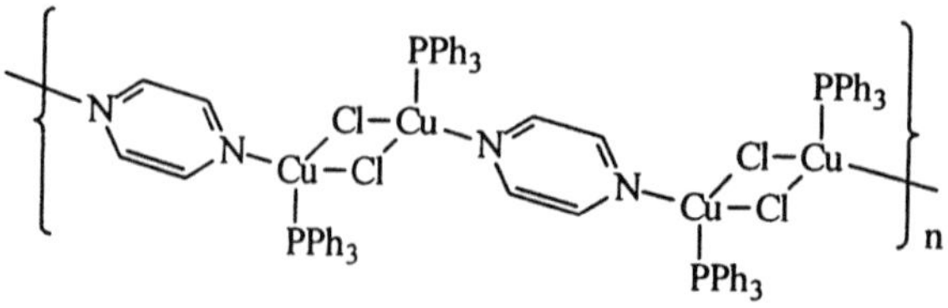

23

E. Organometallic Complexes

Rich and interesting photophysical and photochemical properties of a number of luminescent polynuclear copper(I) acetylide complexes have been reported by us recently [69,70,101–111]. The first series of these compounds is the trinuclear system with one or two μ_3-η^1-bridging acetylide ligands [Cu$_3$(μ-dppm)$_3$(μ_3-η^1-C$\equiv$C-R)$_n$]$^{(3-n)+}$ [n = 2, R = Ph (**24a**), ^{t}Bu (**24b**), C$_6$H$_4$-NO$_2$-4 (**24c**), C$_6$H$_4$-Ph-4 (**24d**), C$_6$H$_4$-OCH$_3$-4 (**24e**), C$_6$H$_4$-NH$_2$-4 (**24f**), ^{n}C$_6$H$_{13}$ (**24g**); n = 1, R = Ph (**25a**), ^{t}Bu (**25b**), C$_6$H$_4$-NO$_2$-4 (**25c**), C$_6$H$_4$-Ph-4 (**25d**), C$_6$H$_4$-OCH$_3$-4 (**25e**), C$_6$H$_4$-NH$_2$-4 (**25f**), ^{n}C$_6$H$_{13}$ (**25g**)] [101–103,111]. Besides, mixed-ligand complexes [Cu$_3$(μ-dppm)$_3$(μ_3-η^1-C$\equiv$C-^{t}Bu)(μ_3-Cl)]$^+$ (**26a**) [101], [Cu$_3$(μ-dppm)$_3$(μ_3-η^1-C$\equiv$C-C$_6$H$_4$-OCH$_3$-4)(μ_3-η^1-C$\equiv$C-C$_6$H$_4$-OCH$_2$CH$_3$-4)]$^+$ (**26b**) [110], and [Cu$_3$(μ-dppm)$_3$(μ_3-η^1-C$\equiv$C-C$_6$H$_4$-OCH$_3$-4)(μ_2-η^1-C$\equiv$C-C$_6$H$_4$-NO$_2$-4)]$^+$ (**26c**) [110] have also been synthesized and characterized crystallographically. In general, the UV-Vis absorption spectra of these complexes show absorption bands at approximately 252–268 nm and 292–328 nm. In view of the similarity in the absorption energies with those of the free dppm ligand and acetylenes, these absorption

24a: R = Ph	**25a**: R = Ph
24b: R = ^{t}Bu	**25b**: R = ^{t}Bu
24c: R = C$_6$H$_4$-NO$_2$-4	**25c**: R = C$_6$H$_4$-NO$_2$-4
24d: R = C$_6$H$_4$-Ph-4	**25d**: R = C$_6$H$_4$-Ph-4
24e: R = C$_6$H$_4$-OCH$_3$-4	**25e**: R = C$_6$H$_4$-OCH$_3$-4
24f: R = C$_6$H$_4$-NH$_2$-4	**25f**: R = C$_6$H$_4$-NH$_2$-4
24g: R = ^{n}C$_6$H$_{13}$	**25g**: R = ^{n}C$_6$H$_{13}$

P$\frown$P = dppm

bands are assigned to ligand-centered π-π*(dppm) and π-π*(acetylide) transitions, respectively. In addition, there are also absorption bands at around 332–404 nm. The energy of this absorption band is in line with the π* orbital energy of the acetylide. For example, the 4-nitrophenylacetylide mono-capped complex **25c** (396 nm) absorbs at lower energy than the 1-octynyl counterpart **25g** (332 nm) [111]. This observation suggests that this low-energy absorption band is a metal-perturbed ligand-centered π-π*(acetylide) and/or an MLCT [d(Cu) $\rightarrow$ π*(acetylide)] transition.

Excitation of the solid samples and fluid solutions of these complexes results in long-lived intense luminescence. The photophysical data of these complexes are listed in Table 7. Some complexes show vibronically structured emission bands (Table 7). Vibrational progressional spacings of around 1350–1600 cm^{-1} and 1800–2000 cm^{-1} have been observed, which are typical of ground state aromatic $\nu(C\mathrel{\vcenter{\hbox{$\cdots$}}}C)$ and acetylide $\nu(C\equiv C)$ stretching frequencies. This suggests the involvement of acetylides in the excited states of these complexes. In general, the complexes with electron-rich acetylides emit at a lower energy. For example, the emission energies of $[Cu_3(\mu\text{-dppm})_3(\mu_3\text{-}\eta^1\text{-}C\equiv C\text{-}R)]^{2+}$ in acetone solution occur in the order R = C_6H_4-OCH_3-4 (483 nm) $\approx$ R = Ph (499 nm) > R = C_6H_4-NH_2-4 (504, 564 nm) > R = ^tBu (640 nm) > R = $^nC_6H_{13}$ (650 nm) [103,111]. This is in line with the increasing donating ability of the acetylide ligand. Therefore, the origin of the emission has been proposed to involve substantial ^{3}LMCT [acetylide $\rightarrow$ Cu$_3$] character. However, in view of the short Cu–Cu distances found in the trinuclear copper(I) complexes, especially in the case of the bicapped acetylide species, a mixing of a metal-centered $3d^94s^1$ state into the lowest lying emissive state is likely. Besides, it has also been found that for the complexes with the same acetylide ligand, the monocapped species emit at a lower energy than the bicapped counterparts [101,103,111]. It is likely that the higher overall positive charge of the monocapped acetylide complexes relative to those of the bicapped species would stabilize the essentially metal-centered LUMO, leading to a lower lying LMCT emissive state. Speaking overall, the lowest lying emissive state could be best described as an admixture of LMCT triplet state and a metal-centered $3d^94s^1$ state modified by copper–copper interaction; the relative degrees of contribution of which depend on the nature of the acetylide ligand as well as the extent of metal–metal interaction. However, for the complexes containing acetylide ligands which are of a lesser degree of electron richness, such as 4-nitrophenylacetylide [110,111], the emission bands are very similar to those of the uncoordinated acetylene units. This is suggestive of an involvement of some ligand-centered π-π*(acetylide) character in the emissive state of these complexes.

On the other hand, a series of related trinuclear copper(I) complexes $[Cu_3\{\mu\text{-}(Ph_2P)_2N\text{-}R\}_3(\mu_3\text{-}\eta^1\text{-}C\equiv C\text{-}R')_2]^+$ [R = $CH_2CH_2CH_3$, R' = C_6H_4-OCH_2CH_3-4 (**27a**), R' = C_6H_4-Ph-4 (**27b**), R' = Ph (**27c**), R' = C_6H_4-NO_2-4 (**27d**); R = Ph, R' = C_6H_4-OCH_2CH_3-4 (**27e**), R' = C_6H_4-Ph-4 (**27f**), R' = Ph,

Table 7 Photophysical Data for **24a–g, 25a–g, 26a–c, 27a–j, 28,** and **29a–h**

Complex	Medium (T K)	Emission wavelength λ_{max} nm (τ_0 μs)	Ref.
24a	Solid (298)	493 (14 $\pm$ 1)	101
	Solid (77)	485, 525 sh	
	Acetone (298)	495 (5.9 $\pm$ 0.5)	103
24b	Solid (298)	450 (0.44 $\pm$ 0.05), 540 (1.7 $\pm$ 0.2)	101
	Solid (77)	450, 530 sh	
	Acetone (298)	444 (0.24 $\pm$ 0.02), 580 sh (16 $\pm$ 1)	103
24c	Solid (298)	420 (0.21)	111
	Solid (77)	630	
	Acetone (298)	438, 469 (1.5)	
24d	Solid (298)	527, 562 sh (5.7)	111
	Solid (77)	532, 574	
	Acetone (298)	530, 570 sh (16.3)	
24e	Solid (298)	450, 482 (63.8)	111
	Solid (77)	450, 478, 490 sh, 525	
	Acetone (298)	481 (11.5)	
24f	Solid (298)	402 (0.3)	111
	Solid (77)	485, 521 sh, 567 sh	
	CH$_2$Cl$_2$ (298)	476, 515 sh (4.0)	
24g	Solid (298)	444 (3.6)	111
	Solid (77)	455	
	Acetone (298)	455, 600 sh (1.0)	
25a	Solid (298)	500 (21 $\pm$ 2)	103
	Solid (77)	492, 530 sh	
	Acetone (298)	499 (6.8 $\pm$ 0.7)	
25b	Solid (298)	627 (14 $\pm$ 1)	103
	Solid (77)	450, 570 sh, 692	
	Acetone (298)	640 (2.6 $\pm$ 0.3)	
25c	Solid (298)	590 (128.0)	111
	Solid (77)	587, 640 sh	
	Acetone (298)	464, 654 (2.1)	
25d	Solid (298)	529, 572 sh (360.0)	111
	Solid (77)	530, 573 sh	
	Acetone (298)	533 (23.9)	
25e	Solid (298)	471, 523 sh, 585 (2.9)	111
	Solid (77)	495, 538 sh, 600 sh	
	Acetone (298)	483 (5.5)	
25f	Solid (298)	418 (3.5)	111
	Solid (77)	508	
	Acetone (298)	504, 564 sh (4.8)	
25g	Solid (298)	601 (24.4)	111
	Solid (77)	540 sh, 640	
	Acetone (298)	650 (1.54)	

Table 7 Continued

Complex	Medium (T K)	Emission wavelength λ_{max} nm (τ_0 µs)	Ref.
26a	Solid (298)	440 sh (<0.01), 535 (33 ± 3)	101
	Solid (77)	440, 572	
	CH₃CN (298)	540 sh (5.3 ± 0.5), 613 (5.4 ± 0.5)	
26b	Solid (298)	475, 502 (17.6)	110
	Solid (77)	467, 515, 569	
	Acetone (298)	479 (5.6)	
26c	Solid (298)	671 (0.2)	110
	Solid (77)	697	
	Acetone (298)	489 (3.6)	
27a	Solid (298)	459 (3.1)	108
	Solid (77)	453 sh, 482	
	Acetone (298)	467 (4.4)	
27b	Solid (298)	521, 558 sh, 615 sh (6.6)	108
	Solid (77)	522, 565, 615 sh	
	Acetone (298)	516, 554, 615 sh (26.1)	
27c	Solid (298)	461 (4.0)	108
	Solid (77)	459, 485, 495, 507	
	Acetone (298)	465 (3.4)	
27d	Solid (298)	400 (4.0)	108
	Solid (77)	400	
	Acetone (298)	454 (1.6)	
27e	Solid (298)	469 (0.3)	108
	Solid (77)	493	
	Acetone (298)	484, 646 (7.0)	
27f	Solid (298)	467, 516, 550 sh (5.5)	108
	Solid (77)	515, 560, 615 sh	
	Acetone (298)	516, 552, 615 sh (35.0)	
27g	Solid (298)	464, 550 sh (5.6)	108
	Solid (77)	462 sh, 487	
	Acetone (298)	461, 633 (4.8)	
27h	Solid (298)	418, 438 sh, 467 (1.2)	108
	Solid (77)	501	
	Acetone (298)	471, 670 (3.4)	
27i	Solid (298)	466, 550 sh (4.8)	108
	Solid (77)	491	
	Acetone (298)	464, 632 (2.5)	
27j	Solid (298)	471, 516, 545, 615 sh (6.3)	108
	Solid (77)	516, 556, 615 sh	
	Acetone (298)	470 sh, 568 (30.0)	

Table 7 Continued

Complex	Medium (T K)	Emission wavelength λ_{max} nm (τ_0 μs)	Ref.
28	Solid (298)	583 (222)	109
	Solid (77)	582	
	CH$_2$Cl$_2$ (298)	596 (40)	
	EtOH/MeOH (4:1 v/v) (77)	579	
29a	Solid (298)	548 (129.0)	111
	Solid (77)	479, 528, 582 sh	
	CH$_2$Cl$_2$ (298)	515, 668 (3.1)	
29b	Solid (298)	450, 480 (7.5, 36.0)	111
	Solid (77)	443, 480	
	CH$_2$Cl$_2$ (298)	672 (2.7)	
29c	Solid (298)	540 (1.3, 4.8)	111
	Solid (77)	537, 600 sh	
	CH$_2$Cl$_2$ (298)	508, 675 (3.1)	
29d	Solid (298)	400 (<0.1)	111
	Solid (77)	618, 675 sh	
	CH$_2$Cl$_2$ (298)	610 (<0.1), 665[a]	
29e	Solid (298)	483 sh (3.7 ± 0.3), 522 (3.7 ± 0.3)	104
	Solid (77)	477, 524	
	CH$_2$Cl$_2$ (298)	420, 520 sh (<0.01), 616 (3.6 ± 0.3)	
29f	Solid (298)	516 (1.3 ± 0.1)	104
	Solid (77)	516	
	CH$_2$Cl$_2$ (298)	420, 510 sh, 606 (0.86 ± 0.09)	
29g	Solid (298)	548 (0.52 ± 0.05)	104
	Solid (77)	535	
	CH$_2$Cl$_2$ (298)	410, 510 sh, 620	
29h	Solid (298)	529 (2.9 ± 0.3)	104
	Solid (77)	521	
	CH$_2$Cl$_2$ (298)	410, 670	

[a] Excitation wavelength = 530 nm.

(27g), R' = $^nC_6H_{13}$ (27h); R = C_6H_4-CH_3-4, R' = C_6H_4-OCH_2CH_3-4 (27i); R = C_6H_4-F-4, R' = C_6H_4-Ph-4 (27j)] have been synthesized and their photophysical properties studied [108]. The complexes displayed long-lived and intense luminescence upon photoexcitation. The photophysical data for these complexes are collected in Table 7. An assignment of an LMCT [acetylide → Cu_3] excited state mixed with copper-centered d-s state for the low-energy emission has been made based on the observed trend in the emission energies of the complexes in the order 27g > 27e > 27h, which is also in line with the increasing electron-donating abilities of the acetylides. The role played by the bridging phosphine ligands has also been studied. It has been found that the low-energy emission for the complexes with the same 4-ethoxyphenylacetylide but different phosphine ligands follows the order 27a > 27i > 27e. This observation is in agreement with the assignment of an excited state with substantial LMCT [acetylide → Cu_3] or LLCT [acetylide → π*(phosphine)] character. However, the possibility of an LLCT excited state is ruled out in view of the small changes in the energies along the series. On the other hand, for the complexes containing biphenylacetylide (27b, 27f, and 27j) the emission spectra reveal very similar vibronically structured bands. The vibrational progressional spacings of about 1400–1500 cm^{-1} are typical of the ν(C≡C) stretching frequency of the aromatic rings. Besides, the emission band of 27d at 77 K occurs at 671 nm which is similar to that for the 4-nitrophenylacetylide ligand. For these complexes with the relatively electron-deficient biphenyl- and 4-nitrophenylacetylide ligands, it is likely that an excited state with large intraligand $^3[\pi \to \pi$*(acetylide)] character is responsible for the low-energy emission.

A hexanuclear copper(I) complex (28) has been synthesized using 1,4-diethynylbenzene [109]. The complex contains two triangular $Cu_3(\mu$-dppm)$_3$ units bridged by an extended conjugated diynyl moiety. The absorption spectrum of complex 28 shows strong absorption bands in the UV region at about 258–324 nm which are assigned as ligand-centered transitions. Vibronically structured absorption bands have also been observed at around 370–416 nm. Progressional spacings of around 1475 to 1500 cm^{-1} which are typical of the ν(C≡C) stretching frequency of phenyl rings in the excited state are observed. Upon excitation with visible light, complex 28 shows long-lived intense orange-yellow emission. The photophysical data of the complex are summarized in Table 7. Both the excitation and emission spectra of complex 28 at 298 K and 77 K show vibronically structured bands with vibrational progressions typical of ν(C≡C) stretches of the aromatic ring, which are suggestive of the involvement of the arylacetylide ligand in the excited state behavior of the complex. The excited state of the complex has been suggested to bear a substantial ^{3}LMCT (acetylide → Cu) character, and probably mixed with a metal-centered d-s triplet state. However, owing to the highly structured emission bands and the significantly long lifetimes, an involvement of a ligand-centered π-π*(acetylide) excited state should not be overlooked.

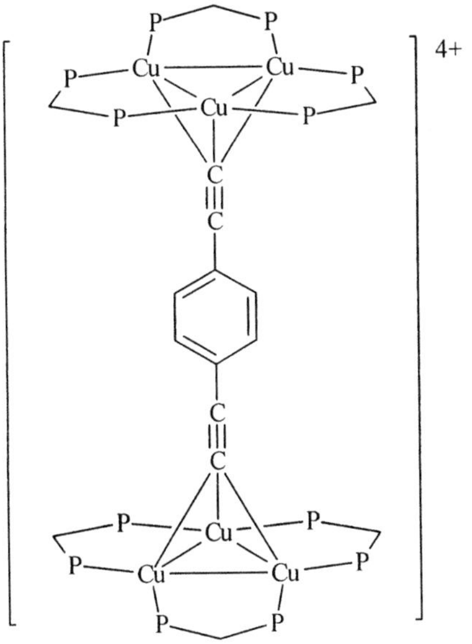

28

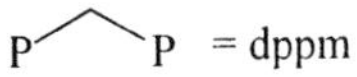

$P\frown P$ = dppm

Another series of copper(I) acetylide complexes that show intriguing luminescence properties are the tetranuclear species with the general formula
$[Cu_4(PR_3)_4(\mu_3-\eta^1-C\equiv C-R')_4]$ [R = Ph, R' = C_6H_4-CH_2CH_3-4 (**29a**), C_6H_4-
OCH_3-4 (**29b**), C_6H_4-Ph-4 (**29c**), C_6H_4-NO_2-4 (**29d**), Ph (**29e**); R = C_6H_4-F-4,
R' = Ph (**29f**); R = C_6H_4-CH_3-4, R' = Ph (**29g**); R = C_6H_4-OCH_3-4, R' = Ph
(**29h**)] [104,111]. The cubane structures of some of these complexes have been
revealed by X-ray crystallographic studies and are similar to the copper(I) halide
complexes with pyridine as ligands. Although the crystal structures of some of
the complexes are not available, it is reasonable to assume that the molecular
structures of this class of complexes are similar in view of their similar spectroscopic properties. The electronic absorption spectra of these tetranuclear cop-
per(I) acetylide complexes show a peak at around 260 nm, with a shoulder at
around 320 nm and long absorption tails extending to 500 nm. The highest energy
absorption bands are similar to those of the free phosphine ligands and are assigned as IL transitions. The complexes display intense luminescence upon photoexcitation. Vibronically structured emission bands are observed for some of
these complexes in the solid state and/or in low-temperature glass. Progressional
spacings of around 1360–1650 cm^{-1} and around 1740–2100 cm^{-1} are observed,
which are typical of ground state $\nu(C\cdots C)$ aromatic and $\nu(C\equiv C)$ acetylide
stretching frequencies, respectively. The photophysical data are collected in Table

7. The emission spectra in CH_2Cl_2 at 298 K of the complexes containing phenylacetylide as ligand are very similar and display dual luminescence at around 420 and 620 nm. Besides, a shoulder at around 520 nm on the high-energy side of the band at 620 nm is also observed. For the complexes with PPh_3 and different acetylides as ligands, similar broad and unstructured low-energy emission bands at about 665–675 nm are also observed. For **29c** and **29d**, the emission bands occur at very low energies and are comparable to those of the free acetylenes. Therefore, the emission of these two cubane-type complexes is assigned to arise from a ligand-centered π-π*(acetylide) emissive state. Comparing the low-energy emission bands of **29a**, **29b**, and **29e** in CH_2Cl_2, a small red shift in energy is observed on going from the phenylacetylide complex to the 4-alkoxyphenylacetylide counterparts. This is in line with the σ-donating properties of the acetylide ligands. Therefore, it is likely that the emissive state of these cubane-type complexes contains some LMCT [acetylide $\rightarrow$ Cu$_4$] character. However, in view of the short Cu–Cu distances [2.5092(5)–2.6636(8) Å] [104,111] observed in these complexes, as well as the relatively small dependence of the emission energies on the nature of the ligands, it is likely that the excited state for the low-energy emission should bear a very high parentage of copper-centered (d-s) character modified by metal–metal interaction delocalized over the cluster core.

29a: R = Ph, R' = C_6H_4-CH_2CH_3-4
29b: R = Ph, R' = C_6H_4-OCH_3-4
29c: R = Ph, R' = C_6H_4-Ph-4
29d: R = Ph, R' = C_6H_4-NO_2-4
29e: R = Ph, R' = Ph
29f: R = C_6H_4-F-4, R' = Ph
29g: R = C_6H_4-CH_3-4, R' = Ph
29h: R = C_6H_4-OCH_3-4, R' = Ph.

On the other hand, reaction of [Cu(CH$_3$CN)$_4$]PF$_6$ with PPh_3 and [{Au(C≡C-C_6H_4-OCH_3-4)}$_\infty$] in CH_2Cl_2 gave a novel complex [Cu$_4$(PPh$_3$)$_4$(μ_3-η^1,η^1,η^2-C≡C-C_6H_4-OCH_3-4)$_3$]$^+$ (**30**) which has been characterized crystallographically [105]. The structure of the complex can be described as a cubane with a vertex missing. The electronic absorption spectrum of the complex shows a high-energy shoulder at around 252 nm and a peak at around 330 nm. Excitation of the complex in the solid state and in fluid solutions resulted in long-lived intense luminescence. The solid sample shows an emission band at 445 nm and a shoulder at about 630 nm ($\tau_0 = 20.7 \pm 1$ μs) at 298 K. The complex emits at 675 nm in acetone ($\tau_0 = 4.0 \pm 0.4$ μs) and CH_2Cl_2 ($\tau_0 = 2.7 \pm 0.3$ μs). In view of the short Cu–Cu distance [2.446(2) Å] and the strong σ-donating properties of the 4-methoxyphenylacetylide moiety, the low-energy emission of the complex is assigned to originate from a ^{3}LMCT [acetylide $\rightarrow$ Cu$_4$] excited state mixed with a metal-centered d-s state modified by copper–copper interaction.

30

The unexpected formation of a novel tetranuclear copper(I) complex [Cu$_4$(μ-dppm)$_4$(μ_4-η^1,η^2-C≡C)]$^{2+}$ (31) has also been reported from our laboratory [106]. The complex was prepared by the reaction of [Cu$_2$(μ-dppm)$_2$(CH$_3$CN)$_2$]$^{2+}$ with trimethylsilylacetylene and n-butyllithium in tetrahydrofuran (THF). X-ray crystallographic studies reveal a novel acetylido ligand bridging the four copper(I) centers with both η^1 and η^2 bonding modes. The electronic absorption spectrum reveals a high-energy intraligand absorption band at 262 nm and a low-energy absorption at around 374 nm. Photoexcitation of the complex results in a strong and long-lived greenish yellow emission. The solid sample emits at 509 nm ($\tau_0 = 9.8$ μs) and 551 nm at 298 and 77 K, respectively. In acetone solution, the emission occurs at 562 nm ($\tau_0 = 16.0$ μs, $\Phi = 0.22$). The origin of the emission has been assigned to be predominantly ^{3}LMCT [(C≡C)$^{2-}$ $\rightarrow$ Cu$_4$] in nature in view of the strong σ-donating properties of the acetylide and the relatively long Cu–Cu contacts.

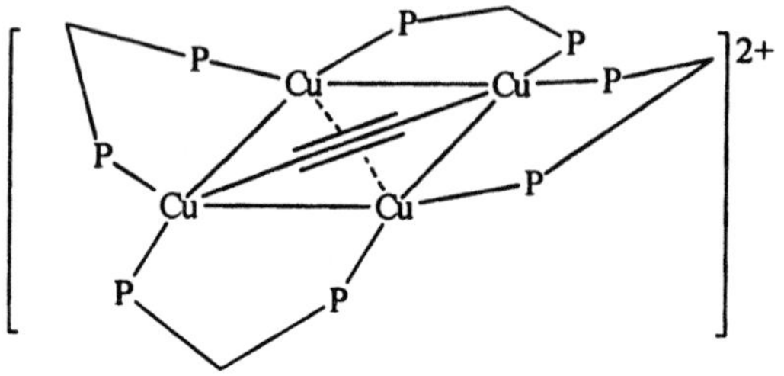

31

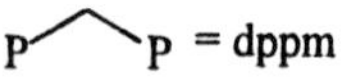

P⌒P = dppm

On the other hand, a dinuclear copper(I) acetylide complex [Cu$_2$(PPh$_2$CH$_3$)$_2$(μ,η^1-C≡C-Ph)$_2$] (**32**) has also been found to possess rich photophysical properties [107]. In CH$_2$Cl$_2$, the complex shows a strong absorption band at about 248 nm with a shoulder at about 315 nm, and a long absorption tail to lower energy. The complex is strongly luminescent in the solid state (at 298 K, λ_{max} = 467 and 509 nm, τ_0 = 87 ± 5 μs) and in fluid solutions (in CH$_2$Cl$_2$, λ_{max} = 529 and 660 nm). At 77 K, the complex shows a vibronically structured emission band with progressional spacings of about 1982 cm^{-1}, typical of ν(C≡C) stretch of the acetylide moiety as observed in the infrared spectrum. Based on the strong σ-donating properties of the acetylide units, together with the observation of a short Cu–Cu distance [2.454(1)Å], the excited state responsible for the long-lived emission is assigned to originate from a ^{3}LMCT [acetylide → Cu$_2$] state, probably mixed with a copper-centered (d-s) state that is modified by the copper–copper interaction in the dimer.

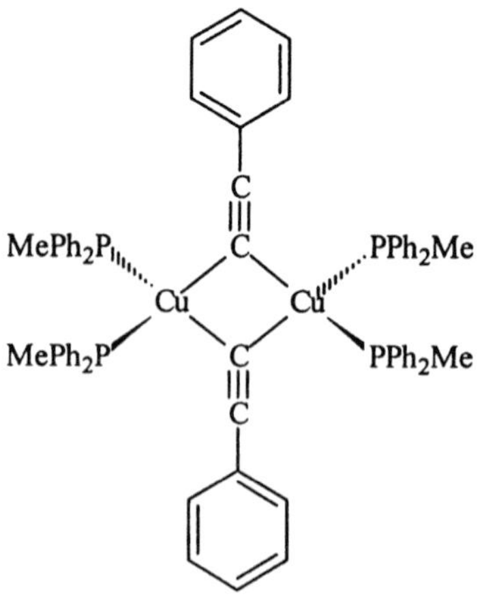

32

These luminescent polynuclear copper(I) acetylide complexes have been found to possess rich photoredox chemistry. Very often the excited complexes are powerful reducing agents. This can be illustrated by the photoreactivities of the phosphorescent states of **25b** [103], **30** [105], and **31** [106]. The emissions of the complexes are quenched in the presence of various structurally related pyridinium ions. The bimolecular quenching rate constants are collected in Table 8. The triplet state energies of the pyridinium ions are too high for any energy transfer reactions to occur. The dependence of the bimolecular quenching rate constants on the reduction potentials of the pyridinium quenchers is suggestive of a quenching mechanism that is oxidative electron transfer in nature. Excited state reduction potentials $E^0[Cu_4^{(n+1)+}/Cu_4^{n+*}]$ of -1.71 V ($n = 1$; $\lambda = 1.36$ eV) and -1.77 V ($n = 2$; $\lambda = 1.36$ eV) vs. SSCE have been estimated for **30*** [105] and **31*** [106], respectively. These highly negative values are indicative of the strong reducing power of these photoexcited complexes.

The photoinduced electron transfer behavior has also been confirmed by nanosecond transient absorption spectroscopy [102,103,108,111]. A representative transient absorption difference spectrum recorded 10 μs after laser flash excitation of **24a** (0.05 mM) and 4-(methoxycarbonyl)-N-methylpyridinium hexafluorophosphate (13 mM) in degassed acetonitrile (0.1 M $^nBu_4NPF_6$) is shown in

Table 8 Bimolecular Quenching Rate Constants for the Oxidative Quenching Reactions of **25b***, **30***, and **31*** with Pyridinium Acceptors

Quencher, Q	$E^0(Q^+/Q^0)/V$ vs. SSCE	Bimolecular quenching rate constant k'_q dm^3 mol^{-1} s^{-1}		
		25b*[a]	**30***[b]	**31***[c]
3,4-Dicyano-N-methylpyridinium	-0.13	6.89×10^8	—	4.05×10^9
2-Chloro-N-methyl-3-nitropyridinium	-0.39	5.18×10^7	—	7.78×10^9
4-Cyano-N-methylpyridinium	-0.67	1.67×10^7	8.80×10^9	5.23×10^9
4-Methoxycarbonyl-N-methylpyridinium	-0.78	$<10^6$	5.46×10^9	2.59×10^9
4-Aminoformyl-N-ethylpyridinium	-0.93	—	5.50×10^8	9.17×10^8
3-Aminoformyl-N-methylpyridinium	-1.14	—	1.55×10^8	9.56×10^7
N-Ethylpyridinium	-1.36	—	1.71×10^7	1.13×10^7
4-Methyl-N-methylpyridinium	-1.49	—	1.22×10^6	1.47×10^6

[a] From Ref. [103].
[b] From Ref. [105].
[c] From Ref. [106].

Fig. 5. The 400-nm absorption band is characteristic of the pyridinyl radical. The reaction involves the reduction of the pyridinium cation to the pyridinyl radical, while the **24a*** is oxidized to the mixed-valence species $[Cu(I)Cu(I)Cu(II)]^{2+}$.

The absorption at around 810 nm (ε = 9940 $dm^3\ mol^{-1}\ cm^{-1}$) has been suggested to arise from the intervalence-transfer transition of the mixed-valence Cu(I)Cu(I)Cu(II) species [102]:

$$Cu(I)Cu(I)Cu(II) + h\nu \rightarrow Cu(I)Cu(II)Cu(I)^*$$

Similar low-energy absorption bands have also been observed for other polynuclear copper(I) acetylide complexes such as **24b** [102], **24e** [111], **25b** [103], **26a** [102], **27c** [108], and **31** [106] with different pyridinium acceptors.

The luminescence properties of organocopper(I) complexes $[Cu_2\{2\text{-}C(SiCH_3)_2C_5H_4N\}_2]$ (**33**) and $[\{Cu[C_6H_2(CH_3)_3\text{-}2,4,6]\}_5]$ (**34**) have also been studied [107]. Solid sample of **33** emits at 520 nm (τ_0 = 12 ± 1 μs) at 298 K. In THF and acetone solutions at 298 K, the emissions of the complex occur at 534 nm (τ_0 = 6.0 ± 0.6 μs) and 515 nm (6.8 ± 0.6 μs), respectively. Complex **34** emits at 644 and 653 nm in the solid state at 298 and 77 K, respectively. The

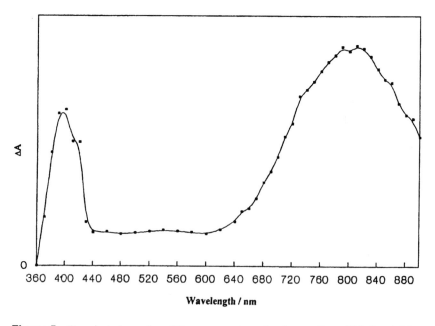

Figure 5 Transient absorption difference spectrum for the reaction of **24a*** and 4-(methoxycarbonyl)-*N*-methylpyridinium in degassed acetonitrile (0.1 mol dm^{-3} $^nBu_4NPF_6$). (Adapted from Ref. [102].)

excited states of these complexes have been suggested to originate from a mixed state of ^{3}LMCT [{C(SiCH$_3$)$_2$C$_5$H$_4$N}$^-$ → Cu$_2$ or mesityl$^-$ → Cu$_5$] and ^{3}MC (d-s) modified by copper–copper interaction in view of the short Cu–Cu distances of 2.412(1) and 2.437(8)–2.469(9) Å for **33** and **34**, respectively. The phosphorescent state of **33** has been found to react with a range of halogenocarbons.

33	**34**

On the other hand, a series of organometallic copper(I) polymers [{Cu(dmb)$_2$Y}$_n$] [Y = BF$_4^-$ (**35a**), NO$_3^-$ (**35b**), ClO$_4^-$ (**35c**)] have also been reported [112]. The solid state emission of **35a** at 77 K occurs at 490 nm. Density functional calculations on the monomeric model complex [Cu(CNH)$_4$]$^+$ reveal that the HOMO is largely composed of Cu(d$_{xy}$, d$_{yz}$, d$_{xz}$) and that the LUMO is π* (CNH) in character. The proposed excited state responsible for the emission is MLCT [Cu → π*(dmb)] in nature.

35a

III. SILVER(I) SYSTEMS

A. Phosphine Complexes

Reaction of $AgCF_3SO_3$ with two-thirds equivalent of dppp gave the trinuclear silver(I) complex $[Ag_3(\mu_3\text{-dppp})_2(CH_3CN)_2(ClO_4)_2]^+$ (36) [113]. X-ray crystallographic studies revealed two coordinating acetonitrile molecules as well as two coordinating perchlorate ions to the silver(I) centers. The two shortest silver–silver(I) distances are 2.943(2) and 3.014(2) Å, indicating some weak metal–metal interaction. The electronic absorption spectrum of the trinuclear silver(I) phosphine complex in CH_2Cl_2 reveals a strong absorption band at 288 nm ($\varepsilon = 2.53 \times 10^4$ dm^3 mol^{-1} cm^{-1}) which has been assigned as a $^1[d_{\sigma*} \rightarrow p_\sigma]$ transition. Upon photoexcitation, the complex emits at 467 nm in the solid state and in CH_2Cl_2 solution at room temperature. The long lifetime of the solid-state emission (11.2 µs) suggests that the emission originates from a phosphorescent state.

Besides, the synthesis and crystal structure of a trinuclear silver(I) phosphine complex $[Ag_3(\mu_3\text{-tppm})_2]^{3+}$ (37) have also been reported [114]. The three silver(I) centers are arranged in a triangular array. It was suggested that the Ag–Ag distances [3.1618(5)–3.2228(9) Å] are too long for any substantial metal–metal interaction to occur. The electronic absorption spectrum of the complex shows no significant absorption beyond 300 nm and is very similar to that of the free ligand. The absence of a low-energy $^1[d_{\sigma*} \rightarrow p_\sigma]$ transition that is commonly observed in polynuclear gold(I) metal complexes has been suggested.

37

The photophysical properties of another trinuclear silver(I) complex $[Ag_3(\mu_3\text{-dppnt})_3]^{3+}$ (38a) have also been studied [115]. The complex shows absorption bands at 258 and 367 nm in CH_2Cl_2. The emission spectrum of the complex in CH_3CN shows an emission band at 550 nm ($\tau_0 = 5.0$ µs) at room temperature. The phosphorescent state has been assigned to be ligand-centered in nature. The X-ray crystal structure of the dinuclear gold(I) counterpart shows that a potassium ion is encapsulated in the macrocyclic cavity, forming the complex $[Au_2K(\mu_3\text{-dppnt})_3]^{3+}$ (38b) [115].

38a

B. Halide Complexes

In 1989, Vogler and co-worker first reported the luminescence properties of poly-nuclear silver(I) complexes. The photophysical properties of tetranuclear silver(I) chloride complexes $[Ag_4Cl_4\{P(OCH_3)_3\}_4]$ (**39a**) and $[Ag_4Cl_4(PPh_3)_4]$ (**39b**) with a cubane structure have been studied [26]. The absorption spectrum of **39a** is featureless and reveals higher intensity absorption at shorter wavelengths. However, **39b** displays a strong absorption band at 249 nm ($\varepsilon = 20{,}000$ dm^3 mol^{-1} cm^{-1}). Both complexes emit at around 480–483 nm at 77 K in the solid state and in toluene glass. A qualitative molecular orbital (MO) diagram for the tetrahedral Ag^I_4 molecule is shown in Fig. 6. In the ground state, all of the d orbitals are filled. However, configuration mixing between 4d and 5s orbitals of the same symmetry results in some net stabilization of the d orbitals, leading to weak Ag–Ag bonding interaction. The emission of the two complexes has been assigned to originate from a silver-centered (d-s) excited state which is strongly modified by silver–silver interaction. The promotion of an electron from an antibonding d orbital to a bonding s orbital significantly increases the bond order between the silver atoms and leads to a strong contraction of the cluster in the excited state. This highly distorted excited state accounts for the large Stokes shift observed between the absorption and the emission band.

On the other hand, the luminescence properties of the cube and chair isomeric forms of $[Ag_4I_4(PPh_3)_4]$ (**40**) have also been reported by Henary and Zink [116]. Solid samples of the cube and chair forms emit at 455 and 418 nm, respectively, at 12 K. The emission has also been assigned to a d-s silver-centered state modified by metal–metal interaction. The lower emission energy of the cube form compared with the chair form can be accounted for by the higher delocalization over the d and s orbitals of the silver(I) centers for the cube isomer. For the cube form, each silver has three neighboring silver centers. However, for the

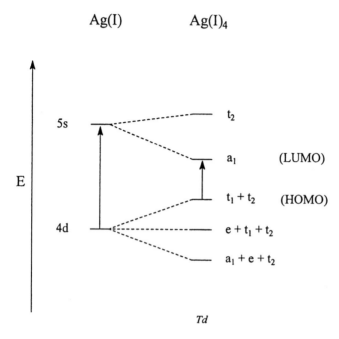

Ag(I)   Ag(I)₄

Figure 6 Qualitative MO diagram for the tetrahedral Ag_4^I molecule **39a, b**. (Adapted from Ref. [26].)

chair isomer, only two silvers have three neighboring silvers and the other two have only two. It is interesting to note that the shortest Ag–Ag distance of the cube form (3.12 Å) is slightly longer than that of the chair form (3.09 Å). This demonstrates that the cluster-centered excited state energy cannot be solely determined from the silver–silver distances in the ground state structure.

40

C. Chalcogenide and Chalcogenolate Complexes

A series of tetranuclear silver(I) chalcogenide complexes $[Ag_4(\mu\text{-dppm})_4(\mu_4\text{-E})]^{2+}$ [E = S (**41a**), Se (**41b**), Te (**41c**)] have been isolated and characterized in our laboratory [117]. The structures of these complexes are analogous to the tetranuclear copper(I) chalcogenides **12a–c** [68–70]. The similarity of the structures of these complexes enables comparison of the photophysical properties of this class of complexes. The UV-Vis absorption spectra of these tetranuclear silver(I) chalcogenides reveal a high-energy intraligand (dppm) absorption at around 246–256 nm and a lower energy absorption in the 350- to 460-nm region. Excitation of the complexes in the solid state and in fluid solutions results in long-lived green to orange luminescence. The photophysical data are summarized in Table 9. Similar to the related copper chalcogenide complexes, the emission of these silver(I) complexes has been assigned to be associated with a ^{3}LMCT [$E^{2-} \rightarrow Ag_4$], mixed with a metal-centered (ds/dp) triplet state. The solid-state emission energy of $[Ag_4(\mu\text{-dppm})_4(\mu_4\text{-E})]^{2+}$ follows the order **41a** (516 nm) > **41b** (527 nm) > **41c** (574 nm). The emission spectra are shown in Fig. 7. This energy trend is in line with the ionization energies of the chalcogens and lends further support to the excited state assignment of large LMCT character. Besides, the small blue shift of around 0.27 eV in emission energy from $[Cu_4(\mu\text{-dppm})_4(\mu_4\text{-E})]^{2+}$ to $[Ag_4(\mu\text{-dppm})_4(\mu_4\text{-E})]^{2+}$ (E = S, Se) disfavors the assignment of an MLCT state, for which the shift should be more significant.

Table 9   Photophysical Data for **41a–c**

Complex	Medium (T K)	Emission wavelength λ_{max} nm (τ_0 μs)	Quantum yield Φ^a
41a	Solid (298)	516 (1.0 ± 0.1)	0.014
	Solid (77)	536	
	Acetone (298)	628 (1.2 ± 0.1)	
	CH$_3$CN (298)	628 (1.5 ± 0.2)	
41b	Solid (298)	527 (0.9 ± 0.1)	<0.001
	Solid (77)	552	
	Acetone (298)	570 (1.3 ± 0.1)	
	CH$_3$CN (298)	572 (3.4 ± 0.3)	
41c	Solid (298)	574 (3.1 ± 0.2)	<0.001
	Solid (77)	588	
	Acetone (298)	615 (1.4 ± 0.1)	
	CH$_3$CN (298)	626 (3.3 ± 0.3)	

[a] In acetone solution (From Ref. [68]).
Source: From Ref. [117].

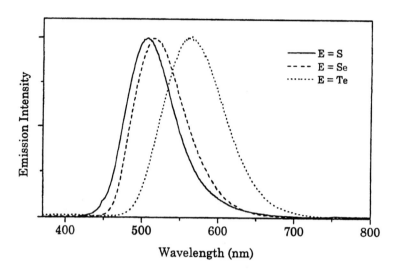

Figure 7 Solid-state emission spectra of **41a**, **41b**, and **41c** at 298 K. (Adapted from Ref. [69].)

The electronic structures of the model complexes [Ag$_4$(μ-H$_2$PCH$_2$PH$_2$)$_4$(μ_4-E)]$^{2+}$ [E = S (**42a**), Se (**42b**), Te (**42c**)] have been studied by Fenske-Hall [118] and ab initio [119] molecular orbital calculations. The MO correlation diagram from Fenske-Hall MO calculations for the model complex **42a** formed by the [Ag$_4$(μ-H$_2$PCH$_2$PH$_2$)$_4$]$^{2+}$ and S fragments are shown in Fig. 8. The bonding pictures between these two fragments for the three highest occupied molecular orbitals are displayed in Fig. 9. The calculation results clearly show that the HOMOs of all three model complexes are composed of Ag(4d)-E(p) bonding character while the LUMOs are almost entirely silver (5s, 5p) in character. The percentage compositions for the frontier orbitals of **42a–c** are collected in Table 10. Accordingly, the lowest lying excited state of this series of silver(I) chalcogenide complexes has been assigned as LMCT [E^{2-} → Ag$_4$]/metal-centered MC (ds/dp) in character. The observation of a decreasing HOMO–LUMO energy gap from the sulfido to the tellurido model complexes is also in accord with the emission energy trend of the complexes. Besides, ab initio MO calculations also give similar results.

Two hexanuclear silver(I) thiolate clusters [Ag$_6$(mtc)$_6$] (**43a**) and [Ag$_6$(dtc)$_6$] (**43b**) have also been found to be luminescent at 77 K [86]. The photophysical data are listed in Table 11. An assignment of IL π-π*(mtc or dtc) is not appropriate as the emission occurs at a low-energy region. Besides, the π* levels for these ligands are too high to justify an MLCT assignment for the lowest lying excited state. In addition, a blue shift of only approximately 0.31 eV in the solid-

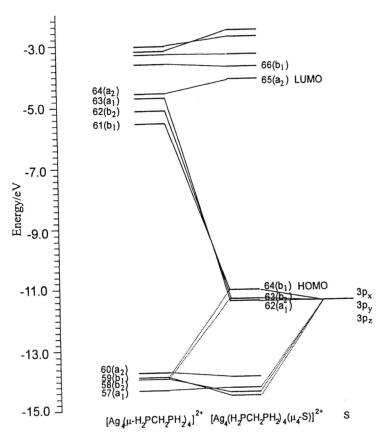

Figure 8 MO correlation diagram from Fenske-Hall MO calculations for the model complex **42a** formed by the $[Ag_4(\mu\text{-}H_2PCH_2PH_2)_4]^{2+}$ and S fragments. (Adapted from Ref. [118].)

state emission of the complexes $[Cu_6(mtc)_6]$ (**13**) to $[Ag_6(mtc)_6]$ also suggests that the excited state is not likely to be MLCT in nature. The excited state of these luminescent hexanuclear silver(I) thiolate complexes has been assigned to be LMCT $[mtc^-/dtc^- \rightarrow Ag_6]$ mixed with a cluster-centered Ag_6 d-s triplet state.

D. Organometallic Complexes

The photophysical properties of a series of trinuclear silver(I) acetylides $[Ag_3(\mu\text{-}P\text{-}P)_3(\mu_3\text{-}\eta^1\text{-}C\equiv C\text{-}R)]^{2+}$ [P-P = dppm, R = Ph (**44a**), $C_6H_4\text{-}OCH_3\text{-}4$ (**44b**), $C_6H_4\text{-}NO_2\text{-}4$ (**44c**); P-P = $(Ph_2P)_2N\text{-}CH_2CH_2CH_3$, R = Ph (**44d**)] [120] and $[Ag_3(\mu\text{-}$

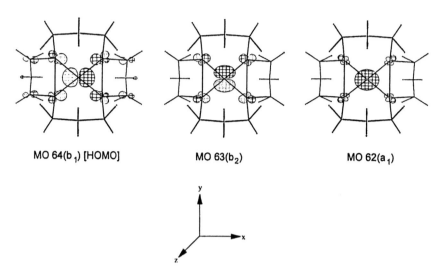

MO 64(b₁) [HOMO]   MO 63(b₂)   MO 62(a₁)

Figure 9 Bonding pictures between $[Ag_4(\mu\text{-}H_2PCH_2PH_2)_4]^{2+}$ and S fragments for the three highest occupied molecular orbitals in **42a**. (Adapted from Ref. [118].)

dppm)$_3$(μ_3-η^1-C≡C-R)$_2$]$^+$ [R = Ph (**44e**) [121], C_6H_4-NO$_2$-4 (**44f**)] [120] have been investigated. The complexes are structurally analogous to the trinuclear copper(I) acetylides. The Ag–Ag distances range from 2.866(2) to 3.4030(6) Å, comparable to the sum of van der Waals radii of silver (3.44 Å) [27]. The photophysical data of these luminescent silver(I) acetylide complexes are collected in Table 12. It has been found that the emission spectra of some of the complexes exhibit

Table 10 Energies and Percent Compositions for the Frontier Orbitals of **42a–c** from Fenske-Hall MO Calculations

Complex	Molecular orbital	Energy eV	E(S, Se, Te)	4Ag	4H$_2$PCH$_2$PH$_2$
			%		
42a	LUMO 65 (a₂)	−4.135	0.00	96.54	3.46
	HOMO 64 (b₁)	−11.080	46.12	42.00	11.88
42b	LUMO 65 (a₂)	−4.521	0.00	96.52	3.48
	HOMO 64 (b₁)	−10.754	50.16	39.26	10.58
42c	LUMO 65 (a₂)	−5.157	0.00	96.30	3.70
	HOMO 64 (b₁)	−10.545	51.64	38.40	9.96

Source: From Ref. [118].

Table 11 Photophysical Data for **43a–b** at 77 K

Complex	Medium	Emission wavelength λ_{max} nm (τ_0 μs)
43a	Solid	644 (<83, >109)
	Toluene	607 (131)
	EtOH	596 (156)
43b	Solid	545 (8.4)
	Toluene	550 (21.3)
	CH$_2$Cl$_2$	547 (13.8)
	EtOH/CH$_2$Cl$_2$	546 (12.9)

Source: From Ref. [86].

vibronic-structured bands at both 77 and 298 K. Progressional spacings of around 1880–2080 cm^{-1} have been observed and are attributed to the ground state stretching frequencies of the acetylide moieties. In some cases, additional vibrational progressions with spacings of around 1450–1600 cm^{-1} have also been observed, which are typical of the stretching frequencies of the aromatic rings in the ground state.

Table 12 Photophysical Data for **44a–f**

Complex	Medium (T K)	Emission wavelength λ_{max} nm (τ_0 μs)	Ref.
44a	Solid (298)	430, 449, 467, 488 sh (93.0)	120
	Solid (77)	428, 462, 469, 495, 515, 536 sh	
	EtOH/MeOH (4:1 v/v) (77)	432, 453, 470, 496 sh	
44b	Solid (298)	464 (5.2)	120
	Solid (77)	443, 464, 482, 508 sh	
	EtOH/MeOH (4:1 v/v) (77)	434, 452, 474, 494, 526 sh	
44c	Solid (298)	526, 546 sh, 615 sh (48.0)	120
	Solid (77)	535, 575	
	EtOH/MeOH (4:1 v/v) (77)	508, 550 sh, 600 sh	
44d	Solid (298)	512 (2.9)	120
	Solid (77)	553	
44e	Solid (298)	440 (0.43)	121
	CH$_2$Cl$_2$ (298)	440 (<0.05)	
44f	Solid (298)	570, 627 sh (<0.1)	120
	Solid (77)	532, 579	
	CHCl$_3$ (77)	534, 598 sh	

The 77 K solid-state emission bands of **44a** and **44b** occur at around 0.29 and 0.33 eV higher in energy than their copper(I) counterparts **25a** and **25e**, respectively. Such an observation rules out the possibility of an MLCT [silver → π^*(acetylide)] excited state in view of the large difference (1.19 eV) between the ionization energies of Cu$^+$(g) and Ag$^+$(g) [122]. Therefore, the emission has been assigned to a ^{3}LMCT [acetylide → Ag$_3$] state, and probably mixed with a silver-centered (ds/dp) triplet state. However, the mixing of a ^{3}IL π-π^* (acetylide) state is also possible, in view of the exceptionally long lifetime and the low-lying π^* orbitals of the less electron-rich acetylides.

A hexanuclear silver(I) acetylide complex [Ag$_3$(μ-dppm)$_3$(μ_3-η^1-C≡C-C$_6$H$_4$-C≡C-4)Ag$_3$(μ-dppm)$_3$]$^{4+}$ (**45**) has also been isolated and its photophysical properties studied [109]. Similar to the case of the copper(I) analog **28**, the UV-Vis absorption spectrum of complex **45** displays strong absorption bands in the UV region at around 258–324 nm, which are assigned as ligand-centered transitions. Vibronically structured absorption bands have also been observed at about 324–364 nm with progressional spacings of 1475–1500 cm^{-1}, which are typical of the v(C⋯C) stretching frequency of the phenyl rings in the excited state. Upon photoexcitation, the complex shows strong green luminescence. The photophysical data are summarized in Table 13. A blue shift in the solid-state emission energy at 77 K on going from the copper(I) complex **28** to the silver(I) complex **45** suggests that the excited states of the complexes should bear a substantial ^{3}LMCT [acetylide → Cu/Ag] character, probably mixed with a metal-centered d-s triplet state. However, an MLCT [Cu/Ag → π^*(acetylide)] would give a similar energy trend, although it is less likely in view of such a small shift (0.27 eV) in the emission energies. Besides, owing to the highly structured emission bands and the significantly long lifetimes, an involvement of a ligand-centered π-π^*(acetylide) excited state is also possible.

On the other hand, a series of silver(I) isocyanide polymeric materials [{Ag(dmb)$_2$Y}$_n$] [Y = BF$_4^-$ (**46a**), PF$_6^-$ (**46b**), NO$_3^-$ (**46c**), CH$_3$CO$_2^-$ (**46d**), ClO$_4^-$ (**46e**)] have been studied [112]. These polymers are emissive at 77 K with

Table 13 Photophysical Data for **45**

Medium (T K)	Emission wavelength λ_{max} nm (τ_0 μs)
Solid (298)	513 (351)
Solid (77)	515
CH$_2$Cl$_2$ (298)	515 (426)
EtOH/MeOH (4:1 v/v) (77)	510

Source: From Ref. [109].

a polyexponential lifetime. In the solid state and in EtOH glass, the emission wavelengths of the complexes are in the range of 467–492 nm and 435–502 nm, respectively. Molecular orbital calculations on the monomeric model complex $[Ag(CNH)_4]^+$ revealed a silver-based HOMO whereas the LUMOs are isocyanide (π^*) in nature. Similar to the analogous copper(I) polymer, $[\{Cu(dmb)_2BF_4\}_n]$ (**35a**) [112], the emission of these silver(I) isocyanide polymers has been suggested to be associated with an MLCT $[Ag \rightarrow \pi^*(dmb)]$ excited state.

However, for the dimeric silver(I) isocyanide complexes $[Ag_2(\mu\text{-}dmb)_2(\mu\text{-}X)_2]$ [X = Cl (**47a**), Br (**47b**), I (**47c**)], their photophysical properties are different [123]. The complexes display absorption bands at approximately 232, 241, and 242 nm ($\varepsilon = 1.8$–2.0×10^4 dm^3 mol^{-1} cm^{-1}) which have been assigned to $^1[d_{\sigma^*} \rightarrow p_\sigma]$ transitions. Lower energy absorption bands at about 260–295 nm ($\varepsilon = 7.8$–14.3×10^2 dm^3 mol^{-1} cm^{-1}) have been assigned to $[Ag_2(4d)/X(p) \rightarrow \pi^*(dmb)]$ transitions. The assignment of this lowest lying transition is supported by EHMO calculations that show that the LUMO are isocyanide-based (π^*) while the high-lying occupied MOs are composed of the interaction between the Ag_2 d_π, d_{π^*}, d_δ, and d_{δ^*} as well as the halide p_x, p_y, and p_z orbitals. Emission of the complexes (77K, ethanol glass) occurs at 450, 460, and 470 nm for **47a**, **47b**, and **47c**, respectively. The origin of the emission has been assigned to a $[Ag_2(4d)/X(p) \rightarrow \pi^*(dmb)]$ triplet excited state.

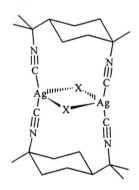

47a: X = Cl
47b: X = Br
47c: X = I

The interesting luminescence properties of $Eu[Ag(CN)_2]_3$ (**48a**) have also been reported by Patterson and co-workers [124]. Excitation of the host ion $[Ag(CN)_2]^-$ leads to the luminescence of the Eu^{3+} ion, indicative of a host-to-guest excited energy transfer. Besides, the luminescence properties of $Tl[Ag(CN)_2]$ (**48b**) have also been studied [125]. It has been found that the Ag–

Ag interaction in the excited state leads to the formation of luminescent exciplexes.

IV. GOLD(I) SYSTEMS

A. Phosphine Complexes

The luminescence properties of the dinuclear gold(I) complex $[Au_2(\mu\text{-dppm})_2]^{2+}$ (**49**) have been reported independently by Fackler [126] and Che [127]. The electronic absorption spectrum of the complex reveals absorption bands at 267 nm ($\varepsilon = 19,870$ dm^3 mol^{-1} cm^{-1}) and 292 nm ($\varepsilon = 29,120$ dm^3 mol^{-1} cm^{-1}). The latter has been assigned to the singlet–singlet $[d_{\sigma*} \to p_\sigma]$ transition. The $d_{\sigma*}$ and p_σ orbitals originate from the $5d_{z^2}$-$5d_{z^2}$ antibonding and $6p_z$-$6p_z$ bonding interaction between the gold(I) centers. The occurrence of the $[d_{\sigma*} \to p_\sigma]$ transition at a lower energy than the mononuclear species is in line with the short Au–Au distance of 2.931(1) Å observed in $[Au_2(\mu\text{-dppm})_2](BF_4)_2$ [128], indicative of the presence of some Au–Au interaction. The complex has been found to be strongly luminescent upon photoexcitation. The emission band occurs at 570 nm ($\tau_0 = 21 \pm 1$ µs) in acetonitrile at 298 K [129].

For a linear P-Au-P unit with a $D_{\infty h}$ symmetry, the relative energies of the occupied 5d orbitals of Au(I) are in the order d_{xz}, $d_{xy} < d_{z^2}$, $d_{yz} < d_{x^2-y^2}$ (z axis taken to be the Au–Au axis, y axis the P-Au-P axis). The lowest unoccupied orbital is $6p_z$ in nature. The formation of (P-Au-P)$_2$ (under D_{2h} symmetry) from two P-Au-P units is associated with the interaction between these two sets of orbitals. Interactions between the two d_{z^2} orbitals and two p_z orbitals result in the formation of d_σ, $d_{\sigma*}$, p_σ, and $p_{\sigma*}$, respectively. Besides, two sets of d_π and $d_{\pi*}$ orbitals are also generated by the interaction between the two d_{xz} and the two d_{yz} orbitals. Similarly, the interactions among the two d_{xy} and two $d_{x^2-y^2}$ orbitals also result in the formation of two sets of d_δ and $d_{\delta*}$ orbitals. Although the energy level of $d_{x^2-y^2}$ is higher than that of d_{z^2} in the P-Au-P building block, the interaction between two d_{z^2} is much more efficient and the splitting between d_σ and $d_{\sigma*}$ is larger than that between d_δ and $d_{\delta*}$ (from $d_{x^2-y^2}$). The nature of the HOMO for the complex **49** is controversial. For example, Mason and co-workers suggested that the HOMO of the complexes $[Au_2(\mu\text{-P-P})_2]^{2+}$ [130] and $[Au_2(\mu\text{-P-P})_3]^{2+}$ [131] is $d_{\sigma*}$ in nature. A similar assignment has also been suggested by an SCF-Xα-SW molecular orbital calculation on the model compound $[Au_2(\mu\text{-H}_2\text{PCH}_2\text{PH}_2)_2]^{2+}$ reported by Fackler and co-workers [126]. However, based on the large Stokes shift between the $^1[d_{\sigma*} \to p_\sigma]$ transition (290 nm) and the emission energy (570 nm), Che and co-workers suggested the lowest lying phosphorescent state of **49** to be $^3[(d_{\delta*})^1(p_\sigma)^1]$ [129]. In other words, the HOMO has been assigned to be $d_{\delta*}$ instead of $d_{\sigma*}$ in character. The absence of an intense absorption for the transition $^1[d_{\delta*} \to p_\sigma]$ in the electronic absorption spectrum has been attributed to symmetry reasons.

On the other hand, a series of dinuclear $[Au_2(\mu\text{-dmpm})_n]^{2+}$ ($n = 2$ (50a) [132], 3 (50b) [133]) and trinuclear $[Au_3(\mu_3\text{-dmmp})_n]^{3+}$ ($n = 2$ (51a) [132], 3 (51b) [133]) have also been reported. X-ray crystallographic studies reveal gold–gold separations of 3.028(2) Å for 50a [130], 3.040(1) and 3.050(1) Å for 50b [134], and 2.981(1) and 2.962(1) Å for 51a [132]. The electronic absorption spectra of 50a and 51a exhibit intense absorption bands at 269 and 315 nm, respectively. These absorption bands are assigned to spin-allowed $^1[d_{\sigma*} \to p_\sigma]$ transitions. A red shift (about 0.67 eV) in the transition energy from the dinuclear to trinuclear species is in line with the decreasing energy gap between the $d_{\sigma*}$ and p_σ orbitals upon addition of a P-Au-P unit. Similarly, the intense absorption bands at 258 and 301 nm for the corresponding three-coordinated species 50b and 51b, respectively, have also been assigned to $^1[d_{\sigma*} \to p_\sigma]$ transitions. The absorption energies also reveal a red shift of around 0.69 eV from the dinuclear to the trinuclear species. In addition, it is interesting to note that an additional phosphine ligand to 50a and 51a in 50b and 51b results in a blue shift in the $^1[d_{\sigma*} \to p_\sigma]$ transition energies of around 0.20 and 0.18 eV, respectively. This can be accounted for by the higher electron density on the gold centers for the three-coordinated species. This additional electron charge has been suggested to reduce the Au–Au interaction and increase the $d_{\sigma*}$ and p_σ energy gap. Besides, the more electron-rich gold centers in the three-coordinated species have also been thought to weaken the $Au[6p_{\sigma(z)}]-P(3d_{\pi*})$ interaction and therefore raises the p_σ orbital energy (z axis taken to be the Au–Au axis, y axis the P-Au-P axis; $P(3d_{\pi*})$ is the empty $3d_{yz}$ orbital of phosphorus in the two-coordinated system).

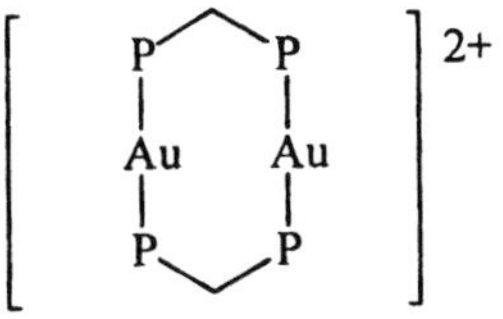

50a

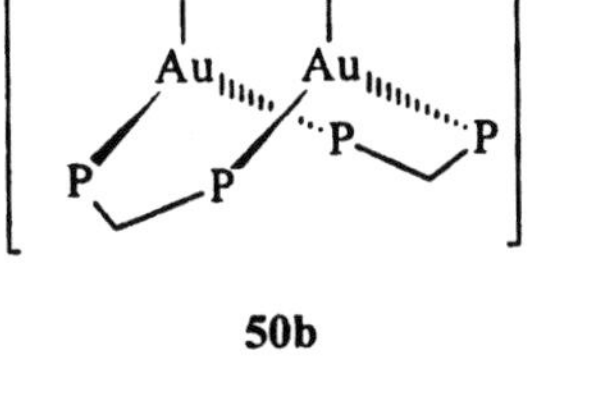

50b

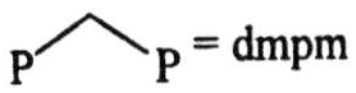

$P\overset{\frown}{\quad}P$ = dmpm

51a

51b

P⌒P⌒P = dmmp

Upon photoexcitation, all these four polynuclear gold(I) phosphine complexes are strongly luminescent. The two-coordinated complexes show dual luminescence in acetonitrile solution. High-energy emission bands at 455 nm (τ_0 = 1.2 ± 0.2 μs) for **50a** and 467 nm (τ_0 = 1.6 ± 0.2 μs) for **51a** are observed. Besides, the complexes also show long-lived emission bands at lower energy. The emission bands of **50a** (555 nm, τ_0 = 2.8 ± 0.2 μs) and **51a** (580 nm, τ_0 = 7.0 ± 0.5 μs) have been assigned to be derived from the $[(d_{\delta*})^1(p_\sigma)^1]$ triplet state. The assignment of a $^3[(d_{\sigma*})^1(p_\sigma)^1]$ state to the low-energy emission is unlikely in view of the large Stokes shifts between the absorption and emission bands. Besides, the comparatively small emission energy difference (about 0.10 eV) between these two complexes also suggests that the lowest lying phosphorescent state is unlikely to be $^3[(d_{\sigma*})^1(p_\sigma)^1]$.

The emission spectra of the three-coordinated complexes **50b** and **51b** in acetonitrile show a single emission band at 588 nm (τ_0 = 0.85 ± 0.10 μs) and 625 nm (τ_0 = 2.2 ± 0.2 μs), respectively. A relatively small red shift (0.13 eV) from the dinuclear to trinuclear complexes is also supportive of a $^3[(d_{\delta*})^1(p_\sigma)^1]$ excited state. The excitation spectra display a maximum at 370 and 390 nm for the dinuclear **50b** and trinuclear **51b** complexes, respectively. It is likely that these excitation maxima are origins of the $^3[(d_{\delta*})^1(p_\sigma)^1]$ emission.

These polynuclear gold(I) phosphine complexes have been found to possess rich photochemistry. The phosphorescence of **49** is found to be quenched by a series of energy acceptors such as *trans*- and *cis*-stilbene, styrene, hept-1-ene, and cyclohexene [129]. The transient absorption difference spectrum of a degassed acetonitrile solution of **49** and *trans*-stilbene displays absorption bands at about 365 and 390 nm, typical of the triplet excited state absorption spectrum of *trans*-stilbene, indicating the energy transfer nature of the quenching mechanism. Besides, the phosphorescent state of **49** has also been found to react with N,N,N',N'-tetramethylbenzene-1,4-diamine (TMPD) with a quenching rate constant of 6.6×10^9 dm^3 mol^{-1} s^{-1} [129]. The transient absorption difference spectrum reveals a peak at 600 nm, characteristic of the cation radical of TMPD. The quenching mechanism is likely to be:

$$[Au_2]^{2+} + h\nu \quad \rightarrow [Au_2]^{2+*}$$
$$[Au_2]^{2+*} + TMPD \rightarrow [Au_2]^+ + TMPD^{\cdot +}$$

However, it has been found that the decay trace for the transient signal of the TMPD$^{\cdot +}$ radical is very long, indicating some irreversibility of the photoinduced electron transfer reaction.

Besides, the photoexcited complex has also been found to react with a series of pyridinium acceptors such as MV^{2+} [129]. The electron transfer nature of the photoreaction mechanism has been established by the appearance of the characteristic MV$^{\cdot +}$ cation radical absorption in the transient absorption difference spectrum. The reaction has been shown to be reversible with a back-electron transfer rate constant of 1.5×10^9 dm^3 mol^{-1} s^{-1}. From the oxidative quenching experiments with a series of structurally related pyridinium acceptors, an excited state reduction potential of $E^0[Au_2^{3+}/Au_2^{2+*}]$ of $-1.6(1)$ V vs. SSCE [RT ln $K\kappa\nu = 0.58(10)$ V vs. SSCE, $\lambda = 0.90(10)$ eV] has been estimated by three-parameter, nonlinear, least-squares fits to the equation:

$$(RT/F)\ln k'_q = (RT/F)\ln K\kappa\nu - (\lambda/4) [1 + (\Delta G^{0\prime}/\lambda)]^2$$

The highly negative value of the excited state reduction potential indicates the strongly reducing characteristic of the excited state of **49**.

In addition, the phosphorescent state of **49** has also been found to react with a series of halogenocarbons [42]. The quenching rate constants follow the order methyl iodide > allyl bromide > chloroform, which indicates that the C-X bond energy governs the quenching of **49*** by the halogenocarbons instead of the reduction potentials $E^0(RX^0/RX^{\cdot -})$. Besides, transient absorption spectroscopic studies reveal that $[Au_2(\mu\text{-dppm})_2X]^{2+}$ is immediately formed at the very early stage of the photoreactions. Therefore, it is suggested that the quenching mechanism involves an initial inner-sphere halogen atom transfer step.

The phosphorescent states of the polynuclear gold(I) complexes **50a, b** and **51a, b** have also been found to react with MV^{2+} [132,133]. The quenching mechanism has been found to be electron transfer in nature. The excited state reduction potentials of these polynuclear gold(I) phosphine complexes have also been estimated by oxidative quenching experiments with a series of structurally related pyridinium acceptors. For example, the bimolecular quenching rate constants for the photoreactions between **50b*** and the pyridinium acceptors are listed in Table 14. A plot of $(RT/F)\ln k'_q$ vs. $E_{1/2}$ values of the pyridinium ions is displayed in Fig. 10. The excited state reduction potentials, RT ln $K\kappa\nu$ values,

Table 14 Bimolecular Quenching Rate Constants for the Oxidative Quenching Reactions of **50b*** with Pyridinium Acceptors

Quencher, Q	$E^0(Q^+/Q^0)$ V vs. SSCE	Bimolecular quenching rate constant k'_q dm^3 mol^{-1} s^{-1}
4-Cyano-N-methylpyridinium	−0.67	8.71×10^9
4-Methoxycarbonyl-N-methylpyridinium	−0.78	2.41×10^9
4-Aminoformyl-N-ethylpyridinium	−0.93	1.80×10^9
3-Aminoformyl-N-methylpyridinium	−1.14	3.17×10^8
N-Ethylpyridinium	−1.36	4.42×10^7
4-Methyl-N-methylpyridinium	−1.49	1.80×10^6

Source: From Ref. [133].

and the reorganization energies for these four polynuclear gold(I) phosphine complexes **50a, b**, and **51a, b** are collected in Table 15. The powerful reducing properties of the phosphorescent states of these complexes are revealed by their highly negative excited state reduction potentials.

Recently, the photolytic cleavage of double-stranded pBR322 plasmid DNA by **51a** has also been demonstrated by our group [135]. Irradiation of an aqueous solution of the complex and the plasmid DNA at $\lambda > 350$ nm in the presence of oxygen causes the photocleavage of the supercoiled form of pBR322 to give the linearized form. It has been found that the strongly reducing and long-lived excited complex **51a*** reacts with oxygen, leading to the formation of superoxide radical:

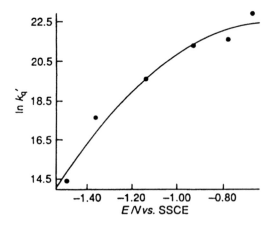

Figure 10 Plot of $(RT/F)\ln k'_q$ vs. $E_{1/2}$ values for the reactions of **50b*** with a series of structurally related pyridinium ions. (Adapted from Ref. [133].)

Table 15 Excited State Reduction Potentials, $RT \ln K\kappa v$ Values, and Reorganization Energies for the Electron Transfer Reactions of **50a–50b, 51a–51b** with Pyridinium Acceptors

Complex	$E^0[Au_m^{(n+1)}/Au_m^{n+}*]/V$ vs. SSCE ($m = 2, 3; n = 2, 3$)	$RT \ln K\kappa v$ V	λ eV	Ref.
50a	−1.7(1)	0.57(10)	0.95(10)	132
50b	−1.6(1)	0.57(10)	1.05(10)	133
51a	−1.6(1)	0.56(10)	0.90(10)	132
51b	−1.7(1)	0.50(10)	1.00(10)	133

$$[Au_3]^{3+}* + O_2 \rightarrow [Au_3]^{4+} + O_2^{\cdot-}$$

Addition of superoxide dismutase to the reaction mixture results in some inhibition of the photocleavage of DNA. This suggests that superoxide radical is responsible for the cleavage. From a series of quenching experiments, it has been found that singlet oxygen formed by the back electron transfer reaction is also responsible for the cleavage of the biomolecule.

$$[Au_3]^{4+} + O_2^{\cdot-} \rightarrow [Au_3]^{3+} + {}^1(O_2)$$

It is also interesting to note that the photocleavage of the plasmid DNA by **51a** can be made to produce either the linearized or the nicked form of the DNA under carefully controlled conditions.

The spectroscopy and photoluminescence properties of another trinuclear gold(I) phosphine complex $[Au_3(\mu_3\text{-dppp})_2]^{3+}$ (**52**) have also been studied [136]. X-ray crystal structure of the complex reveals Au–Au distances of 3.0137(8) and 3.0049(8) Å. The electronic absorption spectrum displays absorption bands at 326 nm ($\varepsilon = 3.16 \times 10^4$ dm^3 mol^{-1} cm^{-1}), which has been assigned to a $^1[d_{\sigma*} \rightarrow p_\sigma]$ transition. The occurrence of this transition at a lower energy than the dinuclear dppm analog **49** is in line with an additional P-Au-P unit and hence a narrowing of the $d_{\sigma*}$-p_σ energy gap. The complex exhibits an emission band at 600 nm ($\tau_0 = 3.7$ μs) with a shoulder at around 500 nm ($\tau_0 = 1.0$ μs). The high-energy emission was suggested to be intraligand phosphorescence. However, the origin of the lower energy emission has been assigned to a $^3[(d_{\delta*})^1(p_\sigma)^1]$ excited state in view of the large Stokes shift and the comparatively similar emission energy with the dinuclear dppm analog **49** (570 nm).

Besides, the triangular gold(I) phosphine complex $[Au_3(\mu_3\text{-tppm})_2Cl]^{2+}$ (**53**) has also been studied [114]. X-ray crystal studies reveal a coordinating chloride ion to one of the gold atoms. Gold–gold separations of 2.9220(8) and 3.0889(8) Å have been observed. Electronic absorption spectrum of the complex in acetonitrile shows intense absorption bands at 271 and 290 nm, which are assigned to $^1[d_{\sigma*} \rightarrow p_\sigma]$ transitions. In addition, there are also broad absorptions at around

300–450 nm, which have been suggested to be $^1[d_{\delta^*}, d_{\pi^*} \to p_\sigma]$ transitions. Upon photoexcitation, the complex in acetonitrile displays long-lived emission at 537 nm (11 μs). The origin of the emission has also been suggested to be derived from a $^3[(d_{\delta^*})^1(p_\sigma)^1]$ excited state. The phosphorescent state of the complex is also highly reducing, as indicated by the highly negative excited state reduction potential $\{E^0\ [Au_3^{3+}/Au_3^{2+*}] = -1.6(2)$ V vs. SSCE, $RT \ln K\kappa\nu = 0.57(10)$ V vs. SSCE, $\lambda = 0.95(10)$ eV$\}$, obtained from oxidative quenching experiments using a series of structurally related pyridinium ions as quenchers.

On the other hand, the spectroscopic and luminescence properties of $[Au_2(\mu\text{-dcpe})_2]^{2+}$ (**54a**) and $[(dcpe)_2Au_2(\mu\text{-dcpe})]^{2+}$ (**54b**) have also been studied by Gray and co-workers [137]. X-ray crystallographic studies reveal a Au–Au separation of 2.936 Å for **54a**, indicative of some Au–Au interaction. However, for **54b**, two of the dcpe ligands are chelating while the remaining bridges the two gold(I) centers resulting in a very long Au–Au separation of 7.0501 Å. The molecule can actually be described as two isolated AuP_3 distorted trigonal planar units. The electronic absorption spectrum of **54a** shows an intense band at 271 nm ($\varepsilon = 10,000$ dm^3 mol^{-1} cm^{-1}) and a weaker shoulder at 320 nm. These absorption bands are assigned to spin-allowed and spin-forbidden $[d_{\sigma^*} \to p_\sigma]$ transitions, respectively. The electronic absorption spectrum of **54b** shows a weak absorption band at 370 nm ($\varepsilon = 600$ dm^3 mol^{-1} cm^{-1}) which has been assigned to a $^1[(d_{xy}, d_{x^2-y^2}) \to p_z]$ transition. Upon photoexcitation, **54a** emits at 489 nm (solid state, 77 K). The emission of **54b** occurs at 501 nm in the solid state and 508 nm [$\tau_0 = 21.1(5)$ μs] in acetonitrile at room temperature. The large Stokes shift between the absorption and the emission of the latter complex implies a highly distorted excited state. This has been rationalized by the strengthening of the Au–P bond in the excited state as the $(d_{xy}, d_{x^2-y^2})$ orbitals are depopulated.

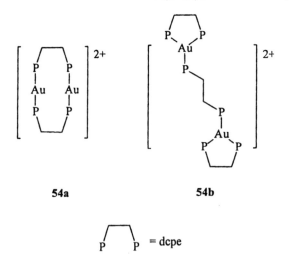

54a   **54b**

$\underset{P}{\diagup}\ \underset{P}{\diagdown}$ = dcpe

The photoredox properties of **54b** have also been studied by oxidative quenching experiments with a series of structurally related pyridinium acceptors [138]. The reorganization energy for the electron transfer reaction has been determined to be 1.66(7) eV. The outer-sphere contribution is estimated to be 0.75 eV. The remaining 0.91 eV suggests that there are substantial inner-sphere changes associated with the formation of Au(I)Au(II) species. This is in line with the observation of a large Stokes shift.

Another dinuclear gold(I) complex $[Au_2(\mu\text{-dpppy})_3]^{2+}$ (**55**) with no Au–Au contact has also been studied [139]. The long Au–Au separation (4.866 Å) suggests that there is no interaction between the two gold(I) centers. The electronic absorption spectrum of the complex is featureless, with a broad absorption band tailing from 250 to 360 nm ($\varepsilon = 8.03 \times 10^4$ dm^3 mol^{-1} cm^{-1} at 260 nm and 3.12 $\times 10^4$ dm^{3}mol^{-1}cm^{-1} at 300 nm). The low-energy absorption at about 300–360 nm has been assigned to a $^1[(d_{xy},d_{x^2-y^2}) \rightarrow p_z]$ transition with the p_z orbital mixed with some π^* character of the phosphine ligand. The emission spectrum of the complex shows a peak at 415 nm ($\tau_0 = 0.3$ μs) and a much more intense emission at 520 nm ($\tau_0 = 1.8$ μs). The higher energy emission has been suggested to probably arise from the metal-perturbed intraligand phosphorescence. However, for the lower energy emission its origin has been assigned to be $^3[(d_{xy}, d_{x^2-y^2})^1(p_z)^1]$ or intraligand in nature. The occurrence of the low-energy emission at a similar energy with the dinuclear Au(I) dppm analog **49** (570 nm) has further suggested that the emission of the polynuclear gold(I) phosphine complexes of this class does not originate from a $^3[(d_{\sigma^*})^1(p_\sigma)^1]$ state.

55

B. Halide Complexes

In 1988, Vogler reported the spectroscopic and luminescence properties of two tetranuclear gold(I) complexes $[Au_4Cl_4(piperidine)_4]$ (**56a**) and $[Au_4(dta)_4]$ (**56b**)

[25]. The four gold(I) centers in these two complexes are arranged in a square array. The piperidine complex **56a** exhibits an absorption band at 305 nm (ε = 770 dm^3 mol^{-1} cm^{-1}) in ethanol. The electronic absorption spectrum of **56b** in CS$_2$ displays an absorption band at 407 nm (ε = 1650 dm^3 mol^{-1} cm^{-1}) and a shoulder at 430 nm (ε = 1400 dm^3 mol^{-1} cm^{-1}). Both complexes are emissive at 77 K in ethanol glass. The emission bands occur at 700 and 743 nm for **56a** and **56b**, respectively. A qualitative molecular orbital diagram for the tetranuclear gold(I) moiety with a D$_{4h}$ symmetry is outlined in Fig. 11. Each gold atom participates with one 5d and one 6s orbital in σ interaction with its closest neighbors. Configuration mixing between the e$_u$ orbitals derived from the filled 5d orbitals and those derived from the empty 6s orbitals results in some metal–metal bonding character. The lowest energy absorption bands observed for these two complexes are assigned to the a$_{2g}$ → a$_{1g}$ transition. The small extinction coefficients are in line with the symmetry-forbidden nature of this transition. Besides, the red shift

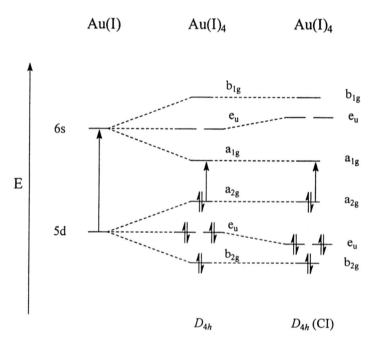

Figure 11 Qualitative molecular orbital diagram for the tetranuclear gold(I) moiety with a D$_{4h}$ symmetry. (Adapted from Ref. [25].)

in the transition energies from **56a** to **56b** can be explained by a shorter metal–metal distance in **56b**, and hence a stronger Au–Au interaction occurring in the latter complex. The emission of the complexes has also been assigned to be derived from a $^3[(a_{2g})^1(a_{1g})^1]$ state.

The photoluminescence behavior of [(TPA)AuCl] (**57a**) and [(TPA-HCl)AuCl] (**57b**) has been reported by Fackler and Schmidbaur [140]. Owing to the small cone angle (102°) of TPA, complex association arisen from gold–gold interaction is possible. In fact, X-ray crystal structures of **57a** and the protonated analog **57b** indicate a dimeric structure with intermolecular Au–Au distances being 3.092(1) and 3.322(1) Å, respectively. These two complexes are luminescent in the solid state at low temperature but not in solution. At 78 K, the solid samples of **57a** and **57b** emit at 674 and 596 nm, respectively. The photoluminescence properties have been suggested to be attributable to the Au–Au interaction. The shorter Au–Au distance observed in **57a** is in line with a smaller HOMO-LUMO gap (d–p) and hence a lower emission energy. EHMO calculations also support the assignment.

57a: X = Cl
58a: X = Br
59a: X = I

The luminescence properties of related complexes [(TPA)AuBr] (**58a**), [(TPA-HBr)AuBr] (**58b**), [(TPA)AuI] (**59a**), and [(TPAH)AuI][AuI₂] (**59b**) have also been reported [141]. The deprotonated complexes **58a** and **59a** show low-energy emission at about 600–700 nm. The solid-state emission energy at 77 K follows the order **57a** (674 nm) < **58a** (647 nm) < **59a** (617 nm). Besides, similar to the case of the chloro analogue, the deprotonated complexes **58a** and **59a** emit at a lower energy than their protonated counterparts **58b** and **59b**, respectively. This low-energy emission has been assigned to originate from a metal-centered triplet state, modified by the Au–Au interaction. In addition, **58a** and **59a** also exhibit a higher energy, structured emission band at around 450–491 nm. This short-lived high-energy fluorescence ($\tau_0 < 10$ ns) has been suggested to originate from an LMCT [halide → gold] singlet excited state.

Balch and Tinti reported the structures and the luminescence properties of the polynuclear gold(I) halide complexes [{(Me$_2$PhP)AuCl}$_3$] (**60**) and [{Me$_2$Ph-P)AuX}$_2$], [X = Cl (**61a**), Br (**61b**), I (**61c**)] [142,143]. X-ray crystal structures show that the Au–Au distances range from 3.091(2) to 3.230(2) Å in these dimeric and trimeric gold(I) halide complexes, indicative of some gold–gold interaction. The solid-state luminescence behavior of these complexes has also been studied. At low temperature, the emission spectra of these complexes show a high-energy emission peak at approximately 360 nm, which has been assigned to intraligand phosphorescence. Besides, there is also a lower energy emission band at about 630–730 nm. The origin of this emission band has been suggested to be derived from a $^3[(d_{\sigma^*})^1(p_\sigma)^1]$ state. The assignment is in agreement with the SCF-Xα-SW calculations carried out on the model complexes [H$_3$PAuX] and [(H$_3$PAuX)$_2$] [143]. However, it has also been found that while the LUMO in the dimeric model complexes are mainly Au(p) character, the Au character in the HOMO decreases from 78% to 36% and the halogen-p character increases from 10% to 58% from X = Cl to I. This suggests that the excited states exhibit increasing LMCT [halide → Au] character upon going from the chloro to the iodo complexes.

Che and co-workers reported the photoluminescence properties of a series of polynuclear gold(I) halide complexes [Au$_2$(μ-dppm)X$_2$] [X = Cl (**62a**), I (**62b**)], [Au$_3$(μ$_3$-tppm)X$_3$] [X = Cl (**63a**), I (**63b**)], and [Au$_3$(μ$_3$-dppp)X$_2$]$^+$ [X = Cl (**64a**), I (**64b**)] with various bridging phosphine ligands [144]. While [Au(PPh$_3$)I] shows no appreciable absorptions in the 300- to 400-nm region, the trinuclear halide complexes (all of them with Au–Au separation < 3.3 Å) exhibit significant absorption bands at wavelengths >300 nm. These absorptions are assigned as $^1[d_{\sigma^*} \rightarrow p_\sigma]$ transitions in view of the short Au–Au distances. All of the complexes are emissive in the solid state and in CH$_2$Cl$_2$ solution. The photophysical data of the complexes are collected in Table 16. The solid samples show dual luminescence at low temperature with the high-energy and low-energy emission bands at around 480–530 nm and 570–700 nm, respectively. The high-energy emission has been assigned to originate from an intraligand and MLCT [Au → π*(phosphine)] mixed state. For the low-energy emission, the emission energy shows a small red shift from the chloride to the iodide analogs. It has been suggested that for the chloride complexes, the origin of the low-energy emission is metal-metal-to-ligand charge transfer ^{3}MMLCT [gold → π*(phosphine)]. However, for the iodide complexes, substantial ^{3}LLCT [iodide → π*(phosphine)] character is also present in the excited state. The assignment is supported by EHMO calculations. It has been proposed that the MMLCT/LLCT triplet state can be thermally populated from the high-energy state by overcoming the crossing barrier, which is estimated to be 945 ± 45 cm^{-1} for **62b**.

Table 16 Electronic Absorption Spectral Data for
62a–b, 63a–b, 64a–b in CH_2Cl_2

Complex	Medium (*T* K)	Emission wavelength λ_{max} nm (τ_0 µs)
62a	Solid (270)	510, 680
	Solid (12)	500
	CH_2Cl_2 (298)	585 (0.5)
62b	Solid (298)	670
	Solid (12)	530, 620 sh
	CH_2Cl_2 (298)	615 (1.1)
63a	Solid (298)	640
	Solid (20)	500
	CH_2Cl_2 (298)	550 (1.1)
63b	Solid (298)	680
	Solid (12)	480, 680
	CH_2Cl_2 (298)	590 (1.2)
64a	Solid (77)	485, 620
	CH_2Cl_2 (298)	650 (1.2)
64b	Solid (12)	575, 665
	CH_2Cl_2 (298)	660 (2.5)

Source: From Ref. [144].

The tetranuclear gold(I) chloride complex [(AuCl)$_4$(dptact)] (**65a**) containing a macrocyclic ligand dptact has also been reported [145]. The tetranuclear units are held together by weak Au–Au interaction [d_{Au-Au} = 3.104(1) Å] forming a two-dimensional polymer. Although the X-ray crystal structures for [(AuBr)$_4$ (dptact)] (**65b**) and [(AuI)$_4$(dptact)] (**65c**) were not available, it was anticipated that they should have similar structures. Besides, it has been suggested that the Au–Au interaction would increase in the order Cl < Br < I based on the theoretical studies on [{(H$_3$P)AuX}$_2$] reported by Pyykkö and co-workers [146]. The polymers **65a–c** exhibit dual luminescence at 77 K. The high-energy emission bands at around 470–530 nm have been assigned to be derived from an intraligand/MLCT [Au → π*(phosphine)] triplet state. A ligand-to-metal charge transfer ^{3}LMCT [X$^-$ → Au] state mixed with a metal-centered ^{3}MC(ds/dp) state has been assigned to be the origin of the low-energy emission at around 600–700 nm. The low-energy emission of the bromide complex (600 nm) occurs at a higher energy than that of the iodide complex (700 nm). This has been rationalized on the grounds that the Au–Au interaction is stronger in the iodide complex and therefore a narrower HOMO-LUMO energy gap results.

65a: X = Cl
65b: X = Br
65c: X = I

C. Thiolate Complexes

Fackler and co-workers reported the luminescence properties of $[Au_2(PPh_3)_2(\mu\text{-}mnt)]$ (**66a**) in 1988 [147]. The complex exhibits an emission band at around 525 nm with a biexponential decay ($\tau_0 = 33$ and 286 μs) at 77 K in CH_3CN. The related complex $[Au_2(\mu\text{-}mnt)_2]^{2-}$ (**66b**) with a Au–Au distance of 2.78 Å has been found to emit at around 510 nm ($\tau_0 = 1.9$ and 40 μs) under similar conditions.

The photophysical properties of a series of mononuclear gold(I) thiolate complexes [L-Au-SR] with phosphine ligands [L = PPh_3, R = Ph (**67a**), C_6H_4-OCH_3-2 (**67b**), C_6H_4-Cl-2 (**67c**); L = TPA, R = Ph (**67d**), C_6H_4-OCH_3-2 (**67e**), C_6H_4-OCH_3-3 (**67f**), C_6H_4-Cl-2 (**67g**), C_6H_4-Cl-3 (**67h**), C_6H_4-Cl-4 (**67i**), $C_6H_3Cl_2$-3,5 (**67j**)] have also been reported [148]. The solid-state emission of these complexes at 77 K occurs in a wide energy range (413–702 nm). The origin of the emission has been ascribed to an LMCT [$RS^- \rightarrow Au$] triplet state. The emission energies of the complexes depend on two factors: the substituent on the phenyl ring of the thiolate ligand, and the extent of the Au–Au interaction, if present. An electron-donating substituent on the thiolate ligand or the presence of Au–Au interaction would cause a lower energy emission.

A series of dinuclear gold(I) thiolate complexes $[Au_2\{\mu\text{-}Ph_2P(CH_2)_nPPh_2\}(S\text{-}C_6H_4\text{-}CH_3\text{-}4)_2]$ [$n = 1–5$ (**68a–e**)] and $[Au_2\{\mu\text{-}Ph_2P(CH_2)_nPPh_2\}\{\mu\text{-}S\text{-}(CH_2)_3\text{-}S)\}]$ [$n = 2 - 5$ (**69a–d**)] have also been studied [149,150]. The low-energy absorption shoulders at approximately 303–357 nm have been suggested to be LMCT transitions $^1[RS^- \rightarrow Au]$. It has been suggested that some Au–Au interaction occurs in the complex $[Au_2(\mu\text{-}dppm)(S\text{-}C_6H_4\text{-}CH_3\text{-}4)_2]$, which gives rise to a lower energy absorption. The emission of these complexes occurs at approximately 485–515 nm, which has been suggested to originate from an LMCT [RS$^-$ → Au] triplet state.

A series of dinuclear gold(I) thiolate complexes [Au$_2${μ-(Ph$_2$P)$_2$NR}(SR')$_2$] [R = C$_6$H$_{11}$, R' = C$_6$H$_4$F-4 (**70a**), C$_6$H$_4$Cl-4 (**70b**), C$_6$H$_4$CH$_3$-4 (**70c**); R = Ph, R' = C$_6$H$_4$CH$_3$-4 (**70d**); R = CH$_2$CH$_2$CH$_3$, R' = C$_6$H$_4$CH$_3$-4 (**70e**); R = CH$_2$CH$_2$CH$_3$, R' = C$_6$H$_4$CH$_3$-4 (**70f**)] have also been investigated [151]. The high-energy absorption bands at around 265–305 nm have been suggested to be ligand-centered (phosphine) transitions. The low-energy absorption bands at around 365–385 nm are absent for the [Au$_2${μ-(Ph$_2$P)$_2$N-R}Cl$_2$] complexes and have been suggested to be LMCT ^{1}[RS$^-$ → Au] transitions. The photophysical data are listed in Table 17. At room temperature, the complexes in the solid state and in fluid solutions display bluish green emission, typical of ligand-centered

Table 17 Photophysical Data for **70a–f**

Complex	Medium (*T* K)	Emission wavelength λ_{max} nm (τ_0 μs)
70a	Solid (298)	519 (0.18, 0.77)
	Solid (77)	540
	CH$_2$Cl$_2$ (298)	503
	CHCl$_3$ (77)	488, 567
70b	Solid (298)	513 (0.64, 3.6)
	Solid (77)	528
	CH$_2$Cl$_2$ (298)	499
	CHCl$_3$ (77)	485, 545
70c	Solid (298)	519, 552 (0.19, 0.81)
	Solid (77)	522, 590
	CH$_2$Cl$_2$ (298)	503
	CHCl$_3$ (77)	498, 607
70d	Solid (298)	506 (0.28)
	Solid (77)	508, 578
	CH$_2$Cl$_2$ (298)	518
	CHCl$_3$ (77)	495, 618
70e	Solid (298)	530 (0.11, 0.45)
	Solid (77)	525, 602
	CH$_2$Cl$_2$ (298)	557
	CHCl$_3$ (77)	495, 602
70f	Solid (298)	517 (0.29, 1.4)
	Solid (77)	520, 609
	CH$_2$Cl$_2$ (298)	540
	CHCl$_3$ (77)	493, 604

Source: From Ref. [151].

emission. However, at 77 K, the emission bands of the solid samples become very unsymmetrical and some samples even reveal two bands. Similar findings are also observed for the emission spectra of the complexes in low-temperature chloroform glass. A high-energy band at similar energy to those observed at room temperature and a low-energy band at around 545–618 nm have been observed. The latter has been proposed to originate from an LMCT [RS$^-$ → Au] triplet state. The assignment is supported by the observation that **70a** and **70b** emit at higher energy than **70c**. The occurrence of the emission of **70a** at a lower energy than that of **70b** has been accounted for by the stronger mesomeric effect of the fluoro substituent which increases the electron density on the S atom through π donation into the 3d orbital. An LLCT [RS$^-$ → phosphine] excited state for the low-energy emission was not favored in view of the similar emission energies for the complexes with the same thiolate but different phosphine ligand.

   Zink and co-workers also reported the emission behavior of the dinuclear gold(I) complex [Au$_2$(PPh$_3$)$_2$(μ-mnt)] (**66a**) [152]. The single-crystal emission at 20 K reveals resolved vibronic structure with the peak maximum at 462 nm. The emission spectrum has been fit using the time-dependent theory of electronic spectroscopy. The vibration frequencies and intensities obtained from resonance Raman spectra have been employed to calculate the distortion of the excited molecule. The high accuracy of the estimation of distortion has been supported by the good fit of the calculated and experimental emission spectra. It has been found that the most significant distortions involve Au-S bonds and C=C bonds of the mnt^{2-} unit. This suggests that the electronic transition giving rise to the emission of the complex is a charge transfer between the gold and mnt^{2-} ligand. As the emission of [Au$_2$(AsPh$_3$)$_2$(μ-mnt)] occurs at a lower energy, the excited state has been suggested to be ^{3}LMCT [mnt^{2-} → Au] in nature. A similar LMCT excited state has also been proposed for the trinuclear thiolate complex [Au$_3$(μ-mnt)$_2$(PEt$_3$)$_2$]$^-$ (**71**) [153].

71

The photophysical properties of a pentanuclear gold(I) thiolate complex $[Au_5(\mu\text{-L})_3(\mu\text{-dppm})_2]^{2+}$ (**72**) (L = quinoline-2-thiolate) have been reported [154]. The X-ray crystal structure reveals Au–Au distances of 2.936(3)–3.351(3) Å. The complex displays electronic absorption bands at 370 nm (ε = 21,900 dm^3 mol^{-1} cm^{-1}) and 385 nm (ε = 18,250 dm^3 mol^{-1} cm^{-1}) which have been assigned as intraligand transitions of the quinoline-2-thiolate ligand L, and probably mixed with gold-centered (ds/dp) and LMCT character. The complex in acetonitrile solution exhibits dual luminescence with emission bands at around 500 nm (τ_0 = 1.9 μs) and 605 nm (τ_0 = 1.6 μs). The high-energy emission has been ascribed to a ligand-centered or an MLCT [Au → dppm/π^*(L)] triplet state. For the low-energy emission, the origin has been assigned to be a metal-centered Au(ds/dp) triplet state and perhaps mixed with some LMCT and ligand-centered character.

72

Another dinuclear gold(I) thiolate complex [{Au(PPh$_3$)$_2$}$_2$(L)] (**73a**) (L = quinoline-8-thiolate) has been found to possess dramatic solvent dependence [155]. The X-ray crystal structure reveals a μ-bridging quinoline-8-thiolate ligand via the S atom. The Au–Au distances are found to be 2.991(2) and 3.081(2) Å (in two asymmetric units). The mononuclear analog [Au(PPh$_3$)$_2$(L)] (**73b**) displays an absorption band at 386 nm, attributed to a spin-allowed LMCT transition. However, the dinuclear complex absorbs at 320 nm in CH$_2$Cl$_2$ which has been assigned to a $^1[d_{\sigma^*} \rightarrow p_\sigma]$ transition modified by Au–Au interaction. Upon addition of acetonitrile, methanol, or ethanol, a new absorption band at around 386 nm appears with the concomitant decrease in the intensity of the 320-nm band. The absorption at 386 nm is similar to that of the mononuclear species, indicative of the absence of Au–Au interaction. The observation has been explained based on the equilibrium in Scheme 2. It has been suggested that the equilibrium constant K_{eq} is larger in CH$_3$CN than in CH$_2$Cl$_2$. The emission spectrum of the dinuclear

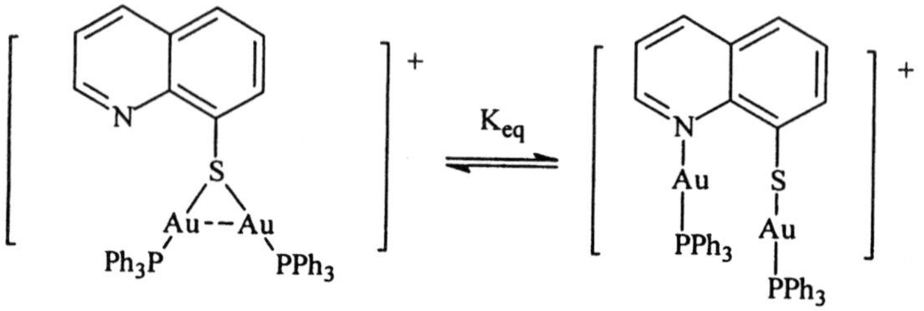

Scheme 2 Equilibrium for the rearrangement of **73a** in solution. (Adapted from Ref. [155].)

complex in CH_2Cl_2 shows a weak emission band at around 466 nm and a more intense band at around 640 nm ($\tau_0 = 26$ μs). The high-energy band is also observed for the mononuclear species. The low-energy band has been suggested to arise from a metal-centered triplet excited state modified by Au–Au interaction, and perhaps with some LMCT character. In the presence of acetonitrile and alcohols, the low-energy emission is quenched. The observation has been ascribed to the disappearance of the Au–Au interaction.

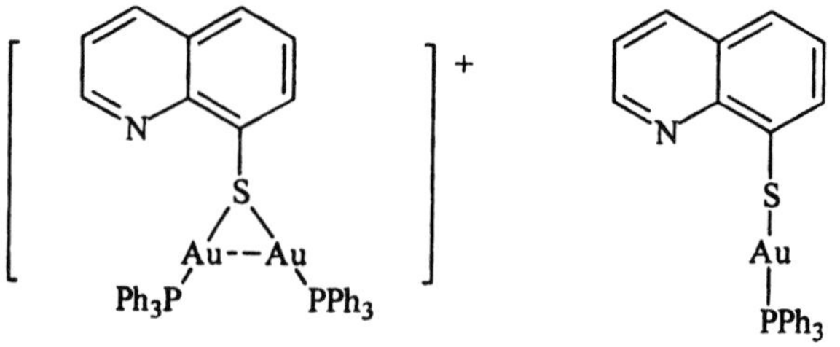

73a

73b

The photophysical properties of a hexanuclear gold(I) thiolate complex $[(AuL)\{AuP(CH_3)_2Ph\}_2]_2$ (**74a**) (H_3L = trithiocyanuric acid) have been described [156]. Four gold(I) centers form a parallelogram with Au–Au distances ranging from 2.964(2) to 2.987(2) Å. The complex exhibits a two-dimensional structure via intermolecular Au–Au interaction [$d_{Au-Au} = 3.130(2)$ Å]. The electronic absorption spectrum of the complex shows a strong absorption band at 320 nm ($\varepsilon = 44{,}870$ dm^3 mol^{-1} cm^{-1}) which is assigned to a spin-allowed [$d_{\sigma*} \rightarrow p_{\sigma}$] transition. This absorption band occurs at a lower energy than that of the trinuclear analog [$\{AuP(CH_3)_2Ph\}_3L$] (**74b**) ($\lambda_{max} = 295$ nm, $\varepsilon = 51{,}320$ dm^3 mol^{-1} cm^{-1}) in which no Au–Au interaction is present. Both complexes emit at about 520–530 nm in the solid state and in CH_2Cl_2 solutions. The excited state has been proposed to be ^{3}LMCT in nature.

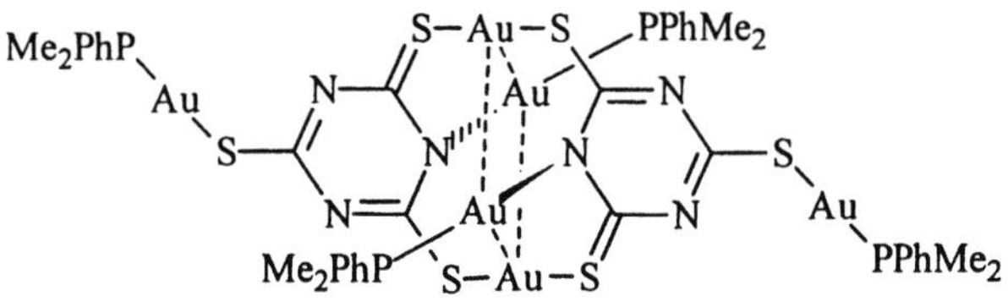

74a

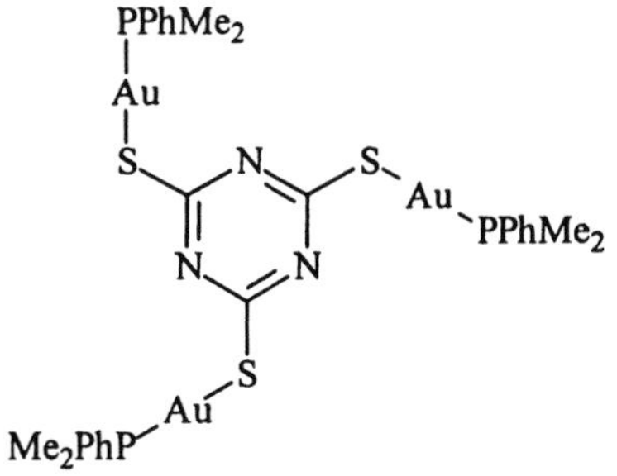

74b

Recently, we reported the luminescence and ion binding properties of dinuclear gold(I) crown ether complexes $[Au_2(\mu\text{-}P\text{-}P)(S\text{-}B15C5)_2]$ [P-P = dppm (**75a**), dcpm (**75b**), HS-B15C5 = 4'-mercaptomonobenzo-15-crown-5) [157]. The UV-Vis absorption spectra of both complexes in CH_2Cl_2/MeOH show spectral changes upon addition of potassium ions. The absence of such changes for the uncrowned complexes $[Au_2(\mu\text{-}P\text{-}P)\{S\text{-}C_6H_4(OCH_3)_2\text{-}3,4\}_2]$ suggests that the spectral changes result from the encapsulation of potassium ion by the benzo-15-crown-5 units of the $[Au_2(\mu\text{-}P\text{-}P)(S\text{-}B15C5)_2]$. It is found that the complexes form a 1:1 adduct with potassium ion with log K values of 3.4 and 4.0, respectively. The emission spectrum of **75a** shows a drop in intensity at around 502 nm, with the concomitant formation of a new long-lived emission band at around 720 nm (0.2 μs) upon addition of potassium ions. However, such a change in emission spectral traces is absent for the uncrowned analog. Besides, despite the well-known binding of sodium ions by benzo-15-crown-5 compounds, similar emission spectral changes are not observed when sodium ions are used instead of potassium ions. This is supportive of the sandwich binding of the potassium ion by the two benzo-15-crown-5 moieties of the same molecule of **75a**. It is likely that the binding of K^+ brings the two gold(I) centers in close proximity to each other, resulting in some weak gold–gold interaction. It has been proposed that the low-energy emission band arises from an LMMCT $[RS^- \rightarrow Au_2]$ excited state.

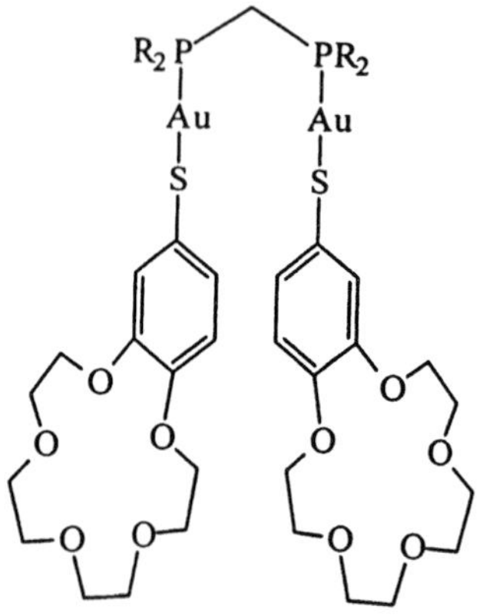

75a: R = Ph
75b: R = C_6H_{11}

D. Polypyridyl Complexes

The photoluminescence properties of [{Au(PPh₃)}₂(μ-bbzim)] (**76a**) and
[{Au(PPh₃)}₄(μ₄-bbzim)]²⁺ (**76b**) have been reported [158]. While the two gold(I)
centers are bridged at the opposite sides of the bbzim ligand, the two gold(I)
centers on each side of **76b** show a Au–Au distance of 3.157(1)–3.222(1) Å.
Both complexes exhibit intense absorption bands at 300–370 nm and the absorption bands for the dinuclear Au(I) complex are vibronically structured. The absorption bands are assigned to intraligand π-π* transitions. Upon photoexcitation,
both complexes display fluorescence at 340–430 nm and ligand-centered phosphorescence at lower energy.

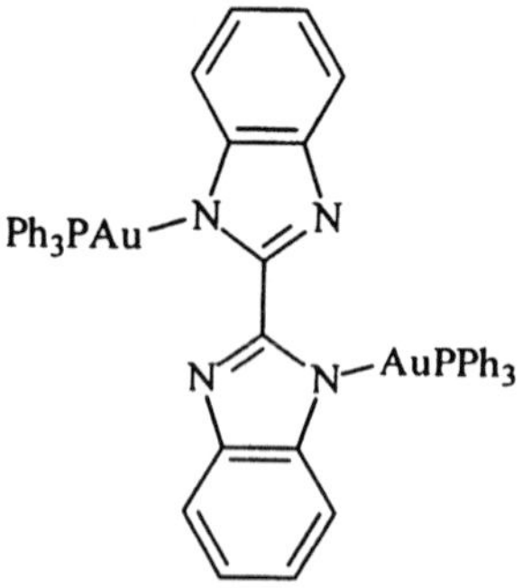

76a

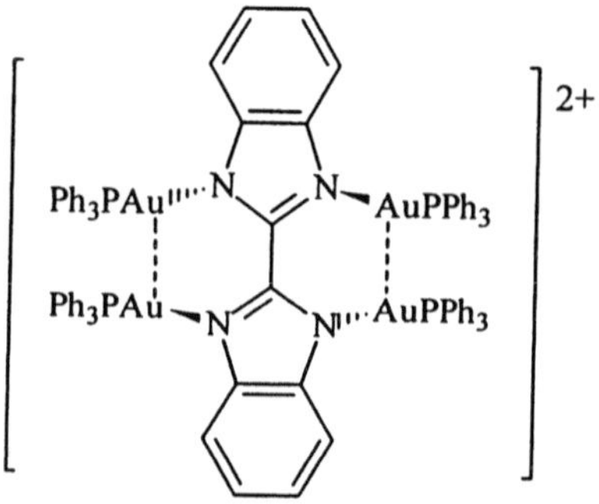

76b

E. Organometallic Complexes

The luminescence properties of organometallic polynuclear gold(I) complexes have been receiving a lot of attention. A series of organogold(I) complexes [Au$_2$(μ-dmpm)R$_2$] [R = CH$_3$ (**77a**), C$_6$H$_4$-OCH$_3$-4 (**77b**)] and [Au$_3$(μ$_3$-dmmp)R$_3$] [R = CH$_3$ (**78a**), C$_6$H$_4$-OCH$_3$-4 (**78b**)] have been studied [159]. The electronic absorption spectra of the complexes are similar. The intense absorption bands at around 269 nm of the dimeric and at around 275 nm of the trimeric species have been suggested to arise from ^{1}[d$_{\sigma*} \rightarrow$ p$_\sigma$] transitions. A red shift from the dinuclear to the trinuclear complexes is in line with a narrowing of the d$_{\sigma*}$–p$_\sigma$ energy gap. The relatively small shift (around 0.1 eV) may be due to the larger Au–Au separations. Photoexcitation of the samples in the solid state and in fluid solutions result in long-lived intense luminescence. The photophysical data are collected in Table 18. The emission energies of the complexes are not sensitive to the nuclearity of the species. This, together with the large Stokes shift observed, suggests that the emissive state is likely to be ^{3}[(d$_{\delta*}$)^1(p$_\sigma$)^1]. For the trinuclear complexes, the possibility of the third gold atom not interacting with the other two has not been ruled out. The phosphorescent state of the complexes has been found to be quenched by pyridinium acceptors. For **78b**, an excited state reduction potential E^o[Au$_3^+$/Au$_3^*$] of −2.0(1) V vs. SSCE (λ = 0.90 eV) has been estimated from oxidative quenching experiments with a series of structurally related pyridinium acceptors. This highly negative value indicates that the photoexcited complex is extremely reducing.

Table 18 Photophysical Data for **77a–b** and **78a–b**

Complex	Medium (T K)	Emission wavelength λ$_{max}$ nm (τ$_0$ μs)
77a	Solid (298)	426, 506
	Solid (77)	486, 575 sh
	CH$_2$Cl$_2$ (298)	424, 485 sh (0.5)
77b	Solid (298)	512
	Solid (77)	502
	CH$_2$Cl$_2$ (298)	432, 500 (7.5), 585 sh (5.0)
78a	Solid (298)	422, 511
	Solid (77)	485, 576
	CH$_2$Cl$_2$ (298)	420, 448 sh (0.7)
78b	Solid (298)	497
	Solid (77)	484
	CH$_2$Cl$_2$ (298)	421, 499 (55), 577 sh (1.8)

Source: From Ref. [159].

The X-ray crystal structures of the organogold(I) complex [Au$_2$(μ-dppm)R$_2$] [R = CH$_3$ (**79a**), Ph (**79b**)] reveal gold–gold distances of 3.251(1) and 3.154(1) Å, respectively [160]. The electronic absorption spectra show an intense absorption band at 290 nm (ε = 8.92 × 10^4 dm^3 mol^{-1} cm^{-1}) and 294 nm (ε = 1.25 × 10^4 dm^3 mol^{-1} cm^{-1}) for the methyl and phenyl complexes, respectively. An absorption tail at 290–350 nm has also been observed. EHMO calculations on the phenyl complex **79b** suggest that the HOMO is basically metal-centered [d$_{\sigma*}$ = 45%, d$_{\delta*}$ = 45%, σ(Au-C) = 22%]. However, the LUMO is mainly the π* of the phenyl moieties. Therefore, the low-energy absorption observed in the dppm series which, unlike the electron-rich dmpm counterparts [159], has been proposed as an MMLCT ^{1}[Au(d$_{\sigma*}$, d$_{\delta*}$) → π*(phenyl or dppm)] transition. The emission spectrum of the methyl complex **79a** in CH$_2$Cl$_2$ shows a broad band at 430–600 nm. For the phenyl analog in CH$_2$Cl$_2$, a broad emission peaking at about 480–490 nm and ranging from 450 to 600 nm was recorded. The solid-state emission spectrum at 77 K shows a vibronically structured band at around 480 nm and a lower energy band at 600 nm. These two bands are suggested to originate from ^{3}[Au(d$_{\sigma*}$,d$_{\delta*}$) → π*(phenyl)] and ^{3}[Au(d$_{\sigma*}$,d$_{\delta*}$) → π*(dppm)] states, respectively.

A series of dinuclear organogold(I) complexes [Au$_2$(μ-dppf)R$_2$] [R = CH$_3$ (**80a**), Ph (**80b**), 1-naphthyl (**80c**), 9-anthryl (**80d**), 1-pyrenyl (**80e**), C≡C-Ph (**80f**), and C≡C-^{t}Bu (**80g**)] containing a diphosphinoferrocene bridging ligand have also been investigated [161]. The UV-Vis absorption spectra are dominated by the π-π* absorption bands of the aromatic groups. For **80f** and **80g**, intense vibronically structured bands at around 270–295 nm with progressional spacings of about 1825 cm^{-1}, typical of ν(C≡C) stretching frequencies in the excited state, have been observed. These absorption bands are proposed to arise from intraligand π-π*(acetylide) transitions.

The complexes are nonemissive in the solid state at 77–298 K. However, in CH$_2$Cl$_2$ solution, the anthryl and pyrenyl complexes show vibronically structured emission bands with progressional spacings of 1200 cm^{-1}, attributable to the aromatic ring stretching mode. The emission has been suggested to be intraligand ^{3}[π-π*] in nature. The emission spectrum of the pyrenyl counterpart is concentration-dependent. It has been found that both excimer formation and exciplex formation with N,N-dimethylaniline occur for this complex. The photoreactivities of **80f** have also been studied. In CH$_2$Cl$_2$, the complex reacts with the solvent leading to the formation of the C-C coupling product Ph-C≡C-C≡C-Ph.

A number of dinuclear gold(I) complexes with diacetylide ligands have also been shown to exhibit rich photophysical properties. The emission spectrum of [(C$_6$H$_4$-OCH$_3$-4)$_3$P-Au-C≡C-Au-P(C$_6$H$_4$-OCH$_3$-4)$_3$] (**81**) reveals a vibronically structured band at 400–500 nm with progressional spacings of about 2100 cm^{-1} [160]. The emissive state is assigned to be intraligand π-π*(acetylide) in origin.

Besides, the complexes $[(Ph)_n(Np)_{3-n}$ P-Au-C$\equiv$C-Au-P(Ph)$_n$(Np)$_{3-n}$] [n = 0 – 3 (**82a–d**)] and [Fc$_2$PhP-Au-C$\equiv$C-Au-PPhFc$_2$] (**83**) have also been studied [162]. The UV-Vis absorption spectra of the complexes with naphthyl-containing phosphines show vibronically structured bands at 296 nm. The absorption has been assigned as $^1[\sigma$(Au-P) $\rightarrow \pi^*$(Np)] transitions. The solid-state emission energies of the complexes at 77 K follow the order **82d** (523 nm) > **82c** (547 nm) > **82b** (556 nm) $\approx$ **82a** (554 nm). This trend is in line with the electron richness around the Au-P bonds and suggests a $^3[\sigma$(Au-P) $\rightarrow \pi^*$(Np)] excited state. The emission of the ferrocene-containing complex is weak and short-lived. This has been suggested to be a consequence of an intramolecular reductive elec-tron transfer quenching of the excited state by the ferrocene unit.

On the other hand, the complex [Au$_3$(μ-dppm)$_2$(C$\equiv$C-Ph)$_2$][Au(C$\equiv$C-Ph)$_2$] (**84**) has also been shown to possess rich photophysical behavior [163]. X-ray crystal studies reveal gold–gold separations of 3.167(2) and 3.083(2) Å. The UV-Vis absorption spectrum shows a high-energy band at 276 nm (ε = 6.9 $\times$ 10^4 dm^3 mol^{-1} cm^{-1}) and a lower energy intense absorption at about 315–375 nm (ε = 10^4–10^3 dm^3 mol^{-1} cm^{-1}). The high-energy absorption has been assigned to be ligand-centered π-π^*(acetylide) transition. The low-energy absorption band has been tentatively suggested to be a $^1[d_{\delta^*} \rightarrow p_\sigma]$ transition, perhaps mixed with some intraligand character. In acetonitrile, the complex displays a high-energy emission at 425 nm (τ_0 = 0.45 μs) and a low-energy band at 600 nm (τ_0 = 8.7 μs). The high-energy band originates from the anion as [^{n}Bu$_4$N][Au(C$\equiv$C-Ph)$_2$] has also been shown to emit at similar energy, which is suggested to be ligand-centered in nature. For the low-energy emission, it has been proposed that the emissive state is $^3[(d_{\delta^*})^1(p_\sigma)^1]$.

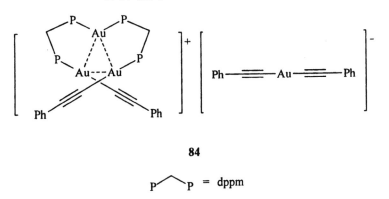

84

P$\frown$P = dppm

X-ray crystal structure of [Au$_2$(μ-dppe)(C$\equiv$C-Ph)$_2$] (**85**) reveals the ab-sence of intramolecular Au–Au interaction [164]. However, two [Au$_2$(μ-dppe)(C$\equiv$C-Ph)$_2$] units are found to interact with each other as indicated by an intermolecular Au–Au separation of 3.153(2) Å. The electronic absorption spectrum reveals a ligand-centered absorption band at 284 nm (ε = 4.9 × 10⁴ dm³ mol⁻¹ cm⁻¹). The solid sample emits at 550 nm at 298 K, whereas in CH_2Cl_2 solution the complex shows a ligand-centered emission at 420 nm. The lower energy emission observed for the complex in the solid state has been suggested to be derived from a $(d_{\delta*})^1(p_\sigma)^1$ triplet excited state.

86

A gold(I) acetylide polymer [{$Au_2(\mu$-dpppy)(C≡C-Ph)_2$}_n$] (**86**) shows a gold–gold distance of 3.252(1) Å [165]. The solid sample emits at 500 nm at 77 K. The emission energy is higher than that of **85** (550 nm) under similar conditions. On the assumption that the excited state is $^3[(d_{\delta*})^1(p_\sigma)^1]$ or $^3[(d_{\sigma*})^1(p_\sigma)^1]$ in nature, the blue shift of emission energy from **85** to **86** has been rationalized by the longer Au–Au distance observed in the latter species.

A series of dinuclear and tetranuclear gold(I) acetylide complexes [$Au_2(\mu$-dppb)(C≡C-R)_2$] [R = $^nC_6H_{13}$ (**87a**), Ph (**87b**), C_6H_4-OCH_3-4 (**87c**)] and [$Au_4(\mu_4$-tppb)(C≡C-R)_4$] [R = $^nC_6H_{13}$ (**88a**), Ph (**88b**), C_6H_4-OCH_3-4 (**88c**)] have been studied [166]. X-ray crystallographic studies on **88b** reveal an intramolecular gold–gold separation of 3.1541(4) Å. The electronic absorption bands of these complexes at around 250–300 nm have been assigned to be $^1[\sigma(Au-P) \rightarrow \pi^*(Ph_{bridge})]$ transitions. The higher transition energies for the dppb complexes than the tppb analogs have been accounted for by the more electron-deficient "bridging" phenyl ring of tppb. The photophysical data are listed in Table 19. In general, **87c** and **88c** emit at a lower energy than the phenylacetylide counterparts. This observation is in line with the more electron-rich OCH_3 moiety on the acetylide which reduces the extent of metal-to-ligand back-π-donation to the acetylide [$d_\pi(Au) \rightarrow \pi^*(acetylide)$]. This leads to an increased $d_\pi(Au)$-3d(P) overlap and therefore a higher $\sigma(Au-P)$ orbital energy.

A series of gold(I) acetylides [(C_6H_4-CH_3-4)_3P-Au-(BL)-Au-P(C_6H_4-CH_3-4)_3] [H_2BL = 1,4-diethynylbenzene (**89a**), 9,10-diethynylanthracene (**89b**)] and [$Au_2(\mu$-P-P)(C≡C-R)_2$] (P-P = dppn, R = C_6H_4-Ph-4 (**90a**), C_6H_4-OCH_3-4 (**90b**), $^nC_6H_{13}$ (**90c**); P-P = dcpn, R = C_6H_4-Ph-4 (**91**); P-P = dmpm, R = Ph (**92a**), C_6H_4-OCH_3-4 (**92b**)] and [$Au_3(\mu_3$-dmmp)(C≡C-R)_3$] [R = Ph (**93a**),

Table 19   Photophysical Data for **87a–c** and **88a–c**

Complex	Medium (*T* K)	Emission wavelength λ_{max} nm (τ_0 μs)[a]
87a	Solid (298)	510 (153.8)
	Solid (77)	427, 537
	CH₂Cl₂ (298)	392 (0.67); 587[b] (0.93)
87b	Solid (298)	507 (5.32)
	Solid (77)	483
	CH₂Cl₂ (298)	395, 411 sh, 476 sh, 601 sh; 628[b] (0.27)
87c	Solid (298)	486; 563[c] (5.06)
	Solid (77)	483; 556[c]
	CH₂Cl₂ (298)	395, 462 sh, 541 (2.66)
88a	Solid (298)	405, 602 (1.28)
	Solid (77)	586
	CH₂Cl₂ (298)	409, 577 (0.71)
88b	Solid (298)	599; 611[c] (0.57)
	Solid (77)	584; 611[c]
	CH₂Cl₂ (298)	510, 538 sh (0.46)
88c	Solid (298)	415; 628[c] (1.85)
	Solid (77)	414; 618[c]
	CH₂Cl₂ (298)	447, 601 (0.47)

[a] Excitation wavelength = 350 nm.
[b] Excitation wavelength = 500 nm.
[c] Excitation wavelength = 450 nm.
Source: From Ref. [166].

C_6H_4-OCH_3-4 (**93b**)] with bridging phosphine or acetylide ligands have been reported [167]. The UV-Vis absorption spectra of **89a** and **89b** show vibronically structured bands at around 294–328 nm (progressional spacings = 1980 cm^{-1}) and at around 410–464 nm (progressional spacings = 1400 cm^{-1}), respectively. The absorption bands are assigned to intraligand π-π*(acetylide) or ^{1}[σ(Au-P) $\rightarrow \pi$*(acetylide)] transitions. For the dppn complexes, the absorption bands occur at around 290–310 nm and a weaker absorption at around 400 nm with tailing to around 500 nm is also observed. These bands are likely to be ^{1}[σ(Au-P) $\rightarrow \pi$*(dppn)] in origin. For the dinuclear complexes with dmpm ligand and trinuclear complexes with dmmp ligand, there are strong absorption bands at around 252–290 nm, attributable to π-π*(acetylide) or [σ(Au-P) $\rightarrow \pi$*(acetylide)] transitions. Besides, absorption shoulders at around 320–332 nm are also observed. These low-energy absorption bands are absent in the mononuclear analogs and have been assigned to arise from a ^{1}[$d_{\sigma^*} \rightarrow p_\sigma$] transition. An MMLCT ^{1}[d_{σ^*}(Au-Au)$\rightarrow \pi$*(C$\equiv$C-R)] assignment is also possible in view of the red shift in the absorption energy from the mononuclear to the trinuclear species.

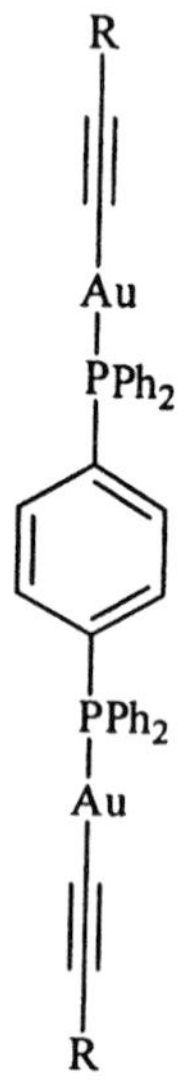

R
‖‖
Au
|
PPh₂

PPh₂
|
Au
‖‖
R

R   R
\X/
Au–Au
Ph₂P   PPh₂

Ph₂P   PPh₂
Au–Au
/\
R   R

87a: R = $^nC_6H_{13}$
87b: R = Ph
87c: R = C_6H_4-OCH_3-4

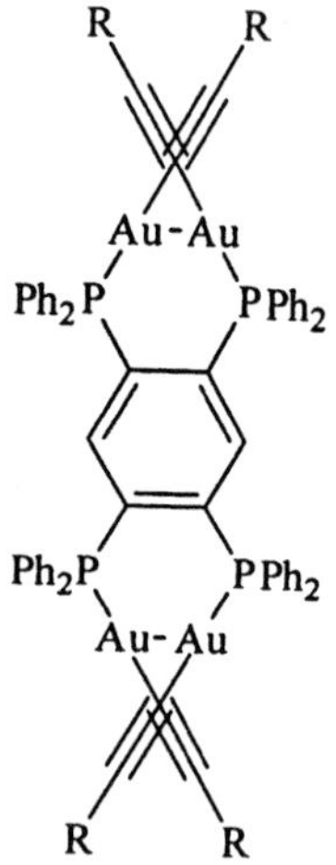

88a: R = $^nC_6H_{13}$
88b: R = Ph
88c: R = C_6H_4-OCH_3-4

The solid-state emission of **89a** appears as a vibronically structured band at around 533 nm. It has been assigned to arise from a ligand-centered π-π*(acetylide) or [σ(Au-P) $\rightarrow$ π*(acetylide)] triplet state. A blue shift of emission energy from this dinuclear complex to the mononuclear species [Au(PPh₃)(C≡CPh)] (λ_{max} = 459 nm under similar conditions) is in accord with a lower lying π* orbital of the bridging acetylide ligand. The emissive properties of **89b** are similar.

For the dppn complexes, the solid-state emission occurs at similar energy (around 571–655 nm). The excited state has also been suggested to be ^{3}[σ(Au-P) $\rightarrow$ π*(naphthyl)]. The assignment is supported by the fact that the dcpn analog **91** emits at a much lower energy (707 nm). The electron-donating cyclohexyl rings of dcpn would increase the electron density of the phosphorus atoms and render the σ(Au-P) electron pair more donating.

The solid-state emission of the acetylide complexes **92a, 92b, 93a**, and **93b** occurs at 490, 521, 538, and 539 nm, respectively. The emission energy shows a red shift from the dimer to the trimer but is not sensitive to the methoxy substituent on the phenylacetylide ligand. In view of this, together with the large Stokes shifts observed, the excited state has been assigned to be ^{3}[$(d_{\delta*})^1(p_\sigma)^1$] in origin. The photochemical properties of these gold(I) acetylides have also been studied. For example, the phosphorescent state of **93a** has been found to be quenched by the electron acceptor 4-methoxycarbonyl-N-methylpyridinium ion, with a quenching rate constant of $= 4.98 \times 10^9$ dm^3 mol^{-1} s^{-1}. The electron transfer mechanism of this photoinduced reaction, as well as that between **90b*** and MV^{2+}, has also been confirmed by nanosecond transient absorption spectroscopy.

Apart from phosphine ligands, gold(I) acetylides with isocyanide ligands have also been reported to be luminescent. For example, mononuclear [Au{CN-C$_6$H$_3$-(CH$_3$)$_2$-2,6}(C≡C-Ph)] (**94**) and dinuclear [Au$_2$(μ-L)(C≡C-Ph)$_2$] [L = tmb (**95a**), dmb (**95b**)] exhibit rich photophysical properties [168]. X-ray crystallographic studies show that the mononuclear complex **94** shows an intermolecular Au–Au separation of 3.329(4) Å. Although **95a** is in an anti configuration, intermolecular Au–Au distance of 3.565(2) Å is observed. The intramolecular Au–Au distance for **95b** is 3.485(3) Å. These figures suggest some very weak Au–Au interaction existing in these complexes. The complexes display intense absorption band at around 240 nm, which has been assigned to a ^{1}[d$_{z^2}$ → π*(isocyanide)] transition. Lower energy absorption bands at around 273 and 288 nm are typical of intraligand π-π*(acetylide) transitions. The emission spectra of the complexes show a structured band at around 420 nm that is assigned to arise from a ligand-centered π-π*(acetylide) triplet state. However, the solid-state emission spectra of the complexes at room temperature reveal a broad band at about 550 nm. This emission band has been assigned to be derived from a metal-centered ^{3}[(d$_{\delta*}$)^1(p$_\sigma$)^1] excited state. For **94**, an excited state reduction potential E^0[Au$^+$/Au*] of $-1.62(1)$ V vs. SSCE (λ = 1.07 eV) has been estimated from oxidative quenching experiments with a series of structurally related pyridinium acceptors.

A tetranuclear gold(I) phenylacetylide complex [Au$_4$(μ$_4$-dptact)(C≡C-Ph)$_4$] (**96**) containing a tetraazamacrocycle cavity has also been studied [169]. The electronic absorption spectrum shows an intraligand π-π*(acetylide) absorption at around 260–310 nm. The emission spectrum of the complex in CH$_2$Cl$_2$ solution at room temperature shows emission bands at around 425 and 560 nm. The higher energy band is ligand-centered in nature. However, this emission band has also been suggested to arise from the Au-PPh$_3$ moiety. On the other hand, it has been suggested that stacking of Au-C≡C-Ph units in solution gives rise to the low-energy emission. In the presence of alkali metal ions, the high-energy emission is strongly enhanced. It has been suggested that inter- and/or intramolecular interactions between the Au(C≡C-Ph) moieties, which provide facile nonradiative decay pathway of the intraligand excited state, are likely to be prohibited in the presence of alkali metal ions. Besides, in the presence of Cu$^+$ or Ag$^+$ the macrocyclic complex enhances the emission at around 550–600 nm. The enhancement in the emission intensity has been ascribed to the formation of a 1:1 heterometallic complex.

A series of gold(I) acetylides with the formula [{-Au-C≡C-Ar-C≡C-Au-L-}$_n$], where L = diphosphine or bis(isocyanide), have been reported to be luminescent [170]. The emission spectra of the complexes [(CH$_3$)$_3$P-Au-C≡C-(C$_6$H$_2$-R$_2$-2,5)-C≡C-Au-P(CH$_3$)$_3$] [R = H (**97a**), CH$_3$ (**97b**)] in CH$_2$Cl$_2$ reveal a high-energy emission band at about 415 nm, attributable to a π-π*(acetylide)/ [σ(Au-P) → π*(acetylide)] state. However, in the solid state, both complexes show an emission band at about 540 nm. An assignment of a $^3[(d_{\delta*})^1(p_\sigma)^1]$ excited state has been suggested to account for the red shift in emission energy, based on the short intermolecular Au–Au separation [3.1361(9) Å] observed in **97b**. However, the complex [{C$_6$H$_3$-(CH$_3$)$_2$-2,6}-N≡C-Au-C≡C-(C$_6$H$_4$-NO$_2$-4)] (**98**) emits at 633 nm in the solid state and at 503 nm in CH$_2$Cl$_2$ solution. The occurrence of such a low-energy emission has been ascribed to the electron-withdrawing NO$_2$ group, which lowers the π* orbital energy of the acetylide. The intermolecular Au–Au separation (3.923 Å) excludes the possibility of any Au–Au interaction. However, the crystal structure reveals that there are some stacking interactions between the phenyl rings of the acetylide and the isocyanide of the two closest molecules. This intermolecular π-π stacking has been suggested to account for the exceptionally low solid-state emission energy.

Balch and co-workers reported the interesting and intriguing solvent-induced luminescence properties of the organometallic complex [Au$_3$(μ-CH$_3$N=COCH$_3$)$_3$] (**99a**) [171]. Strong and long-lived yellow luminescence is observed when a solvent such as chloroform is in contact with the solid sample of the complex that has been irradiated with UV light. On the other hand, upon steady-state photoexcitation, the complex in chloroform solution emits at 422 nm. The solid sample shows a structured and long-lived emission band at 446 nm (τ$_0$ ≈ 1 ms) and a much longer lived emission band at 552 nm (triexponential decay, τ$_0$ = 1.4, 4.4, and 31 s). X-ray crystallographic studies reveal a triangular array of gold centers with a Au–Au separation of 3.308(2) Å. The triangular units also show some stacking interactions resulting in an intermolecular gold–gold distance of 3.346(1) Å. A related complex [Au$_3$(μ-C$_6$H$_5$CH$_2$N=COCH$_3$)$_3$] (**99b**) with Au–Au separation > 3.6 Å does not show similar solvent-induced luminescence. It has been proposed that the supramolecular aggregation is important for the liquid contact–triggered luminescence behavior of [Au$_3$(μ-CH$_3$N=COCH$_3$)$_3$].

99a: R = H
99b: R = C$_6$H$_5$

On the other hand, gold(I) isocyanides have also been shown to possess rich photophysical properties. The first report on luminescent gold(I) isocyanides appeared in 1989 [172]. The complex [Au$_2$(μ-dmb)(CN)$_2$] (**100a**) has an intramolecular gold–gold separation of 3.536 Å [172]. Besides, intermolecular gold–gold distances have been found to be 3.488 and 3.723 Å. The UV-Vis absorption spectrum shows two vibronically structured bands at about 200–260 nm with a weaker absorption tailing from 260 to 350 nm. The intense absorption bands have been suggested to be MLCT ^{1}[Au → π*(dmb)] transitions. The solid sample displays an emission band at 456 nm (τ_0 = 0.59 μs) and a weaker broad emission band at around 600–700 nm. In CH$_2$Cl$_2$, only a high-energy band at 458 nm (τ_0 = 0.13 μs) is observed. Besides, the solid-state emission spectrum of a related complex [Au$_2$(μ-tmb)(CN)$_2$] (**100b**) with a Au–Au separation of 3.21(2) Å also reveals an emission band at 428 nm [173].

The electronic absorption spectra of [Au$_2$(μ-dmb)$_2$]$^{2+}$ (**101a**) and [Au$_2$(μ-tmb)$_2$]$^{2+}$ (**101b**) have also been compared [173]. Both complexes display a strong absorption band at around 240–243 nm which has been assigned as an MLCT ^{1}[Au(d$_{z^2}$) → π*(isocyanide)] transition. Besides, there is also an additional absorption band at around 253 nm for [Au$_2$(μ-tmb)$_2$]$^{2+}$. Such a low-energy absorption has been suggested to arise from a ^{1}[d$_{\sigma*}$ → p$_\sigma$] transition in view of the smaller bite distance of the tmb ligand which can bring the gold(I) centers closer to each other.

The crystal structure of [Au$_2$(μ-tmb)Cl$_2$] (**102**) shows an anti configuration [174]. However, although there is no intramolecular Au–Au interaction, an intermolecular gold–gold separation of 3.3063 Å has been observed. The UV-Vis absorption spectrum of a solid sample of the complex in KBr shows an absorption band at 292 nm. Such an absorption band is also observed in the electronic absorption spectrum at 274 nm recorded in CH$_3$CN. Therefore, it has been suggested that the transition is a metal-centered d-p transition, instead of a ^{1}[d$_{\sigma*}$ → p$_\sigma$] transition due to gold–gold interaction. The solid sample has been found to emit at 417 nm (τ_0 = 0.70 ± 0.03 μs).

The photoluminescence properties of Tl[Au(CN)$_2$] (**103**) and related complexes have been extensively studied by Nagle, Patterson, and co-workers [175]. Covalent Tl–Au interactions in **103** have been suggested and supported by electronic structure calculations. Yersin and co-workers also reported the pressure dependence emissive behavior of M[Au(CN)$_2$] [M = K (**104a**), Cs (**104b**)], and Cs$_2$Na[Au(CN)$_2$]$_3$ (**104c**) [176]. For example, the emission of single crystals of **104a** shifts from 397 nm to 521 nm when the applied hydrostatic pressure is increased from 0 to 30 kbar [176]. This observation has been related to the quasi-one-dimensional metal–metal interactions and the quasi-two-dimensional formation of electronic energy bands.

V. CONCLUSION

Since the first report on the luminescence properties of polynuclear metal complexes of a d^{10} electronic configuration appeared in the literature, there has been fast-growing interest on the photophysical and photochemical studies of this class of compounds. From the examples mentioned throughout this chapter, it is obvious that there are a huge variety of ligands that can form highly luminescent complexes with copper(I), silver(I), and gold(I). Besides, the bonding modes of different ligands to these metal centers have also been shown to be highly diversified. Although there have already been a number of known d^{10} metal aggregates and clusters in the literature, it is not surprising to find that new complexes with novel molecular structures will continue to appear. An important characteristic of these complexes is that their luminescence properties are not limited to only one or two types of ligands. Ligands ranging from strong electron donors such as alkyls and thiolates, all the way to π acceptors such as unsaturated polypyridines, have been shown to form different complexes with remarkably different excited state properties. In other words, production of novel luminescent polynuclear copper(I), silver(I), and gold(I) complexes is almost unlimited.

On the other hand, it is also apparent that apart from fundamental spectroscopic, photophysical, and photochemical studies, as well as molecular and electronic structure investigations, the long-lived luminescence of d^{10} metal complexes also allows this class of compounds to be used as a luminescent probe and sensory material. For example, studies on the interaction of copper(I) [94] and gold(I) [135] complexes with DNA have been reported recently. The attachment of organic receptor type molecules such as crown ethers [157,177,178] and cyclams [145,169] to these metal centers has also been carried out. The binding of different cations to these complexes has been shown to be specific and selective. Studies along these lines will definitely provide a new generation of luminescent sensory materials.

In conclusion, we believe that the photophysical and photochemical studies on polynuclear d^{10} complexes are at a stage of rapid expansion. With the highly flexible molecular geometry and characteristic emissive properties, it is conceivable that luminescent polynuclear copper(I), silver(I), and gold(I) complexes will play a unique and crucial role in inorganic photochemistry.

ACKNOWLEDGMENT

We acknowledge the Research Grants Council and the University of Hong Kong for financial support.

ABBREVIATIONS

bbzim^{2-} = 2,2'-bibenzimidazolate
bpm = 2,2'-bipyrimidine
bpy = 2,2'-bipyridine
^{t}Bu$_2$bpy = 4,4'-di-t-butyl-2,2'-bipyridine
dcdpq = 6,7-dichloro-2,3-bis(2-pyridyl)quinoxaline
dcpe = 1,2-bis(dicyclohexylphosphino)ethane
dcpm = bis(dicyclohexylphosphino)methane
dcpn = 1,8-bis(dicyclohexylphosphino)naphthalene
dmb = 1,8-diisocyano-p-menthane
dmdpq = 6,7-dimethyl-2,3-bis(2-pyridyl)quinoxaline
dmmp = bis(dimethylphosphinomethyl)methylphosphine
dmp = 2,9-dimethyl-1,10-phenanthroline
dmpm = bis(dimethylphosphino)methane
dmpp = 1-phenyl-3,4-dimethylphosphole
DENC = N,N-diethylnicotinamide
dpam = bis(diphenylarsino)methane
dpb = 2,3-bis(2-pyridyl)benzo[g]quinoxaline
dpit^{2-} = N,N'-(2,2'-diphenyl)-bis(1,3-diphenyl-4-iminomethyl-5-thiopyrazole)
dpmp = 2-(diphenylmethyl)pyridine
dpp = 2,3-bis(2-pyridyl)pyrazine
dppb = 1,4-bis(diphenylphosphino)benzene
dppe = 1,2-bis(diphenylphosphino)ethane
dppf = 1,1'-bis(diphenylphosphino)ferrocene
dppm = bis(diphenylphosphino)methane
dppn = 1,8-bis(diphenylphosphino)naphthalene
dppnt = 2,7-bis(diphenylphosphino)-1,8-naphthyridine
dppp = bis(diphenylphosphinomethyl)phenylphosphine
dpppy = 2,6-bis(diphenylphosphino)pyridine
dppy = 2-diphenylphosphinopyridine
dpq = 2,3-bis(2-pyridyl)quinoxaline
dptact = 1,4,8,11-tetra(diphenylphosphinomethyl)-1,4,8,11-tetraazacyclotetradecane
dta$^-$ = dithioacetate
dtc$^-$ = di-n-propyldithiocarbamate
dtpm = bis[bis(4-methylphenyl)phosphino]methane
Fc = ferrocene
form$^-$ = N,N'-di-p-tolylformamidinate
IL = intraligand
IT = intervalence transfer

LC	= ligand-centered
LMCT	= ligand-to-metal charge transfer
MC	= metal-centered
MLCT	= metal-to-ligand charge transfer
mnt^{2-}	= 1,1-dicyanoethylene-2,2-dithiolate
mtc^-	= di-n-propylmonothiocarbamate
MV^{2+}	= methyl viologen
N_p	= 1-naphthyl
NS^{2-}	= S-methylhydrazinecarbodithioate Schiff base
POP^{2-}	= pyrophosphite
SAd^-	= adamantane thiolate
SSCE	= saturated sodium chloride calomel electrode
tmb	= 2,5-diisocyano-2,5-dimethylhexane
TMPD	= N,N,N',N'-tetramethylbenzene-1,4-diamine
TPA	= 1,3,5-triaza-7-phosphaadamantane
tppb	= 1,2,4,5-tetrakis(diphenylphosphino)benzene
tppm	= tris(diphenylphosphino)methane

REFERENCES

1. Balzani, V.; Scandola, F. *Supramolecular Photochemistry*, Ellis-Horwood: Chichester, 1991.
2. Horváth, O.; Stevenson, K.L. *Charge Transfer Photochemistry of Coordination Compounds*, VCH: New York, 1993.
3. Roundhill, D.M. *Photochemistry and Photophysics of Metal Complexes*, Plenum: New York, 1994.
4. Kalyanasundaram, K. *Photochemistry of Polypyridine and Porphyrin Complexes*, Academic Press: London, 1992.
5. Juris, A.; Balzani, V.; Barigelletti, F.; Campagna, S.; Belser, P.; von Zelewsky, A. *Coord. Chem. Rev.* **1988**, *84*, 85.
6. Petersen, J.D.; Gahan, S.L.; Rasmussen, S.C.; Ronco, S.E. *Coord. Chem. Rev.* **1994**, *132*, 15.
7. Sauvage, J.P.; Collin, J.P.; Chambron, J.C.; Guillerez, S.; Coudret, C.; Balzani, V.; Barigelletti, F.; De Cola, L.; Flamigni, L. *Chem. Rev.* **1994**, *94*, 993.
8. Balzani, V.; Juris, A.; Venturi, M.; Campagna, S.; Serroni, S. *Chem. Rev.* **1996**, *96*, 759.
9. Balzani, V.; Campagna, S.; Denti, G.; Juris, A.; Serroni, S.; Venturi, M. *Acc. Chem. Res.* **1998**, *31*, 26.
10. Che, C.M.; Butler, L.G.; Gray, H.B. *J. Am. Chem. Soc.* **1981**, *103*, 7796. Roundhill, D. M.; Gray, H.B.; Che, C.M. *Acc. Chem. Res.* **1989**, *22*, 55.
11. Smith, D.C.; Gray, H.B. *Coord. Chem. Rev.* **1990**, *100*, 169.
12. Ziolo, R.F.; Lipton, S.; Dori, Z. *J. Chem. Soc., Chem. Commun.* **1970**, 1124.
13. Caspar, J.V. *J. Am. Chem. Soc.* **1985**, *107*, 6718.

14. Harvey, P.D.; Gray, H.B. *J. Am. Chem. Soc.* **1988**, *110*, 2145.
15. Crosby, G.A.; Highland, R.G.; Truesdell, K.A. *Coord. Chem. Rev.* **1985**, *64*, 41.
16. Kutal, C. *Coord. Chem. Rev.* **1990**, *99*, 213.
17. Ford, P.C.; Vogler, A. *Acc. Chem. Res.* **1993**, *26*, 220.
18. Gade, L.H. *Angew. Chem. Int. Ed. Engl.* **1997**, *36*, 1171.
19. Pyykkö, P. *Chem. Rev.* **1997**, *97*, 597.
20. Schmidbaur, H. *Gold Bull.* **1990**, *23*, 11.
21. Mehrotra, P.K.; Hoffmann, R. *Inorg. Chem.* **1978**, *17*, 2187.
22. Jiang, Y.; Alvarez, S.; Hoffmann, R. *Inorg. Chem.* **1985**, *24*, 749.
23. Merz, Jr., K.M.; Hoffmann, R. *Inorg. Chem.* **1988**, *27*, 2120.
24. Vogler, A.; Kunkely, H. *J. Am. Chem. Soc.* **1986**, *108*, 7211.
25. Vogler, A.; Kunkely, H. *Chem. Phys. Lett.* **1988**, *150*, 135.
26. Vogler, A.; Kunkely, H. *Chem. Phys. Lett.* **1989**, *158*, 74.
27. Bondi, A. *J. Phys. Chem.* **1964**, *68*, 441.
28. Beck, J.; Strähle, J. *Angew. Chem. Int. Ed. Engl.* **1985**, *24*, 409.
29. Cotton, F.A.; Feng, X.; Matusz, M.; Poli, R. *J. Am. Chem. Soc.* **1988**, *110*, 7077.
30. Meyer, E.M.; Gambarotta, S.; Floriani, C.; Chiesi-Villa, A.; Guastini, C. *Organometallics* **1989**, *8*, 1067.
31. Hathaway, B.J. *Comprehensive Coordination Chemistry*, Vol. 5, Wilkinson, G.; Gillard, R.D.; McCleverty, J.A. eds.; Pergamon: Oxford, 1987, pp. 533–774.
32. van Koten, G.; James, S.L.; Jastrzebski, J.T.B.H. *Comprehensive Organometallic Chemistry II*, Vol. 3, Abel, E.W.; Stone, F.G.A.; Wilkinson, G., Eds.; Pergamon: Oxford, 1995, pp. 57–133.
33. Lancashire, R.J. *Comprehensive Coordination Chemistry*, Vol. 5, Wilkinson, G.; Gillard, R.D.; McCleverty, J.A., Eds.; Pergamon: Oxford, 1987, pp. 775–859.
34. Puddephatt, R.J. *Comprehensive Coordination Chemistry*, Vol. 5, Wilkinson, G.; Gillard, R.D.; McCleverty, J.A., Eds.; Pergamon: Oxford, 1987, pp. 861–923.
35. Grohmann, A.; Schmidbaur, H. *Comprehensive Organometallic Chemistry II*, Vol. 3, Abel, E.W.; Stone, F.G.A.; Wilkinson, G., Eds.; Pergamon: Oxford, 1995, pp. 1–56.
36. Kölmel, C.; Ahlrichs, R. *J. Phys. Chem.* **1990**, *94*, 5536.
37. Abraham, S.P.; Samuelson, A.G.; Chandrasekhar, J. *Inorg. Chem.* **1993**, *32*, 6107.
38. Jansen, M. *Angew. Chem. Int. Ed. Engl.* **1987**, *26*, 1098.
39. Singh, K.; Long, J.R.; Stavropoulos, P. *J. Am. Chem. Soc.* **1997**, *119*, 2942.
40. Siemeling, U.; Vorfeld, U.; Neumann, B.; Stammler, H.G. *Chem. Commun.* **1997**, 1723.
41. Segers, D.P.; DeArmond, M.K.; Grutsch, P.A.; Kutal, C. *Inorg. Chem.* **1984**, *23*, 2874.
42. Li, D.; Che, C.M.; Kwong, H.L.; Yam, V.W.W. *J. Chem. Soc., Dalton Trans.* **1992**, 3325.
43. Diez, J.; Gamasa, M.P.; Gimeno, J.; Tiripicchio, A.; Camellini, M.T. *J. Chem. Soc., Dalton Trans.* **1987**, 1275.
44. Li, D.; Che, C.M.; Wong, W.T.; Shieh, S.J.; Peng, S.M. *J. Chem. Soc., Dalton Trans.* **1993**, 653.
45. Hardt, H.D.; Pierre, A. *Inorg. Chim. Acta* **1977**, *25*, L59.
46. Kyle, K.R.; DiBenedetto, J.; Ford, P.C. *J. Chem. Soc., Chem. Commun.* **1989**, 714.

47. Kyle, K.R.; Ford, P.C. *J. Am. Chem. Soc.* **1989**, *111*, 5005.
48. Kyle, K.R.; Palke, W.E.; Ford, P.C. *Coord. Chem. Rev.* **1990**, *97*, 35.
49. Kyle, K.R.; Ryu, C.K.; DiBenedetto, J.A.; Ford, P.C. *J. Am. Chem. Soc.* **1991**, *113*, 2954.
50. Ryu, C.K.; Kyle, K.R.; Ford, P.C. *Inorg. Chem.* **1991**, *30*, 3982.
51. Vitale, M.; Palke, W.E.; Ford, P.C. *J. Phys. Chem.* **1992**, *96*, 8329.
52. Ryu, C.K.; Vitale, M.; Ford, P.C. *Inorg. Chem.* **1993**, *32*, 869.
53. Døssing, A.; Ryu, C.K.; Kudo, S.; Ford, P.C. *J. Am. Chem. Soc.* **1993**, *115*, 5132.
54. Vitale, M.; Ryu, C.K.; Palke, W.E.; Ford, P.C. *Inorg. Chem.* **1994**, *33*, 561.
55. Tran, D.; Ryu, C.K.; Ford, P.C. *Inorg. Chem.* **1994**, *33*, 5957.
56. Ford, P.C. *Coord. Chem. Rev.* **1994**, *132*, 129.
57. Lindsay, E.; Ford, P.C. *Inorg. Chim. Acta* **1996**, *242*, 51.
58. Tran, D.; Bourassa, J.L.; Ford, P.C. *Inorg. Chem.* **1997**, *36*, 439.
59. Henary, M.; Zink, J.I. *J. Am. Chem. Soc.* **1989**, *111*, 7407.
60. Lai, D.C.; Zink, J.I. *Inorg. Chem.* **1993**, *32*, 2594.
61. Li, D.; Yip, H.K.; Che, C.M.; Zhou, Z.Y.; Mak, T.C.W.; Liu, S.T. *J. Chem. Soc., Dalton Trans.* **1992**, 2445.
62. Provencher, R.; Harvey, P.D. *Inorg. Chem.* **1996**, *35*, 2235.
63. Ho, D.M.; Bau, R. *Inorg. Chem.* **1983**, *22*, 4079.
64. Harvey, P.D.; Drouin, M.; Zhang, T. *Inorg. Chem.* **1997**, *36*, 4998.
65. Yam, V.W.W.; Lee, W.K.; Lai, T.F. *J. Chem. Soc., Chem. Commun.* **1993**, 1571.
66. Yam, V.W.W.; Lo, K.K.W.; Cheung, K.K. *Inorg. Chem.* **1996**, *35*, 3459.
67. Yam, V.W.W.; Lo, K.K.W.; Wang, C.R.; Cheung, K.K. *J. Phys. Chem. A* **1997**, *101*, 4666.
68. Yam, V.W.W.; Lo, K.K.W. *Comments Inorg. Chem.* **1997**, *19*, 209.
69. Yam, V.W.W. *J. Photochem. Photobiol. A, Chem.* **1997**, *106*, 75.
70. Yam, V.W.W.; Lo, K.K.W.; Fung, W.K.M.; Wang, C.R. *Coord. Chem. Rev.* **1998**, *171*, 17.
71. Wang, C.R.; Lo, K.K.W.; Fung, W.K.M.; Yam, V.W.W., *Chem. Phys. Lett.* **1998**, *296*, 505.
72. Yam, V.W.W.; Lam, A.C.H., unpublished results.
73. Schugar, H.J.; Ou, C.; Thich, J.A.; Potenza, J.A.; Lalancette, R.A.; Furey, Jr., W. *J. Am. Chem. Soc.* **1976**, *98*, 3047. Miskowski, V.M.; Thich, J.A.; Solomon, R.; Schugar, H.J. *J. Am. Chem. Soc.* **1976**, *98*, 8344.
74. Robin, M.B.; Day, P. *Adv. Inorg. Chem. Radiochem.* **1967**, *10*, 247.
75. Gagné, R.R.; Koval, C.A.; Smith, T.J. *J. Am. Chem. Soc.* **1977**, *99*, 8367.
76. Gagné, R.R.; Koval, C.A.; Smith, T.J.; Cimolino, M.C. *J. Am. Chem. Soc.* **1979**, *101*, 4571.
77. Scott, B.; Willett, R.; Porter, L.; Williams, J. *Inorg. Chem.* **1992**, *31*, 2483.
78. Barr, M.E.; Smith, P.H.; Antholine, W.E.; Spencer, B. *J. Chem. Soc., Chem. Commun.* **1993**, 1649.
79. Farrar, J.A.; Grinter, R.; Neese, F.; Nelson, J.; Thomson, A.J. *J. Chem. Soc., Dalton Trans.* **1997**, 4083.
80. Slutter, C.E.; Sanders, D.; Wittung, P.; Malmström, B.G.; Aasa, R.; Richards, J.H.; Gray, H.B.; Fee, J.A. *Biochemistry* **1996**, *35*, 3387.
81. Houser, R.P.; Young, Jr., V.G.; Tolman, W.B. *J. Am. Chem. Soc.* **1996**, *118*, 2101.

82. Wallace-Williams, S.E.; James, C.A.; de Vries, S.; Saraste, M.; Lappalainen, P.; van der Oost, J.; Fabian, M.; Palmer, G.; Woodruff, W.H. *J. Am. Chem. Soc.* **1996**, *118*, 3986.

83. Beinert, H. *Eur. J. Biochem.* **1997**, *245*, 521.

84. Luchinat, C.; Soriano, A.; Djinovic-Carugo, K.; Saraste, M.; Malmström, B.G.; Bertini, I. *J. Am. Chem. Soc.* **1997**, *119*, 11023.

85. Hay, M.T.; Ang, M.C.; Gamelin, D.R.; Solomon, E.I.; Antholine, W.E.; Ralle, M.; Blackburn, N.J.; Massey, P.D.; Wang, X.; Kwon, A.H.; Lu, Y. *Inorg. Chem.* **1998**, *37*, 191.

86. Sabin, F.; Ryu, C.K.; Ford, P.C.; Vogler, A. *Inorg. Chem.* **1992**, *31*, 1941.

87. Knotter, D.M.; Blasse, G.; van Vliet, J.P.M.; van Koten, G. *Inorg. Chem.* **1992**, *31*, 2196.

88. Liu, C.W.; Stubbs, T.; Staples, R.J.; Fackler, Jr., J.P. *J. Am. Chem. Soc.* **1995**, *117*, 9778.

89. Chan, C.K.; Cheung, K.K.; Che, C.M. *Chem. Commun.* **1996**, 227.

90. Rasmussen, J.C.; Toftlund, H.; Nivorzhkin, A.N.; Bourassa, J.; Ford, P.C. *Inorg. Chim. Acta* **1996**, *251*, 291.

91. Fujisawa, K.; Imai, S.; Kitajima, N.; Moro-oka, Y. *Inorg. Chem.* **1998**, *37*, 168.

92. McMillin, D.R.; Buckner, M.T.; Ahn, B.T. *Inorg. Chem.* **1977**, *16*, 943.

93. McMillin, D.R.; Kirchhoff, J.R.; Goodwin, K.W. *Coord. Chem. Rev.* **1985**, *64*, 83.

94. McMillin, D.R.; Liu, F.; Meadows, K.A.; Aldridge, T. K.; Hudson, B.P. *Coord. Chem. Rev.* **1994**, *132*, 105.

95. Vogler, C.; Hausen, H.D.; Kaim, W.; Kohlmann, S.; Kramer, H.E.A.; Rieker, J. *Angew. Chem. Int. Ed. Engl.* **1989**, *28*, 1659.

96. Armaroli, N.; Balzani, V.; Barigelletti, F.; De Cola, L.; Sauvage, J.P.; Hemmert, C. *J. Am. Chem. Soc.* **1991**, *113*, 4033.

97. Yam, V.W.W.; Lo, K.K.W. *J. Chem. Soc. Dalton Trans.* **1995**, 499.

98. Harvey, P.D. *Inorg. Chem.* **1995**, *34*, 2019.

99. Brown, I.D.; Dunitz, J.D. *Acta Crystallogr.* **1961**, *14*, 480.

100. Henary, M.; Wootton, J.L.; Khan, S.I.; Zink, J.I. *Inorg. Chem.* **1997**, *36*, 796.

101. Yam, V.W.W.; Lee, W.K.; Lai, T.F. *Organometallics* **1993**, *12*, 2383.

102. Yam, V.W.W.; Lee, W.K.; Yeung, P.K.Y.; Phillips, D. *J. Phys. Chem.* **1994**, *98*, 7545.

103. Yam, V.W.W.; Lee, W.K.; Cheung, K.K.; Crystall, B.; Phillips, D. *J. Chem. Soc., Dalton Trans.* **1996**, 3283.

104. Yam, V.W.W.; Lee, W.K.; Cheung, K.K. *J. Chem. Soc., Dalton Trans.* **1996**, 2335.

105. Yam, V.W.W.; Choi, S.W.K.; Chan, C.L.; Cheung, K. K. *Chem. Commun.* **1996**, 2067.

106. Yam, V.W.W.; Fung, W.K.M.; Cheung, K.K. *Angew. Chem. Int. Ed. Engl.* **1996**, *35*, 1100.

107. Yam, V.W.W.; Lee, W.K.; Cheung, K.K.; Lee, H. K; Leung, W.P. *J. Chem. Soc., Dalton Trans.* **1996**, 2889.

108. Yam, V.W.W.; Fung, W.K.M.; Wong, M.T. *Organometallics* **1997**, *16*, 1772.

109. Yam, V.W.W.; Fung, W.K.M.; Cheung, K.K. *Chem. Commun.* **1997**, 963.

110. Yam, V.W.W.; Fung, W.K.M.; Cheung, K.K. *Organometallics* **1998**, *17*, 3293.

111. Yam, V.W.W.; Fung, W.K.M.; Cheung, K.K. *J. Cluster Science*, **1999**, in press.

112. Fortin, D.; Drouin, M.; Turcotte, M.; Harvey, P.D. *J. Am. Chem. Soc.* **1997**, *119*, 531.
113. Che, C.M.; Yip, H.K.; Li, D.; Peng, S.M.; Lee, G. H.; Wang, Y.M.; Liu, S.T. *J. Chem. Soc., Chem. Commun.* **1991**, 1615.
114. Che, C.M.; Yip, H.K.; Yam, V.W.W.; Cheung, P.Y.; Lai, T.F.; Shieh, S.J.; Peng, S.M. *J. Chem. Soc., Dalton Trans.* **1992**, 427.
115. Uang, R.H.; Chan, C.K.; Peng, S.M.; Che, C.M. *J. Chem. Soc., Chem. Commun.* **1994**, 2561.
116. Henary, M.; Zink, J.I. *Inorg. Chem.* **1991**, *30*, 3111.
117. Yam, V.W.W.; Lo, K.K.W.; Wang, C.R.; Cheung, K.K. *Inorg. Chem.* **1996**, *35*, 5116.
118. Wang, C.R.; Lo, K.K.W.; Yam, V.W.W. *J. Chem. Soc., Dalton Trans.* **1997**, 227.
119. Wang, C.R.; Lo, K.K.W.; Yam, V.W.W. *Chem. Phys. Lett.* **1996**, *262*, 91.
120. Yam, V.W.W.; Fung, W.K.M.; Cheung, K.K. *Organometallics* **1997**, *16*, 2032.
121. Wang, C.F.; Peng, S.M.; Chan, C.K.; Che, C.M. *Polyhedron* **1996**, *15*, 1853.
122. Moore, C.E. *Natl. Stand. Ref. Data Series (U.S. Natl. Bur. Stand.)* **1971**, *NSRDS-NBS, 35*, pp. 51, 116.
123. Piché, D.; Harvey, P.D. *Can. J. Chem.* **1994**, *72*, 705.
124. Assefa, Z.; Shankle, G.; Patterson, H.H.; Reynolds, R. *Inorg. Chem.* **1994**, *33*, 2187.
125. Omary, M.A.; Patterson, H.H. *Inorg. Chem.* **1998**, *37*, 1060.
126. King, C.; Wang, J.C.; Khan, Md. N.I.; Fackler, Jr., J.P. *Inorg. Chem.* **1989**, *28*, 2145.
127. Che, C.M.; Kwong, H.L.; Yam, V.W.W.; Cho, K.C. *J. Chem. Soc., Chem. Commun.* **1989**, 885.
128. Porter, L.C.; Khan, Md. N.I.; King, C.; Fackler, Jr., J.P. *Acta Crystallogr., Sect. C* **1989**, *45*, 947.
129. Che, C.M.; Kwong, H.L.; Poon, C.K.; Yam, V.W.W. *J. Chem. Soc., Dalton Trans.* **1990**, 3215.
130. Jaw, H.R.C.; Savas, M.M.; Rogers, R.D.; Mason, W.R. *Inorg. Chem.* **1989**, *28*, 1028.
131. Jaw, H.R.C.; Savas, M.M.; Mason, W.R. *Inorg. Chem.* **1989**, *28*, 4366.
132. Yam, V.W.W.; Lai, T.F.; Che, C.M. *J. Chem. Soc., Dalton Trans.* **1990**, 3747.
133. Yam, V.W.W.; Lee, W.K. *J. Chem. Soc., Dalton Trans.* **1993**, 2097.
134. Bensch, W.; Prelati, M.; Ludwig, W. *J. Chem. Soc., Chem. Commun.* **1986**, 1762.
135. Yam, V.W.W.; Choi, S.W.K.; Lo, K.K.W.; Dung, W.F.; Kong, R.Y.C. *J. Chem. Soc., Chem. Commun.* **1994**, 2379.
136. Li, D.; Che, C.M.; Peng, S.M.; Liu, S.T.; Zhou, Z.Y.; Mak, T.C.W. *J. Chem. Soc., Dalton Trans.* **1993**, 189.
137. McCleskey, T.M.; Gray, H.B. *Inorg. Chem.* **1992**, *31*, 1733.
138. McCleskey, T.M.; Winkler, J.R.; Gray, H.B. *Inorg. Chim. Acta* **1994**, *225*, 319.
139. Shieh, S.J.; Li, D.; Peng, S.M.; Che, C.M. *J. Chem. Soc., Dalton Trans.* **1993**, 195.
140. Assefa, Z.; McBurnett, B.G.; Staples, R.J.; Fackler, Jr., J.P.; Assmann, B.; Angermaier, K.; Schmidbaur, H. *Inorg. Chem.* **1995**, *34*, 75.
141. Assefa, Z.; McBurnett, B.G.; Staples, R.J.; Fackler, Jr., J.P. *Inorg. Chem.* **1995**, *34*, 4965.

142. Toronto, D.V.; Weissbart, B.; Tinti, D.S.; Balch, A.L. *Inorg. Chem.* **1996**, *35*, 2484.
143. Weissbart, B.; Toronto, D.V.; Balch, A.L.; Tinti, D.S. *Inorg. Chem.* **1996**, *35*, 2490.
144. Xiao, H.; Weng, Y.X.; Wong, W.T.; Mak, T.C.W.; Che, C.M. *J. Chem. Soc., Dalton Trans.* **1997**, 221.
145. Tzeng, B.C.; Cheung, K.K.; Che, C.M. *Chem. Commun.* **1996**, 1681.
146. Pyykkö, P.; Li, J.; Runberg, N. *Chem. Phys. Lett.* **1994**, *218*, 133.
147. Khan, Md. N.I.; Fackler, Jr., J.P.; King, C.; Wang, J.C.; Wang, S. *Inorg. Chem.* **1988**, *27*, 1672.
148. Forward, J.M.; Bohmann, D.; Fackler, Jr., J.P.; Staples, R.J. *Inorg. Chem.* **1995**, *34*, 6330.
149. Narayanaswamy, R.; Young, M.A.; Parkhurst, E.; Ouellette, M.; Kerr, M.E.; Ho, D.M.; Elder, R.C.; Bruce, A.E.; Bruce, M.R.M. *Inorg. Chem.* **1993**, *32*, 2506.
150. Jones, W.B.; Yuan, J.; Narayanaswamy, R.; Young, M.A.; Elder, R.C.; Bruce, A.E.; Bruce, M.R.M. *Inorg. Chem.* **1995**, *34*, 1996.
151. Yam, V.W.W.; Chan, C.L.; Cheung, K.K. *J. Chem. Soc., Dalton Trans.* **1996**, 4019.
152. Hanna, S.D.; Zink, J.I. *Inorg. Chem.* **1996**, *35*, 297.
153. Hanna, S.D.; Khan, S.I.; Zink, J.I. *Inorg. Chem.* **1996**, *35*, 5813.
154. Tzeng, B.C.; Che, C.M.; Peng, S.M. *J. Chem. Soc., Dalton Trans.* **1996**, 1769.
155. Tzeng, B.C.; Chan, C.K.; Cheung, K.K.; Che, C.M.; Peng, S.M. *Chem. Commun.* **1997**, 135.
156. Tzeng, B.C.; Che, C.M.; Peng, S.M. *Chem. Commun.* **1997**, 1771.
157. Yam, V.W.W., Li, C.K.; Chan, C.L. *Angew. Chem. Int. Ed.* **1998**, *37*, 2857.
158. Tzeng, B.C.; Li, D.; Peng, S.M.; Che, C.M. *J. Chem. Soc., Dalton Trans.* **1993**, 2365.
159. Yam, V.W.W.; Choi, S.W.K. *J. Chem. Soc., Dalton Trans.* **1994**, 2057.
160. Hong, X.; Cheung, K.K.; Guo, C.X.; Che, C.M. *J. Chem. Soc., Dalton Trans.* **1994**, 1867.
161. Yam, V.W.W.; Choi, S.W.K.; Cheung, K.K. *J. Chem. Soc., Dalton Trans.* **1996**, 3411.
162. Müller, T.E.; Choi, S.W.K.; Mingos, D.M.P.; Murphy, D.; Williams, D.J.; Yam, V.W.W. *J. Organomet. Chem.* **1994**, *484*, 209.
163. Che, C.M.; Yip, H.K.; Lo, W.C.; Peng, S.M. *Polyhedron* **1994**, *13*, 887.
164. Li, D.; Hong, X.; Che, C.M.; Lo, W.C.; Peng, S.M. *J. Chem. Soc., Dalton Trans.* **1993**, 2929.
165. Shieh, S.J.; Hong, X.; Peng, S.M.; Che, C.M. *J. Chem. Soc., Dalton Trans.* **1994**, 3067.
166. Yam, V.W.W.; Choi, S.W.K.; Cheung, K.K. *Organometallics* **1996**, *15*, 1734.
167. Yam, V.W.W.; Choi, S.W.K. *J. Chem. Soc., Dalton Trans.* **1996**, 4227.
168. Xiao, H.; Cheung, K.K.; Che, C.M. *J. Chem. Soc., Dalton Trans.* **1996**, 3699.
169. Tzeng, B.C.; Lo, W.C.; Che, C.M.; Peng, S.M. *Chem. Commun.* **1996**, 181.
170. Irwin, M.J.; Vittal, J.J.; Puddephatt, R.J. *Organometallics* **1997**, *16*, 3541.
171. Vickery, J.C.; Olmstead, M.M.; Fung, E.Y.; Balch, A.L. *Angew. Chem. Int. Ed. Engl.* **1997**, *36*, 1179.
172. Che, C.M.; Wong, W.T.; Lai, T.F.; Kwong, H.L. *J. Chem. Soc., Chem. Commun.* **1989**, 243.

173. Che, C.M.; Yip, H.K.; Wong, W.T.; Lai, T.F. *Inorg. Chim. Acta* **1992**, *197*, 177.
174. Perreault, D.; Drouin, M.; Michel, A.; Harvey, P.D. *Inorg. Chem.* **1991**, *30*, 2.
175. Assefa, Z.; DeStefano, F.; Garepapaghi, M.A.; LaCasce, Jr., J.H.; Ouellete, S.; Corson, M.R.; Nagle, J.K.; Patterson, H.H. *Inorg. Chem.* **1991**, *30*, 2868.
176. Yersin, H.; Riedl, U. *Inorg. Chem.* **1995**, *34*, 1642.
177. Yam, V.W.W.; Lo, K.K.W.; Cheung, K.K. *Inorg. Chem.* **1995**, *34*, 4013.
178. Yam, V.W.W.; Pui, Y.L.; Li, W.P.; Lo, K.K.W.; Cheung, K.K. *J. Chem. Soc., Dalton Trans.* **1998**, 3615.

3

Electron Transfer Within Synthetic Polypeptides and De Novo Designed Proteins

Michael Y. Ogawa
Bowling Green State University,
Bowling Green, Ohio

I. INTRODUCTION

Biological electron transfer (ET) reactions often occur over long distances (>10 Å) between cofactors that are embedded within a complicated protein matrix [1–4]. It has therefore become a goal of current research to investigate the putative role of proteins, and specific protein structures, in propagating long-range electron donor–acceptor interactions [5–12]. Recent studies of both native and chemically modified ET proteins have produced conflicting information concerning the ability of the protein matrix to mediate long-range electronic coupling. For example, Dutton and co-workers [13,14] have compared the available ET data obtained from the bacterial photosynthetic reaction centers to conclude that the native protein matrix behaves much as an isotropic solvent that presents a homogeneous barrier to electron tunneling. This conclusion was based on the observation that the rates of activationless ET reactions occurring in the reaction center follow a uniform distance dependence, as measured by their through-space edge-to-edge separations. The similarity of such behavior obtained for reactions that traverse different regions of this protein complex suggests that the donor–acceptor coupling strengths are not strongly affected by their specific structural environments. Quantitation of these observations further indicates that the photosynthetic protein medium is a relatively inefficient conduit for mediating long-range donor–acceptor interactions and must therefore play a minor role in promoting the high efficiency of these reactions. An opposing view of protein-based ET mechanisms is held by Gray and co-workers [15,16] who have conducted numerous investigations of ruthenium-modified metalloproteins to show that these proteins do indeed display considerable heterogeneity in their ET properties. Thus, several situations have been reported in which protein-based ET reactions can occur across comparable through-space distances, but with rate constants that vary by as much as several orders of magnitude [15–17]. Such observations have been explained in terms of a tunneling pathway model in which long-range donor–acceptor interactions are mediated by optimal combinations of covalent, hydrogen bond, and through-space interactions that exist in a particular region of a given protein [18,19]. In contrast to the conclusions of Dutton and co-workers [13,14], this work appears to demonstrate that different protein structures can indeed display different barriers to electron tunneling. If true, these observations could have important implications for the future design of synthetic peptide systems for use in bioelectronics, therapeutics, and energy conversion schemes. Thus, the ongoing debate concerning the fundamental nature of protein-based electron transfer has caused numerous studies to be performed in an effort to learn more about the properties of native, chemically modified, and de novo designed ET systems. The latter approach to this problem offers a controlled manner in which to investigate this problem through the synthesis of well-defined, peptide-based donor–acceptor systems that can adopt the various structural motifs found in native proteins. After a brief introduction to electron transfer theory and experimental methods, this chapter will review some of the recent work conducted on model ET systems whose range of structural complexity extends from relatively simple peptide-bridged donor–acceptor compounds to organized peptide assemblies having well-defined tertiary and quaternary protein structures.

The redox-active cofactors that participate in biological ET reactions are usually weakly coupled. Thus, a golden rule expression is often used to describe the factors that affect the rates of these nonadiabatic ET reactions:

$$k_{et} = (4\pi^2/h)(\text{FCWD})|H_{DA}(r)|^2 \tag{1}$$

In Eq. (1), FCWD is the Franck-Condon weighted density of states that describes the nuclear contribution to k_{et}, and $H_{DA}(r)$ is the electronic coupling matrix element that defines the interaction strength occurring between the reactant and product wavefunctions at the transition state. In the classical limit, the Franck-

Condon term can be described by Eq. (2), which predicts a parabolic dependence of k_{et} on the driving force ($-\Delta G^0$) for electron transfer at a given nuclear reorganization energy (λ):

$$FCWD = (4\pi\lambda kT)^{-1/2} \exp[-[(\Delta G^0 + \lambda)^2/4\lambda kT]] \tag{2}$$

This formulation predicted the existence of the now familiar inverted Marcus region in which the ET rate constants are seen to decrease with increasing driving force under conditions in which $-\Delta G^0 > \lambda$. Numerous examples of this behavior have been reported for both small-molecule systems [20–24] and metalloprotein-based ET reactions [16,25–27].

In simple electron donor–acceptor complexes, the intervening chemical spacer is expected to provide a uniform barrier to electron tunneling from which the magnitude of H_{DA} is expected to decrease exponentially with increasing donor–acceptor separation [Eq. (3)]:

$$H_{DA}(r) = H_0 \exp[-\beta(r - r_0)/2] \tag{3}$$

Here r is the separation distance, r_0 the sum of the van der Waals radii of the donor and acceptor sites, H_0 the electronic coupling strength at van der Waals contact, and β the distance attenuation factor, which can be used to parameterize the efficiency of the donor–acceptor coupling. Thus, under conditions of activationless electron transfer ($-\Delta G^0 = \lambda$), the maximum ET rate constant, k_{max}, is expected to obey Eq. (4):

$$k_{max} = k_0 \exp[-\beta(r - r_0)] \tag{4}$$

Using this approach, values of $\beta = 0.6-1.2$ Å^{-1} have been reported for covalently bound donor–acceptor systems [20,28–31]. However, many of the reported data have not been properly corrected for reorganization effects and may thus represent only an upper limit to the distance dependence of H_{DA}. A recent comparison of activationless ET rate constants obtained from several different covalent donor–acceptor compounds yields a value of $\beta = 0.7$ Å^{-1} [13].

Seminal work by Closs, Miller, and co-workers [20] obtained the rate constants for intramolecular ET reactions occurring within a series of compounds in which the donor and acceptor sites were covalently attached to one another by rigid hydrocarbon spacers composed of fused cyclohexane rings. These studies showed that the magnitude of electronic coupling was not only influenced by the distance between their donor and acceptor sites according to Eq. (3), but also by the stereochemical geometry in which these sites were attached to the hydrocarbon spacer. Indeed, very similar values for H_{DA} were observed for two stereoisomers, D27aa (**1**) vs. D27ee (**2**) of a decalin-bridged compound that contained the same number of covalent bonds separating their donor and acceptor sites. This near-similarity in coupling strength was observed to occur despite an approximately 6 Å difference in the through-space (center-to-center) donor–acceptor separation in these compounds. This observation provided important evidence to suggest that the long-range donor–acceptor coupling was propagated by the number of intervening covalent bonds that linked the redox sites, and not by through-space or through-solvent interactions [20,32]. This work provided an important example of how the systematic study of a series of well-defined donor–acceptor systems can provide important information regarding the mechanism of long-range ET reactions. However, when compared to the chemically discrete small-molecule systems just described, the situation found in natural ET proteins presents a much more complicated problem to study. Redox-active cofactors are often embedded within a complicated protein matrix whose structural diversity can present a large number of putative chemical routes through which long-range donor–acceptor coupling can occur. Thus, computational methods designed to study the nature of protein-based ET reactions are confronted with the difficulties associated with considering large macromolecular systems, which necessitates the development of approximate methods to help simplify this problem [14, 18,33–35]. In light of these advances, it has become a challenge for the synthetic chemist to create new systems that are specifically designed to test the various theories of protein-based ET reactions. Indeed, the ongoing controversies in this field have made it of particular interest to study systems that can probe the ability, if any, of diverse protein structures to affect the magnitudes of long-range electronic coupling. An attractive approach to this problem involves the use of well-defined model compounds that retain many of the structural features found in native proteins. A particular advantage of this work is that these relatively small, chemically discrete donor–acceptor compounds can be designed to pursue systematic studies of the effects of structure in mediating long-range ET reactions. This chapter will review some recent studies on the ET properties of synthetic polypeptides and de novo designed proteins.

II. MEASUREMENT OF PHOTOINDUCED ELECTRON TRANSFER RATES

A. Photoinduced Electron Transfer Reactions

Laser flash photolysis offers a convenient method to study the kinetics of excited state ET reactions [36]. This technique uses a short pulse of light to generate a high-energy excited state that can either donate (or accept) an electron to (or from) a neighboring redox partner. A particular advantage of flash photolysis is that its time resolution is limited, in principle, only by the width of the laser pulse and the unperturbed lifetime of the excited state species.

As shown in Fig. 1a, optical excitation of a unimolecular ET compound **3** can be used to produce the excited state species **4** in which *D is the photoinduced electron donor, A is the ground state electron acceptor, and ''Bridge'' denotes the

(a) (b)

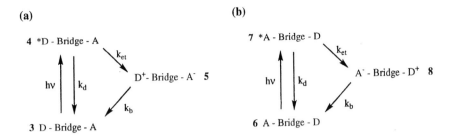

Figure 1 Reaction schemes for photoinduced ET studies in the presence of (a) a photoinduced electron donor and (b) a photoinduced electron acceptor.

intervening chemical bridge that serves to separate these sites from one another. A similar situation obtains in Fig. 1b except that in **7**, *A is now a photoinduced electron acceptor and D is a ground state electron donor. In both situations the decay of the excited state occurs by a process that includes contributions arising from the intrinsic radiative and nonradiative decay processes of the species, having the rate constant $k_d = k_r + k_{nr}$, in addition to the photoinduced ET reaction. Thus, emission lifetime measurements can be used to determine the ET rate constant (k_{et}) according to Eq. (5):

$$k_{et} = k_{obs} - k_d \tag{5}$$

where k_{obs} is the decay rate observed for the donor–acceptor complex and k_d is the rate constant for the intrinsic decay mechanism as measured for an appropriate model compound. In most cases, this is taken to be the identical fluorophore that lacks the bridged redox partner. In practice, it should be noted that suitable control experiments should always be performed to confirm that the emission quenching process is in fact due to an ET reaction and not a related energy-transfer mechanism.

Transient absorption spectroscopy offers another, somewhat more direct method to determine the rate constants for photoinduced ET reactions. This technique can be used when the products of an ET reaction (**5** or **8**) can be spectroscopically distinguished from its excited state reactants (**4** or **7**). Thus, under optimal conditions the rates of both the forward photoinduced ET reaction (k_{et}) and its thermal back reaction (k_b) can be studied for a given donor–acceptor complex by studying the magnitude of these spectroscopic changes as a function of time. However, regardless of the method used to determine ET rate constants, a concentration dependence of k_{et} should be determined to verify that the electron transfer is indeed due to an intramolecular event and not a bimolecular reaction.

B. Ground State Electron Transfer Reactions by Pulse Radiolysis

Two limitations to the study of photoinduced ET reactions are as follows: (1) they are necessarily confined to the study of donor–acceptor systems having access to a photoexcited state and (2) they can only be conducted in cases where the ET event occurs rapidly on the time scale of the excited state decay. Thus, to help overcome these problems the electron pulse radiolysis technique has been used as a complementary tool for the study of ground state ET reactions that can occur over a more extended time domain [37,38]. This method uses a rapid pulse of electrons to generate a population of solvated radicals that can be used to initiate a subsequent intramolecular ET reaction within a donor–acceptor compound. Thus, pulse radiolysis of an aqueous solution is known to create a population of hydrated electrons (e_{aq}^-), hydroxyl radicals (OH•), and hydrogen atoms (H•). Other products such as hydrogen peroxide and hydronium ion can also be formed but are easily controlled by the appropriate use of buffers. An advantage of this technique is that each of the generated radicals can be selectively converted to other radicals and/or eliminated by scavengers, which gives the user exquisite control over the creation of either a strongly reducing or a strongly oxidizing environment in which to initiate subsequent intramolecular ET reactions. In the example shown in Fig. 2a, pulse radiolysis is used to produce a population of hydrated electrons (e_{aq}^-), which are powerful reducing agents ($E^0 \approx -2.8$ V) that can proceed to nondiscriminately attack the bifunctional donor–acceptor complex **9**. This second-order reaction creates a *nonequilibrium* distribution of the two singly reduced species **10** and **11** which can then follow the thermodynamic path back to equilibrium via an intramolecular electron transfer. The kinetics of the ET process can be followed by time-resolved absorption measurements that monitor either the disappearance of the donor species (D⁻) in **10** or the appearance of the reduced acceptor (A⁻) in **11**. Figure 2b shows a complementary series of reactions in which pulse radiolysis is used to create an initial population of radical oxidants, such as CO₃·⁻ ($E^0 = +1.5$ V) and N₃·⁻ ($E^0 = +1.3$ V), which undergo a reaction

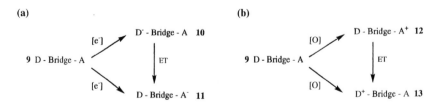

Figure 2 Reaction schemes for intramolecular ET processes initiated by (a) reductive pulse radiolysis and (b) oxidative pulse radiolysis.

to produce a nonequilibrium distribution of the two singly oxidized species **12** and **13**. Again, intramolecular electron transfer returns the system to equilibrium. In examining Fig. 2, it is noted that the temporal resolution of this experiment is limited at fast times by the rate of the initial diffusional reaction between the generated radical species and the donor–acceptor complex. The longer time limit is defined by the intrinsic lifetimes of ground state reactants, which is usually much longer than those of typical photoexcited states, and the rates of the related intermolecular ET reaction. Thus, pulse radiolysis is often used to study slower ET reactions than can be examined by direct laser flash photolysis.

C. Ground State Electron Transfer Reactions by the Laser Flash-Quench Method

Ground state ET reactions can also be studied using a laser "flash-quench" technique that employs an exogenous ET reagent to form the precursor compound for a subsequent intramolecular, ground state ET reaction [39,40]. As shown in Fig. 3, the oxidative flash-quench method begins with the photoexcitation of a ground state complex **14** in the presence of a large (approximately 10^2-fold) excess of the exogenous oxidant Q. The resulting intermolecular quenching reaction rapidly produces the oxidized electron acceptor (A^+) within complex **16**. The newly formed A^+ species can then oxidize the proximal electron donating site (D) by intracomplex electron transfer to produce **17**, in a process that can be monitored by transient absorption spectroscopy. An advantage of this method as compared to pulse radiolysis is that it is a relatively inexpensive technique for studying ground state ET reactions and requires only the presence of a photosensitizer whose excited state lifetime can be relatively short (<100 ns).

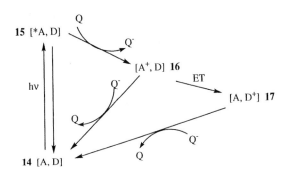

Figure 3   Reaction scheme for studying ground-state ET reactions by the oxidative flash-quench method.

III. DONOR–ACCEPTOR COMPOUNDS WITH COVALENT PEPTIDE BRIDGES

A. Electron Transfer Across Polyproline Bridges

During the mid-1980s, Isied and co-workers began a systematic investigation of the ET properties of peptide-bridged donor–acceptor systems. Several series of unimolecular donor–acceptor complexes were prepared in which small polypeptide chains were used to provide a covalent link between kinetically inert, redox-active transition metal ions. The first series of compounds studied were of the type $[SO_4(NH_3)_4Ru^{II}(iso)\text{-}(AA)_n\text{-}Co^{III}(NH_3)_5]^{2+}$ in which $n = 0$ (**18**), 1, and 2, and where $(AA)_1$ = Gly (**19**), Phe (**20**), or Pro (**21**), and $(AA)_2$ = Phe-Phe (**22**), Phe-Gly (**23**), or Leu-Gly (**24**) [41]. Here, the iso linker (iso = isonicotinic acid) was used to attach the ruthenium(II) electron donor to the N-terminal residue of the bridging ligand via an amide bond, and the cobalt(III) acceptor was directly coordinated to the C-terminal residue through a carboxalato linkage. The ET process was thermodynamically endoergic ($\Delta G^0 = +0.54$ eV), being driven by the rapid aquation of the reduced Co(II) complex. Not surprisingly, the rate constants for these reactions were quite small ($k_{et} = 1.24 \times 10^{-2}$ s^{-1} for **18**) and underwent a substantial decrease upon the introduction of successive amino acids to the intervening peptide bridge. In addition, the ET rate constants appeared to be relatively insensitive to the specific amino acid composition of the bridge and did not follow the distance dependence expected from Eq. (4). Rather, the rate constants decreased by $>10^2$-fold between the $n = 0$–1 compounds, but only <10-fold upon introduction of the second amino acid to the bridge. Significant differences were observed for the activation parameters ($\Delta H^{\ddagger}$ and $\Delta S^{\ddagger}$) obtained for the different series of mono- and dipeptides studied. Together, these results were interpreted in terms of an ET mechanism in which the rate-determining step involved a conformational rearrangement of the peptide bridge to bring the donor and acceptor sites in close spatial proximity to one another. The importance of this early work was that it emphasized the need to construct model systems having structurally rigid peptide bridges to better investigate the ET properties of specific protein structures.

In subsequent work, Isied and co-workers began to use peptide spacers that consisted of more rigid oligoproline chains [42], as these peptides were previously shown to form relatively stable helical structures at small chain lengths. This unusual characteristic of oligoprolines is due to the cyclic nature of the amino acid side chain, which restricts rotation about the peptide bond to produce a cis/trans conformational equilibrium that is slow on the NMR time scale. Thus, homooligoprolines can exist within two different types of solvent-stabilized helical structures (Fig. 4). In nonpolar solvents the compact, all-cis poly(L-proline)-I secondary structure is favored, which exists as a right-handed helix having 3-1/3 residues per turn and a through-space translation of 1.85 Å per residue.

trans   cis

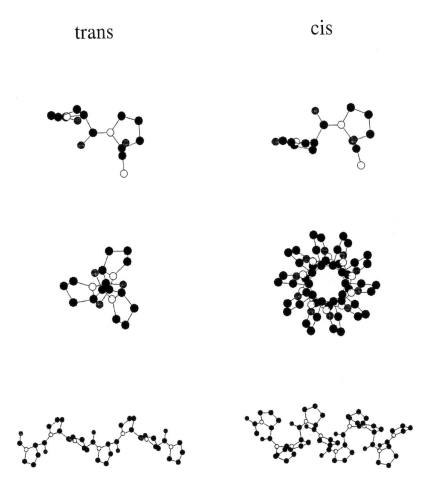

Figure 4 Views of the polyproline structures in the all-trans (i.e., polyproline II) and all-cis (i.e., polyproline I) conformations: (Top) proline dipeptides, (middle) down the helical axis, (bottom) perpendicular to the helical axis.

In contrast, polar media encourages the formation of the all-trans poly(L-proline)-II structure, which is an extended left-handed helix having a repeat length of three amino acids and a translation length of 3.12 Å per residue. It should be noted that both polyproline structures exist as solvent-stabilized conformations which, unlike the α helix, do *not* rely on the formation of long-range, interresidue hydrogen bonds for their stability. Thus proline *trimers* [43] and *tetramers* [44] crystallized from aqueous solution have been shown to adopt the conformational properties of the proline II helix. As an illustration of the solvent-dependent nature of proline conformations, Schanze and Sauer [45] studied photoinduced ET reactions occurring between a ruthenium polypyridyl donor and a quinone acceptor attached to the ends of a polypeptide spacer composed of zero to four proline residues (25–29) as shown in Fig. 5. Due to the limited solubility of these compounds in aqueous solvent, the photophysical studies were conducted in CH_2Cl_2 where the equilibrium of proline conformations was weighted more heavily to the cis isomer. The resulting population of multiple peptide conformations produced multiphasic ET kinetics. Whereas an overall $>10^3$-fold decrease in ET rate constant was observed as the length of the proline bridge was increased from zero to four residues, an unambiguous analysis of this distance dependence was not possible due to the complicated kinetics.

An early study of proline-bridged donor–acceptor compounds conducted in aqueous solvent involved compounds (30–34) of the type shown in Fig. 6a where the electron-donating Os(II) species was generated in situ by reductive pulse radiolysis [46]. The subsequent intramolecular ET reaction proceeded with a modestly exoergic driving force of $\Delta G^0 = -0.17$ eV under solvent conditions in which the *trans*-polyproline II structure was favored. The intramolecular ET rate constants decreased from $k_{et} = 2.7 \times 10^2$ s^{-1} for 31 to $k_{et} < 9 \times 10^{-2}$ s^{-1} for 34. However, in a manner similar to that described above, the magnitude of k_{et} again did not show a simple exponential decrease with increasing peptide length. Rather, the rate constants decreased by $>10^2$-fold between 31 and 32, >8-fold between 32 and 33, and only >4-fold between 33 and 34. Apparently, the mechanisms of the slower reactions were again complicated by the equilibration of different conformational forms of the proline bridge, perhaps mediated by the $Os^{II/III}$ self-exchange which occurs rapidly on the time scale of the slow ET reactions. Nevertheless, these results did allow an important comparison to be made between the rate constants observed for the various dipeptide-bridged

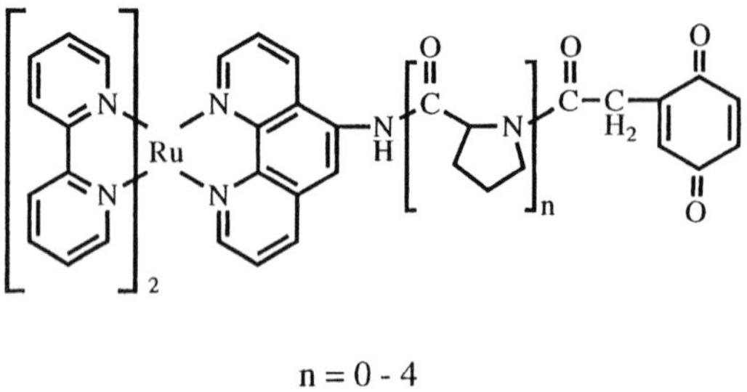

n = 0 - 4

(25 - 29)

Figure 5 Proline-bridged donor–acceptor complexes with organic redox centers.

species studied by these workers. It was found that when the donor and acceptor sites were joined by the more flexible spacers (i.e., Phe_2 and Gly_2, respectively) the resulting ET reaction occurred with rate constants that were almost an order of magnitude larger than that observed across the Pro_2 bridge ($k_{et} = 0.74$ s^{-1}). Since all three compounds contained exactly the same through-bond distances separating their donor and acceptor sites, the observed differences in rate were thought to reflect a somewhat closer through-space approach in the more flexible systems. In addition, the observation of single exponential kinetics in the more rigid donor–acceptor system placed a probable lower limit to the cis/trans conformational equilibrium time of $\tau > 1.4$ s.

Further investigation of the distance dependence of the rates of electron transfer occurring across proline bridges required that systems be developed that could display rate constants that were larger than that of the peptide conformational equilibrium. Thus, another series of complexes was prepared in which the pentamminecobalt(III) site was replaced by the more oxidizing pentamminruthenium(III) acceptor (**35–39**) [47,48]. The ET reaction for these systems proceeded with much larger rate constants due to the larger thermodynamic driving force for the reaction ($\Delta G^0 = -0.25$ eV) and the lower inner-sphere reorganization energy of the pentammineruthenium(III) acceptor. In these studies, pulse radiolysis techniques were used to show that LRET reactions can occur rapidly on the cis/trans conformational time scale in cases where the redox centers were separated by as many as four proline residues ($d_{M-M} \approx 21$ Å). The unimolecular ET rate constants in these systems decreased from a value of $k_{et} = 3.6 \times 10^6$ s^{-1} for **36** to $k_{et} \approx 50$ s^{-1} for **39**. Analysis of the activation parameters for these reactions demonstrated that solvent reorganization effects have a significant effect in governing the distance dependence of intramolecular ET rates and that the electronic factor in these systems follows an attenuation factor of $\beta = 0.68$ Å^{-1} using the through-space distance estimate of 3.12 Å per residue. Unfortunately, accurate values for $\Delta H^{\ddagger}$ and $\Delta S^{\ddagger}$ could not be obtained for the $n = 4$ compound (**39**), which was not included in the analysis. Interestingly, Schanze and Cabana [49] observed a nearly identical distance dependence of the activation parameters for k_{et} occurring between a rhenium bipyridyl electron acceptor and a (dimethylamino)benzoate donor across zero to two proline residues.

In later work, two series of high-driving-force, proline-bridged ET compounds were prepared having the form $[(bpy)_2Ru^{II}L\text{-}(Pro)_n\text{-}M_2^{III}(NH_3)_5]^{m+}$, where bpy = 2,2′-bipyridine, L = 4-carboxy-4′-methyl-2,2′-bipyridine, and M_2 was either $Co^{III}(NH_3)_5$, or $apyRu^{III}(NH_3)_5$, in which apy = 4-aminopyridine (Fig. 6c and 6d) [50,51]. In these systems, ET reactions were initiated by reductive pulse radiolysis which generated the radical anion $[(bpy)_2Ru^{II}L]^{-\cdot}$ as the electron donor. It was expected that the presence of this highly reducing ($E^0_{1/2} = -1.2$ V vs. NHE) ligand-centered radical would increase the magnitude of k_{et}, allowing the study of ET reactions occurring over longer distances. The first series of com-

(a)

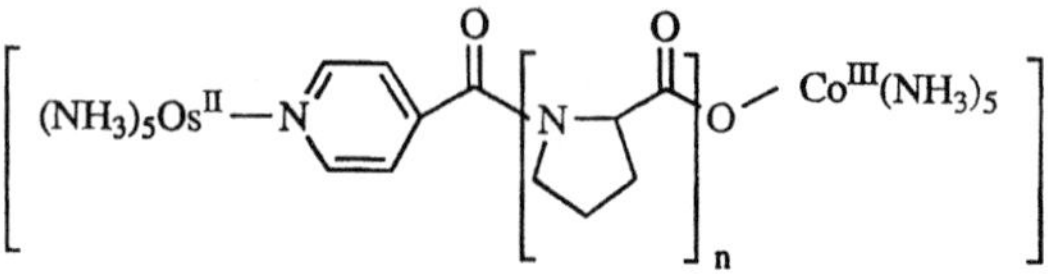

n = 0, 1, 2, 3, 4

(30 - 34)

(b)

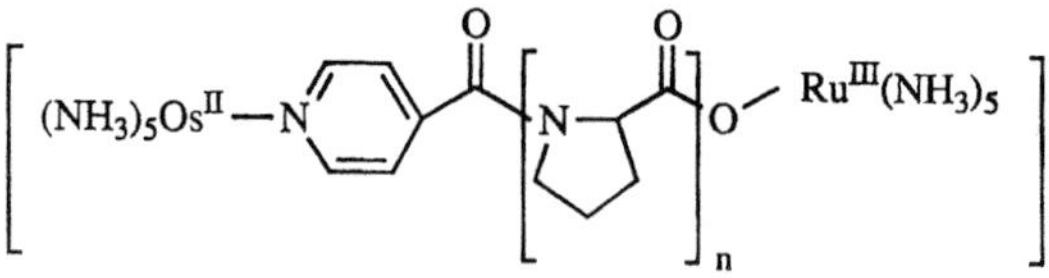

n = 0, 1, 2, 3, 4

(**35** - **39**)

(c)

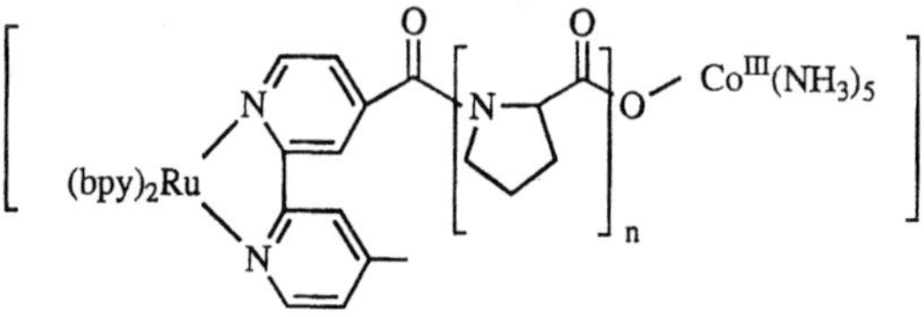

n = 0, 1, 2, 3, 4, 5, 6

(**40** - **45**)

(d)

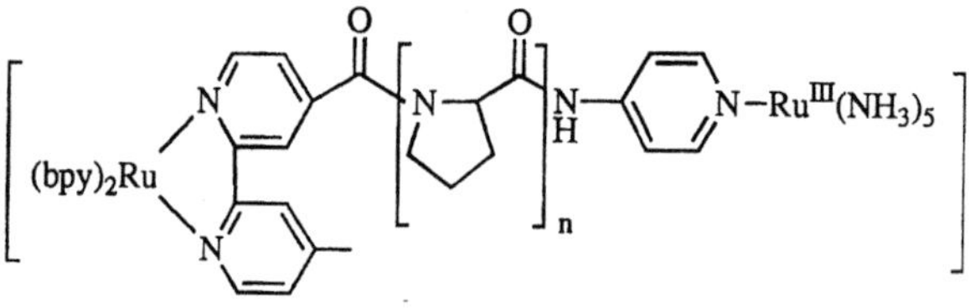

n = 5, 6, 7, 8, 9

(46 - 50)

Figure 6 Proline-bridged donor–acceptor complexes having metal-based redox sites.

pounds studied was $[(bpy)_2Ru^{II}L\text{-}(Pro)_n\text{-}Co^{III}(NH_3)_5]$, $n = 1\text{–}6$ (**40–45**), in which the driving force for electron transfer was $\Delta G^0 \approx -1.1$ eV. As predicted, the intramolecular ET reactions observed for the entire series of compounds occurred on a much faster time scale. Single exponential ET kinetics were observed having rate constants ranging from $k_{et} > 5 \times 10^8\ s^{-1}$ for **40** to $k_{et} = 8.9 \times 10^3\ s^{-1}$ for

45. An interesting result of this study is shown in Fig. 7 where the distance dependence of these rates can be divided into two separate regions. The shorter metallopeptides (**40–42**) define a region of k_{et} values which decay strongly with distance, indicative of a relatively weak donor–acceptor interaction. In contrast, the compounds that contain longer oligoproline bridges (**43–45**) appear to comprise a region of stronger electronic coupling in which the ET rates are much less sensitive to the donor–acceptor separation. Analysis of these data according to Eq. (4), using the appropriate through-space distance estimates yields the empirical values of $\beta = 1.3$ Å^{-1} and 0.29 Å^{-1} for each of these two regions, respectively [50]. It should be noted that these values only represent upper limits to the actual distance attenuation factor for electronic coupling as these ET reactions were not free energy–optimized and occur below the top of the Marcus curve (vide infra). This point is especially valid for the shorter compounds in which solvent reorganization effects would be expected to make a larger contribution to the distance dependence of k_{et}. However, despite this caveat the authors noted that the demarcation between the putative regions of weak and strong electronic coupling occurred at the point where the polyproline-II bridge completes its first helical turn. Thus, these results provided a tantalizing suggestion that secondary protein structures may indeed play an important role in mediating long-range ET reactions.

In order to address the concern that the region of slowly decaying ET rates may be due to the effects of a rate-limiting conformational motion, the authors prepared a second series of compounds in which the -Co^{III}(NH$_3$)$_5$ electron ac-

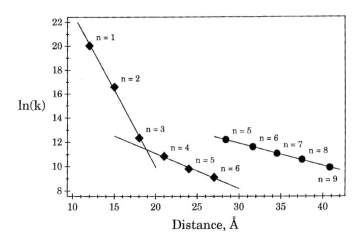

Figure 7 Plot of ln k_{et} vs. through-space metal-to-metal distance at high-driving forces for **40–45** (♦) and **46–50** (●).

ceptor was replaced by the more oxidizing apyRu^{III}(NH$_3$)$_5$ species and $n = 5–9$ (**46–50**) [51]. This substitution was designed to simultaneously raise the thermo-dynamic driving force for the reaction and lower its inner-sphere reorganization barrier. These effects were expected to produce even faster rates of electron trans-fer *if* the ET mechanisms did not involve a rate-limiting conformational motion of the peptide bridge. As shown in Fig. 7, larger values of k_{et} were indeed observed at comparable metal-to-metal distances for this new series of compounds and, significantly, the distance dependence of these rates was nearly identical to that previously observed for the region of strong electronic coupling within the series of Ru-Co compounds. Thus, the distance dependence of k_{et} in both systems likely reflects a modulation of the electronic coupling between the donor and acceptor sites, and is not a result of conformational gating. The authors note that in **50**, electron transfer occurs over a metal-to-metal distance of >40 Å, which is com-parable to the diameter of cytochrome *c*.

A similar weak distance dependence of k_{et} at long bridge lengths has been independently observed by several other research groups for different polypro-line-based systems. In particular, Bobrowski and co-workers [52–54] used pulse radiolysis to study the reaction:

H-Trp$^{\bullet}$-(Pro)$_n$-Tyr-OH $\rightarrow$ H-Trp[H]-(Pro)$_n$-Tyr$^{\bullet}$ [O]-OH

in which the proton-coupled ET process produced an oxidized tyrosine radical, and $n = 0–5$ (**51–56**). A combination of multinuclear NMR, circular dichroism, and molecular modeling methods was used to describe the conformational proper-ties of the peptides (Fig. 8a). In a manner similar to that previously discussed by Sneddon and Brooks [55], the shorter bridged peptides (**51–53**) were found to exist within a set of rapidly converting rotamers in which the flexibility of the aromatic tyrosine and tryptophan side chains allowed them to be separated by short edge-to-edge separations. In contrast, the longer chain peptides (**54–56**) were found to reside within a more rigid polyproline II type of structure. The observed distance dependence of k_{et} in the above systems was analyzed in terms of a model that accounts for the specific conformational properties of the peptides. Thus, a through-space mechanism was shown to be operational in the short-bridged systems, and a through-bond mechanism was observed for longer proline bridges. Interestingly, these latter compounds yielded an empirical value of $\beta = 0.25$ Å^{-1}. The authors further suggested that the bimodal distance dependence observed by Isied and co-workers [50] could be rationalized by similar consider-ations of peptide conformation for the shorter peptide bridges.

In closely related work, Farraggi, Klapper, and co-workers [56–58] used pulse radiolysis to study the ET reaction discussed above, in addition to one in which the relative positions of the electron donor and acceptor were reversed in the amino acid sequence. Experiments using 1-*N*-methyltryptophan in place of tryptophan showed that the net proton transfer was not the rate-limiting step in

a

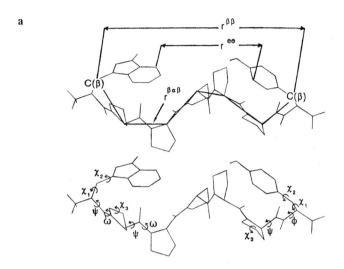

b

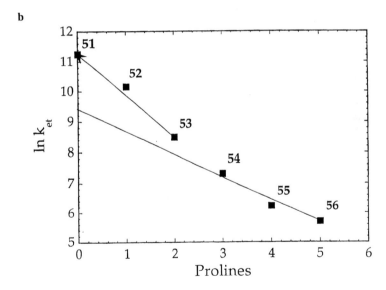

Figure 8   (a) Structural and conformational parameters for H-Trp-(Pro)$_n$-Tyr-OH compounds (**51–56**). (From Ref. [54].) (b) Distance dependence of ET rate constants.

this ET process. These workers further showed that the observed distance attenuation factor for the nonactivationless ET rates was $\beta = 0.33$ Å^{-1} when the Trp residue was located toward the N terminal of the peptide, but $\beta = 0.2$ Å^{-1} when the direction of the electron transfer was reversed. Finally, Tamiaki and co-workers [59–61] have been studying photoinduced ET processes in peptide-bridged donor–acceptor compounds. In one system a zinc(II) tetraphenylporphyrin moiety and an iron(III) tetraphenylporphyrin were coupled to the ends of an oligoproline bridge ($n = 0$–8) by methylene linkers (**57–65**). Steady-state fluorescence measurements taken in ethanol showed that the observed rates of electron transfer decayed with a value of $\beta = 0.16$ Å^{-1}. This result was independent of whether the ZnTPP was attached to the N or C terminus of the proline bridge.

In summary, the body of work described above indicates that the polyproline II secondary protein structure provides an extremely effective ET pathway that allows electrons to be transferred over very long distances (>40 Å). It is interesting to note that the small distance attenuation factors observed for these systems are of the same order of magnitude as that originally claimed by Barton and co-workers [62] in DNA-based donor–acceptor compounds. However, the specific mechanism for the efficiency of the proline-bridged ET reactions is still not understood.

B.   Electron Transfer Over α-Helical and 3₁₀-Helical Bridges

One limitation to work described above is that many of the donor and acceptor sites of the proline-bridged systems are directly attached to the backbone of the biological spacer. Thus, systematic variation of the helical bridge length produces monotonic changes in both the through-space *and* through-bond donor–acceptor separation. A simple correlation of ET rates with the length of the intervening peptide bridge can therefore offer little insight into whether a through-space or through-bond ET mechanism operates in these systems. Thus, several research groups have studied the ET properties of polypeptides in which the donor and acceptor sites were attached to the amino acid side chains of an α-helical or 3_{10}-helical peptide bridge. Such systems can provide a convenient means to study the relative contributions of through-bond vs. through-space ET mechanisms as systematic changes in the residue length of the α-helical bridge will produce monotonic changes in the through-bond donor–acceptor distances, but *not* the through-space separations. With this situation in mind, a significant synthetic challenge was confronted in the design of donor–acceptor compounds having stable α-helical bridges. The difficulty resides in the fact that α helices are stabilized by long-range interresidue hydrogen bonds, and most small polypeptides are unable to form the necessary number of long-range interactions to form the desired structure. Indeed, it has been reported that "helical" polypeptides composed of <50 amino acid residues exhibit a relatively large degree of conforma-

tional disorder at ambient temperature [63]. The following paragraphs will describe some successful routes used to surmount this difficulty.

In early studies, Sisido and co-workers [64–68] used a host–guest approach
to stabilize several series of synthetic α helices for use in both electron and energy
transfer studies. A variety of short donor–acceptor sequences was introduced into
poly(γ-benzyl-L-glutamate) whose steady-state conformational properties were
assigned to be α-helical by circular dichroism spectroscopy. Intramolecular excimer fluorescence studies conducted on a L-1-pyrenylalanine-substituted
polyglutamate showed that the conformational fluctuation time of the helical
bridge is about 20 ns at 20°C [66]. Further work showed that the efficiency of
long-range energy transfer between the side chain modified amino acids, L-4-bi-
phenylalanine and L-1-naphthalene followed the expected r^{-6} distance dependence if at least four γ-benzylglutamate residues were placed at the C terminus
of the donor–acceptor sequence and >30 glutamate residues were placed at the
N terminus [68]. Apparently, peptides having a smaller number of glutamate
residues experienced a larger degree of conformational flexibility. These results
were used to demonstrate the viability of the host–guest approach to produce
conformationally stable α-helical peptide bridges. Thus, unimolecular ET studies
were conducted on compounds **66–68** (Fig. 9). in which a polyglutamate host
was modified to incorporate a pyrenylalanine photoinduced electron donor and
a dimethylaminophenylalanine acceptor that were separated from one another by
$m = 0-2$ alanine residues. As stated above, an important consequence of placing
the donor and acceptor sites on the amino acid side chains is that the helical
conformation forces the through-space donor–acceptor distance to vary nonuni-

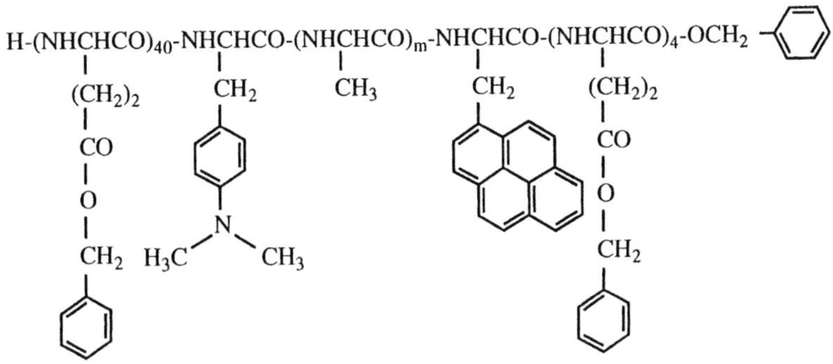

$m = 0, 1, 2$

(66 - 68)

Figure 9 Structure of α-helical donor–acceptor peptides using a host–guest approach.

formly with the residue length of the spacer. Thus, the through-bond separation increases by exactly three σ bonds for each additional alanine residue, whereas this elongation of the peptide bridge produces through-space donor–acceptor distances of $d_{edge-to-edge}$ = 5.4 (**66**), 9.4 (**67**), and 5.5 Å (**68**) for m = 0, 1, and 2, respectively. The results of laser flash photolysis studies conducted in trimethyl phosphate showed that the rate constants for intramolecular electron transfer from the DMA-Phe donor to the PYR-Ala acceptor were k_{et} = 1.9 × 10^7, 6.6 × 10^5, and 2.1 × 10^7 s^{-1} for **66–68**, respectively [67]. Clearly, the rates do not depend on the number of covalent bonds separating the donor and acceptor groups but rather scale with their through-space distances. These results indicate that in such *short α-helical segments* relatively weak through-bond interactions occur between these sites in a manner that can be contrasted to the behavior reported for donor–acceptor compounds containing rigid organic bridges. Analysis of the data obtained for the α-helical systems using the through-space donor–acceptor distances yielded a value of β = 0.93 Å^{-1}. It must be noted, however, that these reactions have not been corrected for reorganization effects, and these results can only provide an upper limit to the electronic distance attenuation factor.

Related research by Kuki and co-workers [69–71] provided additional evidence to suggest that ET rates are not predominately controlled by a through-bond mechanism in short hydrogen-bonded helices. Earlier studies from these [72] and other [73] workers have determined that short sequences rich in α,α-dialkylated amino acids such as α-aminoisobutyric acid (Aib) can exist within stable α helices or 3$_{10}$ helices depending on their chain length and specific Aib content. These structures can be distinguished from one another other by identification of their characteristic hydrogen bonding patterns by ^{1}H NMR. Thus, a series of Aib-rich octamers was prepared in which a β-(1′-naphthyl)-L-alanine fluorophore (Nap) was separated from a p-bromo-L-phenylalanine quencher (Bph) by, respectively, one, two, and three Aib residues (**69–71**). ^{1}H NMR experiments showed that **69** existed as an α helix whereas **70** and **71** formed 3$_{10}$ helices. As seen in Fig. 10, the fluorophore/quencher pair in **71** are arranged along the same face of the 3$_{10}$ helix and come within van der Waals contact of each other. In contrast, these groups are forced to maintain longer through-space distances in the remaining two peptides, **69** and **70**. Both steady-state and time-resolved fluorescence studies were used to determine that the naphthylalanine emission was quenched by the bromophenylalanine moiety by a mechanism that includes both a nominally spin-forbidden singlet-triplet energy transfer process and a remote heavy atom effect (RHAE) which enhances the intersystem crossing rate of the fluorophore. Importantly, this latter effect can be used to measure the degree of electronic interaction occurring between the naphthylalanine fluorophore and the heavy atom quencher. The rates of the bromine-induced singlet-triplet energy-transfer process were therefore analyzed according to Forster theory to

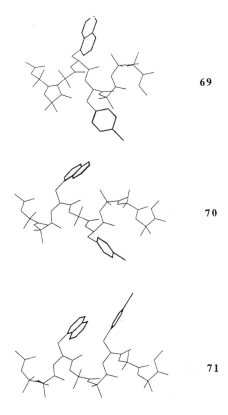

69

70

71

Figure 10 Structures of octameric Aib-containing peptides in their preferred helical conformations as determined by NMR experiments. (From Ref. [71].)

provide an internal measure of the through-space donor–acceptor distance in these systems to yield 4.8, 5.8, and 3.2 Å for **69–71**, respectively. The rate constants for the RHAE process were then measured to be $k_{RHAE} = 0.5 \times 10^6$, 1.1 $\times 10^6$, and 14.1 $\times 10^6$ s^{-1} for these compounds. Interestingly, the values of k_{RHAE} do not simply scale with either the through-bond or the through-space separation between these moieties. As may be expected, the fastest rate for the heavy atom effect was seen for the **71** species in which the fluorophore and quencher are in near–van der Waals contact with one another. However, the smallest rate constant was seen for **69**, which had an intermediate through-space separation. The authors suggested that this behavior lends support to the pathway model for biological ET reactions in which long-range donor–acceptor coupling is mediated through an appropriate combination of covalent, hydrogen bond, and through-space inter-actions.

In recent work, Fox and co-workers [74–76] generated a series of Aib-containing polypeptides (Fig. 11) that incorporated alanine residues that were derivatized to contain either a pendant electron donor (N-dimethyl-p-anilino) or electron acceptor (2-pyrenyl) attached to their methyl side chain. NMR studies that showed that these peptides displayed nuclear Overhauser effect (NOE) inter-actions that were characteristic of an α helix, but not a 3_{10} helix. However, unlike the previously discussed body of work, these donor–acceptor systems were spe-cifically designed to test the hypothesis that the electrostatic properties of α heli-ces can affect the magnitudes of k_{et}. This work was based on previous observa-tions that within an α helix the amide bond dipole moments are arranged in a head-to-tail fashion that is oriented almost parallel to the helical axis [77]. This situation produces a large (10^9 V/m) electrostatic field along the length of the helix that places a net positive charge at the N terminus of the helix and a net negative charge at the C terminus. It was thus speculated that a larger thermody-namic driving force for electron transfer should exit in cases where the electron is transferred in a direction that is antiparallel to the helix dipole (i.e., C → N) than in cases where it is transferred along the helix dipole (i.e., N → C). Thus, two ET polypeptides were prepared (72, 73) having opposite donor–acceptor orientations with respect to the chain direction. In both cases, the donor–acceptor separations were estimated to be about 10 Å. Significantly, the magnitudes of k_{et} were found to be 5–27 times faster in 72 than in 73, with smaller differences being observed in solvents of higher dielectric constant. These observations were in agreement with the predicted behavior to suggest that electrostatic effects can

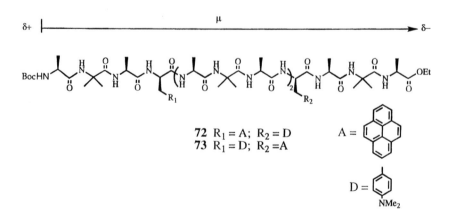

Figure 11   Structures of α-helical donor–acceptor peptides for helix dipole studies.

indeed be used to provide directionality to ET efficiencies in model proteins. However, biexponential decays were consistently observed in the emission life-time experiments which indicated that the ET peptides adopted multiple conformations on the timescale of the ET reaction. Nevertheless, when the helical nature of the peptides were destroyed by the use of chemical denaturants or the incorporation of helix-breaking proline residues into the peptide sequence, the observed ET rate constants for the two systems became nearly identical.

C. Electron Transfer Within Models for β-Pleated Sheets and β Strands

β-Pleated sheets compose an important protein structure motif whose ET properties have only recently begun to be explored in natural proteins [78,79]. These structures consist of an antiparallel (or, less commonly, parallel) array of fully extended peptide chains that are connected by a dense network of interchain hydrogen bonds and hydrophobic side chain contacts. The rational design of β-sheet models is faced with the difficulty of constructing a colinear array of extended peptide chains which have the proper values of the peptide dihedral angles (ϕ and ψ) that define this structure. Thus, relatively few artificial β-pleated sheets have been described in the literature [80–82]. The synthesis of β-sheet ET complexes is made even more difficult by their need to incorporate suitable electron donor and acceptor sites along their lengths. However, based in part on the earlier work of Kelly and co-workers [80], Ogawa and co-workers have prepared two series of metallopeptides that exist as water-soluble mimics of individual β strands [83] and β-pleated sheets that can undergo photoinduced electron transfer [84–86].

The single β-strand peptides (**74–76**) are shown in Fig. 12 in which oligovaline chains were used due to their known propensity to exist within the desired secondary structure. Two-dimensional ^{1}H NMR experiments were performed on each of these compounds at 298 K in both aqueous and methanol solutions to show that they displayed values of $^3J_{\text{NH-C}\alpha} > 7.5$ Hz, and strong sequential (i, $i + 1$) NH-C$_\alpha$H contacts that are diagnostic of the β-strand conformation [87]. Emission lifetime measurements and high-performance liquid chromatography product analysis showed that all three binuclear donor–acceptor compounds underwent photoinduced electron transfer in both aqueous solution at 298 K, as well as in a frozen ethanol-methanol glass at 77 K. Significantly, the compounds displayed single exponential ET kinetics in the frozen glass, indicating that they each existed within a unique donor–acceptor separation. No evidence for multiple conformations was observed. The values of k_{et} observed for **74–76** decreased with increasing donor–acceptor distance and were fit to Eq. (4) to yield a distance attenuation factor of $\beta_{\text{space}} = 1.1 \pm 0.4$ Å^{-1} in both fluid solution and the cryogenic glass. The cause for the possible deviation from a simple exponential distance

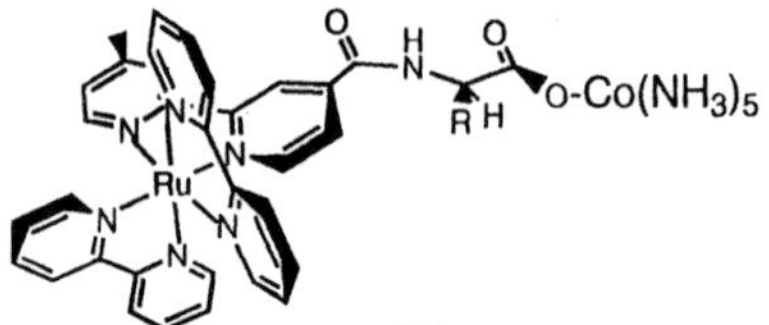

74

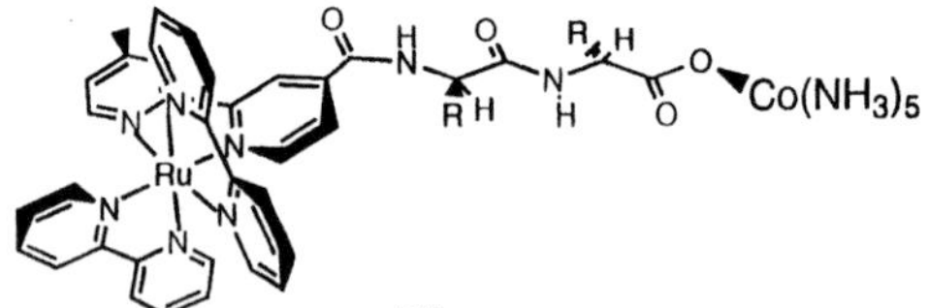

75

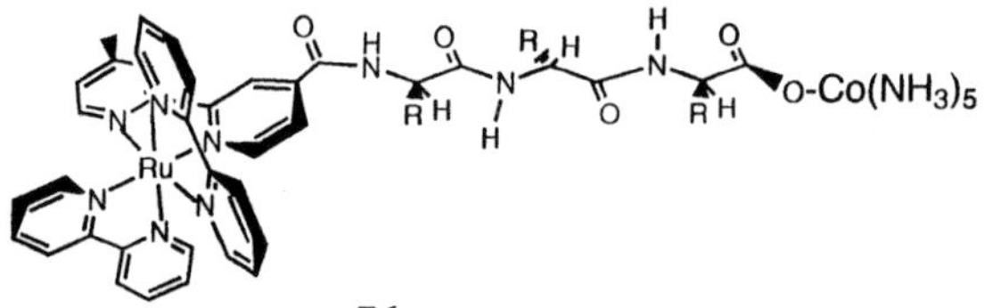

76

Figure 12   Single β-strand donor–acceptor peptides: R = −CH(CH₃)₂.

dependence is not clear. However, the authors note that the similarity in behavior obtained at room temperature and at 77 K indicated that the electronic coupling term dominates the distance dependence of the k_{et}.

The two-stranded β-sheet models (**77, 78**) used a novel ruthenium polypyridyl complex, Ru(bpy)₂L, to bring two valinyl peptide chains in close proximity to one another (Fig. 13), where L = 3,5-dicarboxy-2,2′-bipyridine [88]. In these studies, 2-D NMR was again used to show that the individual peptide chains of **78** possessed the requisite properties of a β strand. Evidence for the assignment of significant β-sheet character to **78** was obtained from the temperature coefficients of the chemical shifts of the amide protons, which indicated that the C-terminal amide of the peptide chain attached to the C-3 position of L was involved in an intramolecular hydrogen bond (Fig. 13). In contrast, the bis(monovaline) peptide, **77**, was too small to form any type of organized structure.

The luminescence of both **77** and **78** were quenched relative to that of the ruthenium metallopeptide which contained no cobalt(III) acceptor. As before, product analysis indicated that the quenching process was due to an ET event.

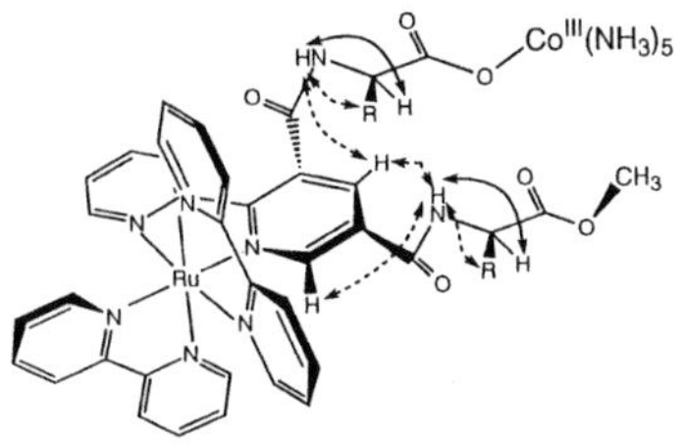

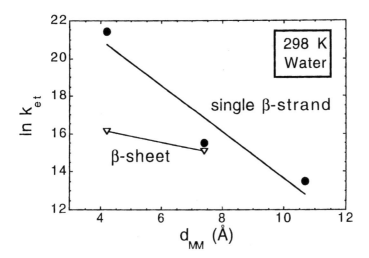

77 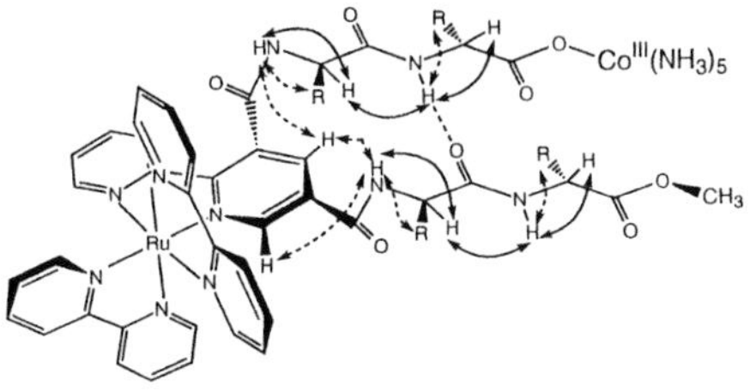   78

Figure 13 Two-stranded β-sheet models. NMR studies indicate that the peptide chains of **78** are joined by an interchain hydrogen bond.

Both steady-state and time-resolved emission data yielded a concentration-independent quenching rate constant of $k_q = 6.9(1) \times 10^6$ s^{-1} and $k_q = 3.8(1) \times 10^6$ s^{-1} for **77** and **78**, respectively. At this point, it is interesting to compare the distance dependence of k_{et} observed for the two-stranded β-sheet mimics (**77, 78**) with that seen for the single β strands having the analogous number of intervening valine spacers (**74, 75**). As shown in Fig. 14, k_{et} for **77** is 20-fold *smaller* than that measured for the analogous β strand (**74**) where each compound has a single

Figure 14 Comparison of ET rate constants for the single-chain β-strand compounds **74–76** (●) and double-chain β-sheet models **77** and **78** (△), respectively.

valinyl residue separating the donor and acceptor sites. This observation can be rationalized by a simple driving force argument since the emission maxima of the two-stranded compounds occur at a much lower energy (λ_{em} = 725 nm vs. 670 nm), which makes the photoinduced ET reaction more endoergic. Interestingly, despite its significantly lower driving force, the value of k_{et} for the two-stranded β-sheet model, **78**, is nearly identical to that of its single-stranded analog, **75**. This nonthermodynamic result is very interesting since NMR measurements show that each of the peptide chains in these compounds have nearly identical conformations with the exception that the two-stranded (n = 2) compound shows evidence for an interchain hydrogen bond. Thus, over an identical number of amino acid residues, and with a substantially lower driving force, it appears that ET occurs very rapidly along the hydrogen-bonded β-sheet mimic (**78**).

IV. ELECTRON TRANSFER IN MULTICOMPONENT SYSTEMS AND ORGANIZED PEPTIDE ASSEMBLIES

A. Multifunctional Electron Transfer Systems

The structural properties of polypeptides in general, and of oligoprolines in particular, allow them to be used as molecular scaffolds upon which multifunctional photochemical arrays can be built. During the last several years, Meyer and co-workers [89–93] have been studying an interesting family of peptide-based arrays that have been shown to generate long-lived, redox-separated states for potential use in energy conversion schemes. The original work produced a functionalized lysine residue (**79**) which contained a donor–chromophore–acceptor triad consisting of a side-chain-coupled ruthenium polypyridyl chromophore, an electron-donating phenothiazine moiety coupled to the amino terminus of the residue, and an electron-accepting paraquat site coupled to the carboxyl terminus (Fig. 15). The redox-separated state was produced from one of two proposed reaction mechanisms shown in Fig. 16 [89]. In the right-hand branch of the scheme, initial photoexcitation of the ruthenium polypyridyl complex produces a metal-to-ligand charge transfer state which is quenched by a rapid (k_1 = 4.6 × 10^7 s^{-1}) one-electron oxidation of a phenothiazine moiety to produce the reduced "Ru(I)" species. This reactive species then undergoes an extremely rapid (k_2 > 2 × 10^{10} s^{-1}) reduction of the paraquat acceptor to produce the redox-separated state, D$^+$-Ru^{II}-A$^-$, in relatively high yield (ϕ = 0.34). It is calculated that this species can be used to store 1.17 eV of energy. Significantly, this high-energy state is relatively long-lived, decaying with a relatively slow first-order rate constant of k_7 = 9.26 × 10^6 s^{-1} (τ = 108 ns). In the alternative reaction mechanism, shown on the left-hand branch of the reaction scheme, photoexcitation of the ruthenium chromophore results in an initial reduction of the paraquat acceptor to form the reactive Ru(III) intermediate. This species then rapidly oxidizes the phenothi-

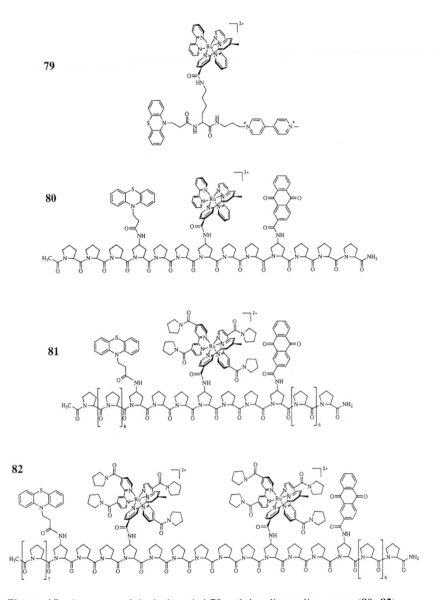

Figure 15 Structures of the lysine triad **79** and the oligoproline arrays (**80–82**).

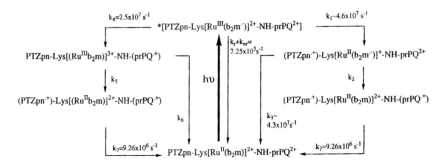

Figure 16 Proposed reaction scheme for the photoprocesses of the lysine triad **79**.

azine donor to produce D^+-Ru^{II}-A^-. However, control experiments performed on model dyads indicate that this proposed route may be much less efficient in producing the redox-separated state due to the rapid back reaction (k_6) arising from specific conformational effects found within the trifunctionalized lysine residue.

The peptide-based strategy for preparing new photochemical arrays has recently been applied to the use of longer oligoproline scaffolds [93]. A solid-phase synthesis was reported in which photochemical triads and tetrads of the general type described above were incorporated into oligoproline helices. In these systems, ruthenium polypyridyl chromophores, phenothiazine donors, and anthraquinone acceptors were covalently attached to *cis*-4-aminoproline residues via amide linkages. These were in turn incorporated into oligoproline helices having a total of 13, 21, or 27 amino acid residues (**80–82**). It was found that the 13-mer photochemical array produced a redox-separated state that stored 1.65 eV of energy with a lifetime of 175 ns. An important facet of these studies was that the choice of solvent was used to produce a conformational change of the oligoproline peptide bridge from the all-trans polyproline II helix in aqueous solution to the all-cis polyproline I helix in acetonitrile. These changes allowed specific variation of the through-space, but not through-bond, separation between the reduced anthraquinone and oxidized phenothiazine in the back ET reaction. Significantly, the back ET reaction in **82** was seen to proceed with an almost identical rate constant when it existed within the polyproline I conformation ($k_{bet} = 5.0 \times 10^5 \text{ s}^{-1}$) as when it was made to form the polyproline II conformation ($k_{bet} = 4.9 \times 10^5 \text{ s}^{-1}$). This similarity in rate constant occurred despite the approximately 14 Å change in the through-space donor–acceptor distance caused by this conformational change. These results provide convincing evidence to show that electron transfer occurs via a through-bond mechanism in these systems. It was further shown that the coupling efficiency of this mechanism is very high, as the

ET rate constants changed from $k_{bet} = 5.9 \times 10^6$ s^{-1} for **81** in which the bridging peptide has 29 intervening covalent bonds to a value of $k_{bet} = 5.0 \times 10^5$ s^{-1} for **82** which has 47 intervening bonds. Interestingly, these results indicate a very small distance attenuation factor for ET rates.

B. Electron Transfer in Model Heme Proteins

In recent years, several groups have begun to develop peptide-based ET systems containing metalloporphyrin redox sites. Aoudia and Rodgers [94] have used electrostatic interactions to prepare a self-assembled model hemeprotein in which a string of four glutamic acid residues terminated by either a tryptophan (**83**) or tyrosine (**84**) residue was electrostatically bound to a tetracationic Pd(II) porphyrin. Laser flash photolysis showed that the triplet state of Pd(II)TMPyP^{4+} decayed by the sum of two exponential terms in the presence of the redox peptide. The fast decay components (1.4 ± 0.15 10^7 s^{-1} for **83** and 6.6 ± 0.2 10^6 s^{-1} for **84**) were independent of peptide concentration and thus assigned to an intracomplex quenching process involving the aromatic residues. The slower component was found to be associated with bimolecular interactions between the free peptide and porphyrin molecules. Evidence that the intracomplex triplet state decay was due to an ET reaction was derived from the observation that an increase of the intramolecular rate constant resulted from the pH-governed increase in the oxidation potential of the amino acid. Interestingly, the intramolecular rate constant reached a limiting value at pH > 8.5, although the driving force continued to increase. These results indicate that local diffusional motion of the peptide may be a necessary precursor to the ET step.

Another route toward the design of model heme proteins involves the use of some of the proteolytic digestion products of cytochrome c [95,96]. These short polypeptides, called *microperoxidases*, are composed of 8, 9, or 11 amino acid residues that retain both their native thioether attachments between the two vinyl groups of the heme macrocycle via Cys-14 and Cys-17, and the axial coordination of His-18 to the iron center. An interesting feature of these peptides is that the coordinatively unsaturated iron site provides a convenient site for ligand complexation. Indeed, Low et al. [97] recently found that the photoinduced, bimolecular oxidization of aquoferric microperoxidase-8 produces the ferryl, Fe^{IV} (O), product at high pH. In other work, Ondrias and co-workers [98] used the open coordination site of microperoxidase-11 to bind a histidine-containing dipeptide (Pro-His) to which Ru(bpy)$_2$(mcbpy) was attached to the N-terminal proline via an amide bond (**85**). This process was found to have an association constant of 1.4×10^4 M^{-1} and give a center-to-center ruthenium-to-iron separation of 12 Å (edge-to-edge distance of 8 Å) as determined by molecular modeling studies. Transient resonance Raman showed the formation of a significant population of reduced heme upon photoexcitation of the complex. However, the transient emission results showed the existence of multiple kinetic components in which a static contribution was assigned to an intracomplex electron transfer between the coordinated RuPro-His donor and the ferric microperoxidase heme. Analysis of the data gave an upper limit for k_{et} of 4.2×10^7 s^{-1}. In a related study [99], these workers prepared another peptide-based ET system in which the Ru(bpy)$_2$(mcbpy) donor was covalently attached to a lysine side chain of microperoxidase-11 (**86**). This approach was taken to simplify the ET kinetics by eliminating any contributions arising from diffusional reactions. Energy minimization studies showed that this system has a center-to-center ruthenium-to-iron separation of 20.4 Å (edge-to-edge distance of 12.2 Å), which apparently assumes a rigid conformation of the lysine side chain. A combination of transient resonance Raman and transient absorption experiments was used to demonstrate the occurrence of a prompt, intramolecular, photoinduced ET reaction occurring between the ruthenium donor and the heme acceptor. The rate constant for this reaction was estimated to be approximately 2×10^7 s^{-1} by static emission quenching experiments.

C. Electron Transfer in De Novo Designed Metalloproteins

Recent advances in the field of de novo protein design have afforded the opportunity to prepare a new generation of synthetic ET proteins that retain much of the structural complexity found in native systems. To this end, Dutton and co-workers have been preparing a series of multihelical peptide bundles that contain native redox sites [100–104]. These protein models have been termed "molecular maquettes," being designed to contain redox-active cofactors that reside within a minimalistic protein structure. The archetype for this approach was constructed from a 62-residue α-helical dimer that can specifically bind heme sites through axial ligation to the buried histidine residues belonging to the individual chains of the dimer. These peptide dimers (**87**) were shown to self-assemble into four-helix bundles that possess an overall twofold symmetry and a probable all-parallel alignment (Fig. 17). The heme sites of this model protein were found to be chemically interactive, displaying negative binding cooperativities and mutually dependent electrochemical potentials. Whereas at the time of writing no ET measurements have been reported for these maquettes, these systems nonetheless represent an elegant example of the design of artificial heme proteins.

Mihara and co-workers have synthesized some interesting donor–acceptor maquettes that could be incorporated into bilayer membranes [105,106]. The first such system (**88**) consisted of a molecular triad in which a ruthenium polypyridyl photosensitizer, an anthraquinone primary oxidant, and a propylviologen terminal acceptor were placed along the lengths of two amphiphilic 21-residue polypeptides that were covalently attached to the ruthenium template. Circular dichroism

SH
|
Ac-NH-Cys-Gly-Gly-Gly-Glu-Leu-Trp-Lys-Leu-**His**-Glu-Glu-Leu-Leu-Lys-Lys

Phe-Glu-Glu-Leu-Leu-Lys-Leu-**His**-Glu-Glu-Arg-Leu-Lys-Lys-Leu-CONH$_2$

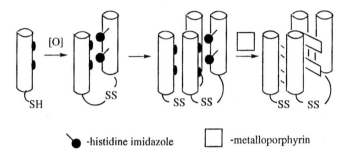

● -histidine imidazole ☐ -metalloporphyrin

Figure 17 Self-assembly of a four-helix bundle, heme-binding maquette.

spectra showed that **88** was 45% α-helical in methanol but became more disordered in H$_2$O/CH$_3$OH mixtures. Unfortunately, when the metalloprotein was incorporated into lecithin vesicles, the circular dichroism spectrum showed that it assumed an unidentified conformational motif and these workers were not able to obtain unambiguous evidence for an intramolecular ET reaction. In a related study, this group used a functionalized manganese(III) tetraphenylporphyrin moiety to nucleate the formation of a four-helix bundle (**89**) in both organic solvents and lecithin vesicles, as identified by circular dichroism and IR spectroscopy. An important feature of **89** was that it contained a flavin group attached to one of its 21-residue polypeptides, which served as a redox shuttle, enhancing the bimolecular rate constant for manganese reduction when exogenous reductants were added to the vesicle solution.

Seminal work by Mutz et al. [107,108] produced the first systematic study of the ET mechanisms of a synthetic metalloprotein (**90**). Based on the original designs of Lieberman and Sasaki [109] and Ghadiri et al. [110], these workers prepared a de novo ET protein in which viologen groups were attached to the carboxyl ends of a three-helix bundle capped by a Co^{III}(bpy)$_3$ electron acceptor (Fig. 18). Both electron pulse radiolysis and laser flash-quench techniques were used to create a prompt reduction of a viologen donor that subsequently transferred an electron to the Co^{III} acceptor via simple first-order kinetics. An important aspect of this system was that a variation of solvent conditions was found

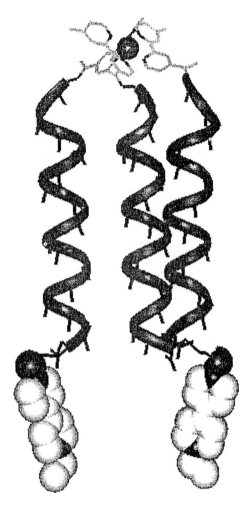

Figure 18 Sketch of the putative structure of maquette **90**. The Co(bpy)$_3$ moiety is located toward the top of the page and the viologen groups are toward the bottom end of the maquette. (From Ref. [108].)

to produce significant changes in the conformational properties of the metalloprotein as determined by circular dichroism spectroscopy. Thus, the protein was found to be 77% α-helical when dissolved in 25% (v/v) trifluorethanol (TFE), but was completely unfolded in 6 M urea. Significantly, the observed intramolecular ET rate constants were also found to vary with solvent, having a maximum value of $k_{et} = 2.0 \pm 0.2 \times 10^3 \text{ s}^{-1}$ in 25% TFE, which decreased to $k_{et} =$

$0.9 \pm 0.1 \times 10^3$ s^{-1} in 6 M urea. This effect could not be attributed to changes occurring within either the thermodynamic driving force or the outer-sphere reorganization energy for the reaction that occurred upon changing the solvent. Rather, the more than twofold decrease in ET rate was attributed to a lowering of the donor–acceptor coupling strength in the denatured protein. In order to determine the source of this loss of coupling strength, a series of energy-transfer experiments was performed on an analogous system in which the viologen redox sites were replaced by naphthalene-1-sulfonic acid (i.e., dansyl) fluorophores, and the cobalt capping group was substituted by a ruthenium polypyridyl quencher (**91**). The results showed that the average through-space donor–acceptor separation in the metalloprotein increased from 17.8 to 18.5 Å when the protein was denatured by changes in solvent. Thus, the smaller degree of electronic interaction in the unfolded peptide was ascribed to an increased donor–acceptor separation. Interestingly, these results are in qualitative agreement with those previously discussed for the α-helical bridged ET systems, arguing that a simple through-bond mechanism cannot be used to describe the ET properties of this α-helical maquette as only its through-space, and not through-bond, distances can be altered by changes in solvent. It is of further interest to note that the observed distance dependence of k_{et} can be analyzed in terms of Eq. (4) to yield a value of $\beta = 1.1$ Å^{-1} for this synthetic redox protein.

An alternate approach to the design of synthetic ET proteins was recently reported by Ogawa and co-workers [111,112] who have designed de novo designed metalloproteins (**92**) based on the "leucine zipper" motif found within the class of bZIP transcription factors (Fig. 19) [113,114]. This native protein structure is formed by the intertwining of two right-handed α helices to form a left-handed supercoil and is stabilized by the close interaction of regularly spaced hydrophobic residues belonging to each strand of the peptide dimer. Numerous studies of synthetic polypeptides have shown that these "coiled-coil" structures can be formed from relatively short peptide sequences if they are based on a 4-3 hydrophobic repeat [115]. In earlier work, Lee et al. [116] attempted to study ET reactions occurring between Trp and Tyr radicals located along a single strand of a synthetic coiled coil (**93**). However, the resulting kinetic data were extremely difficult to interpret due to complications arising from multiple conformational equilibria. In hindsight, the conformational variability of this system may have been predicted from the sequence of this peptide which, while loosely based on the sequence of the GCN4 transcriptase, had multiple amino acid substitutions occurring within the important interfacial region of the putative coiled coil. The work of Ogawa and co-workers modified this approach to use synthetic peptide sequences shown by Hodges and co-workers [115] to exist as stable, two-stranded coiled coils. Thus, the 31-residue polypeptide [NH$_2$-K-(I-E-A-L-E-G-K)$_2$-(I-E-A-L-E-H-K)-(I-E-A-L-E-G-K)-C'-G-OH] was prepared in which a single histidine residue was placed at position 21, which is the most highly solvent-exposed posi-

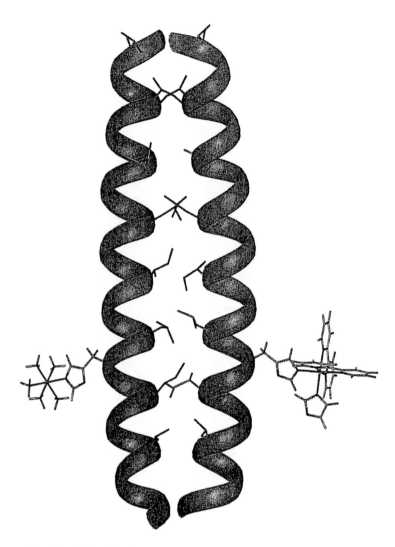

Figure 19   Synthetic coiled-coil metalloprotein (**91**).

tion of the third heptad repeat. This peptide, called H21(31-mer), was therefore designed to contain a convenient binding site for a metal-based redox center.

In recent work, the ET heterodimer [Ru(trpy)(bpy)H21(31-mer)]/[Ru (NH$_3$)$_4$(isn)H21(31-mer)] was prepared (trpy = 2,2': 6',2''-terpyridine; bpy = 2,2'-bipyridine, and isn = isonicotinamide).

A combination of circular dichroism, sodium dodecyl sulfate–polyacrylamide gel electrophoresis, chemical crosslinking, and analytical ultracentrifugation studies showed that both the apo- and metallated derivatives of H21(31-mer) form two-stranded α-helical coiled coils in aqueous solution. Further characterization of these derivatives by EPR spin-label experiments helped to determine its three-dimensional backbone structure. In these studies, a Cys-21 mutant of the 31-mer coiled coil, H21/C21(31-mer), was prepared and labeled with a thiol-specific nitroxide spin label (MTSL = 1-oxyl-2,2,5,5-tetramethyl-Δ^3-pyrroline-3-methyl-methanethiosulfonate) at position 21 of the peptide sequence which is the site of metal substitution in the ET heterodimer. Comparison of the low-temperature, dipolar-broadened spectrum of the spin-labeled dimer with those of magnetically dilute peptide samples yielded a backbone-to-backbone distance that was nearly identical to that of the GCN4 homodimer. Based on these results, computer modeling studies provided an estimate of the metal-to-metal distance in the ET heterodimer of $d_{M-M} > 25$ Å. The electron-transfer properties of this system are now being studied by a combination of laser flash-quench and pulse radiolysis techniques.

V. SUMMARY

Peptide-based ET complexes display a wide range of structural complexity that spans the gamut between appropriately derivatized single amino acid residues and large molecular weight synthetic proteins. The study of these types of systems has demonstrated a widespread diversity in their ET properties. It now appears that different types of secondary protein structures can indeed display different mechanisms for mediating long-range donor–acceptor interactions. It is hoped that future work in this field will offer additional insight into the principles of peptide-based ET reactions that may be applied to the design of new systems with interesting chemical properties.

Note added in proof

Another approach towards the design of synthetic ET proteins was published while this manuscript was in press [117].

ACKNOWLEDGMENT

I thank Ms. Anna Federova for her invaluable assistance in the development of this manuscript. Our work on peptide-based electron transfer complexes has been supported by the National Science Foundation grant no. CHE-9307791.

REFERENCES

1. DeVault, D. *Quantum-Mechanical Tunnelling in Biological Systems, 2nd ed.*; Cambridge University Press: New York, 1984.
2. Johnson, M.K.; King, R.B.; Kurtz, D.M. Jr.; Norton, M.L.; Scott, R.A., Eds. *Electron-Transfer in Biology and the Solid State*; Advances in Chemistry Series 226; American Chemical Society: Washington, DC, 1990.
3. Bolton, R.R.; Mataga, N.; McLendon, G.L., Eds. *Electron-Transfer in Inorganic, Organic and Biological Systems*; Advances in Chemistry Series 228; American Chemical Society: Washington, DC, 1991.
4. Deisenhofer, J.; Norris, J.R. *The Photosynthetic Reaction Center*, Academic Press: San Diego, 1993; Vol 1.
5. Bowler, B.E.; Raphael, A.L.; Gray, H.B. *Prog. Inorg. Chem.* **1990**, *38*, 259.
6. Nocek, J.M.; Zhou, J.S.; De Forest, S.; Priyadarshy, S.; Beratan, D.N.; Onuchic, J.N.; Hoffman, B.M. *Chem. Rev.* **1996**, *96*, 2459.
7. Isied, S.S. in *Electron Transfer Reactions Inorganic, Organometallic, and Biological Applications*; Isied, S.S., Ed.; Advances in Chemistry Series 253; American Chemical Society: Washington, DC, 1997, p. 331.
8. McLendon, G.; Hake, R. *Chem. Rev.* **1992**, *92*, 481.
9. Farver, O; Pecht, I. *J. Biol. Inorg. Chem.* **1997**, *2*, 387.
10. Durham, B.; Fairris, J.L.; McLean, M.; Millett, F.; Scott, J.R.; Sligar, S.G.; Willie, A. *J. Bioenerg. Biomembr.* **1995**, *27*, 331.
11. Ullmann, G.M.; Knapp, E.W.; Kostic, N.M. *J. Am. Chem. Soc.* **1997**, *119*, 42.
12. Tollin, G. *J. Bioenerg. Biomembr.* **1995**, *27*, 303.
13. Moser, C.C.; Keske, J.M.; Warncke, K.; Farid, R.S.; Dutton, P.L. *Nature* **1992**, *355*, 796.
14. Moser, C.C.; Page, C.C.; Farid, R.; Dutton, P.L. *J. Bioenerg. Biomembr.* **1995**, *27*, 263.
15. Winkler, J.R.; Gray, H.B. *J. Biol. Inorg. Chem.* **1997**, *2*, 399.
16. Winkler, J.R.; Gray, H.B. *Chem. Rev.* **1992**, *92*, 369.
17. Moreira, I.; Sun, J.; Cho, M.O.-K.; Wishart, J.F.; Isied, S.S. *J. Am. Chem. Soc.* **1994**, *116*, 8396.
18. Beratan, D.N.; Betts, J.N.; Onuchic, J.N. *Science* **1991**, *252*, 1285.
19. Curry, W.B.; Grabe, M.D.; Kurnikov, I.V.; Skourtis, S.S.; Beratan, D.N.; Regan, J.J.; Aquino, A.J.A.; Beroza, P.; Onuchic, J.N. *J. Bioenerg. Biomembr.* **1995**, *27*, 285.
20. Closs, G.L.; Calcaterra, L.T.; Green, N.J.; Penfield, K.W.; Miller, J.R. *J. Phys. Chem.* **1986**, *90*, 3673.
21. Gould, I.R.; Ege, D.; Mattes, S.L.; Farid, S. *J. Am. Chem. Soc.* **1987**, *109*, 3794.
22. Ohno, T.; Yoshimura, A.; Mataga, N. *J. Phys. Chem.* **1986**, *90*, 3295.
23. MacQueen, D.B.; Schanze, K.S. *J. Am. Chem. Soc.* **1991**, *113*, 7470.
24. Chen, P.; Mecklenburg, S.L.; Meyer, T.J. *J. Phys. Chem.* **1993**, *97*, 13126.
25. McLendon, G.; Miller, J.R. *J. Am. Chem. Soc.* **1985**, *107*, 7811.
26. Scott, J.R.; McLean, M.; Sligar, S.G.; Durham, B.; Millett, F. *J. Am. Chem. Soc.* **1994**, *116*, 7356.

27. Turro, C.; Zaleski, J.M.; Karabatsos, Y.M.; Nocera, D.G. *J. Am. Chem. Soc.* **1996**, *118*, 6060.

28. Oevering, H.; Paddon-Row, M.N.; Heppener, M.; Oliver, A.M.; Cotsaris, E.; Verhoeven, J.W.; Hush, N.S. *J. Am. Chem. Soc.* **1987**, *109*, 3258.

29. Knapp, S.; Dhar, T.G.M.; Albaneze, J.; Gentemann, S.; Potenza, J.A.; Holten, D.; Schugar, H.J. *J. Am. Chem. Soc.* **1991**, *113*, 4010.

30. Finckh, P.; Heitele, H.; Volk, M.; Michel-Beyerle, M.E. *J. Phys. Chem.* **1988**, *92*, 6584.

31. Stein, C.A.; Lewis, N.A.; Seitz, G. *J. Am. Chem. Soc.* **1982**, *104*, 2596.

32. Closs, G.L.; Miller, J.R. *Science*, **1988**, *240*, 440.

33. Kuki, A.; Wolynes, P.G. *Science* **1987**, *236*, 1647.

34. Broo, A.; Larsson, S. *Int. J. Quant. Chem.: Quantum Biol. Symp.* **1989**, *16*, 185.

35. Siddarth, P.; Marcus, R.A. *J. Phys. Chem.* **1993**, *97*, 13078.

36. Bensasson, R.V.; Land, E.J.; Truscott, T.G. *Flash Photolysis and Pulse Radiolysis: Contributions to the Chemistry and Biology of Medicine*; Pergamon Press, New York, 1983.

37. Klapper, M.H.; Faraggi, M. *Quart. Rev. Biophys.* **1979**, *12*, 465.

38. Wishart, J.F.; Nocera, D.G., Eds. *Photochemistry and Radiation Chemistry, Complementary Methods for the Study of Electron-Transfer*; ACS Advances in Chemistry Series 254; American Chemical Society: Washington, DC, 1998.

39. Chang, I-J.; Gray, H.B.; Winkler, J.R. *J. Am. Chem. Soc.* **1991**, *113*, 7056.

40. Mines, G.A.; Bjerrum, M.J.; Hill, M.G.; Casimiro, D. R.; Chang, I.-J.; Winkler, J.R.; Gray, H.B. *J. Am. Chem. Soc.* **1996**, *118*, 1961.

41. Isied, S.S.; Vassilian, A. *J. Am. Chem. Soc.* **1984**, *106*, 1726.

42. Isied, S.S.; Vassilian, A. *J. Am. Chem. Soc.* **1984**, *106*, 1732.

43. Kartha, G.; Ashida, T.; Kakudo, M. *Acta Cryst.* **1974**, *B30*, 1861.

44. Matsuzaki, T. *Acta Cryst.* **1974**, *B30*, 1029.

45. Schanze, K.S.; Sauer, K. *J. Am. Chem. Soc.* **1988**, *110*, 1180.

46. Isied, S.S.; Vassilian, A.; Magnuson, R.H.; Schwarz, H.A. *J. Am. Chem. Soc.* **1985**, *107*, 7432.

47. Isied, S.S.; Vassilian, A.; Wishart, J.F.; Creutz, C.; Schwarz, H.A.; Sutin, N. *J. Am. Chem. Soc.* **1988**, *110*, 635.

48. Vassilian, A.; Wishart, J.F.; van Hemelryck, B.; Schwarz, H.; Isied, S.S. *J. Am. Chem. Soc.* **1990**, *112*, 7278.

49. Schanze, K.S.; Cabana, L.A. *J. Phys. Chem.* **1990**, *94*, 2740.

50. Ogawa, M.Y.; Wishart, J.F.; Young, Z.; Miller, J.R.; Isied, S.S. *J. Phys. Chem.* **1993**, *97*, 11456.

51. Ogawa, M.Y.; Moreira, I.; Wishart, J.F.; Isied, S.S. *Chem. Phys.* **1993**, *176*, 589.

52. Bobrowski, K.; Holcman, J.; Poznanski, J.; Ciurak, M.; Wierzchowski, K.L. *J. Phys. Chem.* **1992**, *96*, 10036.

53. Bobrowski, K.; Wierzchowski, K.L.; Holcman, J.; Ciurak, M. *Stud. Biophys.* **1987**, *122*, 23.

54. Bobrowski, K.; Poznanski, J.; Wierzchowski, K.L. in *Photochemistry and Radiation Chemistry, Complementary Methods for the Study of Electron-Transfer*; Wishart, J. F.; Nocera, D.G., Eds.; ACS Advances in Chemistry Series 254; American Chemical Society: Washington, DC, 1998, p. 131.

55. Sneddon, S.F.; Brooks, C.L. III *J. Am. Chem. Soc.* **1992**, *114*, 8220.
56. Faraggi, M.; DeFilippis, M.R.; Klapper, M.H. *J. Am. Chem. Soc.* **1989**, *111*, 5141.
57. DeFelippis, M.R.; Faraggi, M.; Klapper, M.H. *J. Am. Chem. Soc.* **1990**, *112*, 5640.
58. Mishra, A.K.; Chandrasekar, R.; Faraggi, M.; Klapper, M.H. *J. Am. Chem. Soc.* **1994**, *116*, 1414.
59. Tamiaki, H.; Maruyama, K. *J. Chem. Soc. Perkin Trans.* **1991**, 817.
60. Tamiaki, H.; Nomura, K.; Maruyama, K. *Bull. Chem. Soc. Jpn.* **1993**, *66*, 3062.
61. Tamiaki, H.; Nomura, K.; Maruyama, K. *Bull. Chem. Soc. Jpn.* **1994**, *67*, 1863.
62. Murphy, C.J.; Arkin, M.R.; Jenkins, Y.; Ghatlia, N. D.; Bossman, S.H.; Turro, N.J.; Barton, J.K. *Science* **1993**, *262*, 1025.
63. Scholtz, J.M.; Qian, H.; York, E.J.; Stewart, J.M.; Baldwin, R.L. *Biopolymers*, **1991**, *31*, 1463.
64. Sisido, M.; Tanaka, R.; Inai, Y.; Imanishi, Y. *J. Am. Chem. Soc.* **1989**, *111*, 6790.
65. Inai, Y.; Sisido, M.; Imanishi, J. *J. Phys. Chem.* **1990**, *94*, 6237.
66. Inai, Y.; Sisido, M.; Imanishi, Y. *J. Phys. Chem.* **1990**, *94*, 8365.
67. Inai, Y.; Sisido, M.; Imanishi, Y. *J. Phys. Chem.* **1991**, *95*, 3847.
68. Kuragaki, M; Sisido, M. *J. Phys. Chem.* **1996**, *100*, 16019.
69. Basu, G.; Kubasik, M.; Anglos, D.; Secor, B.; Kuki, A. *J. Am. Chem. Soc.* **1990**, *112*, 9410.
70. Basu, G.; Anglos, D.; Kuki, A. *Biochemistry* **1993**, *32*, 3067.
71. Basu, G.; Kubasik, M.; Anglos, D.; Kuki, A. *J. Phys. Chem.* **1993**, *97*, 3956.
72. Basu, G.; Bagchi, K.; Kuki, A. *Biopolymers* **1991**, *31*, 1763.
73. Toniolo, C.; Benedetti, E. *Macromolecules* **1991**, *24*, 4004.
74. Gallopini, E.; Fox, M.A. *J. Am. Chem. Soc.* **1996**, *118*, 2299.
75. Knorr, A.; Galoppini, E.; Fox, M.A. *J. Phys. Org. Chem.* **1997**, *10*, 484.
76. Fox, M.A.; Galoppini, E. *J. Am. Chem. Soc.* **1997**, *119*, 5277.
77. Brant, D.A. *Macromolecules* **1968**, *1*, 291.
78. Langen, R.; Chang, I-J.; Germanas, J.P.; Richards, J. H.; Winkler, J.R.; Gray, H.B. *Science* **1995**, *268*, 1733.
79. Regan, J.J.; di Bilio, A.J.; Langen, R.; Skov, L.K.; Winkler, J.R.; Gray, H.B.; Onuchic, J.N. *Chem. Biol.* **1995**, *2*, 489.
80. Tsang, K.Y.; Diaz, H.; Graciani, N.; Kelly, J.W. *J. Am. Chem. Soc.* **1994**, *116*, 3988.
81. Nowick, J.S.; Smith, E.M.; Noronha, G. *J. Org. Chem.* **1995**, *60*, 7386.
82. Kemp, D.S.; Li, Z.Q. *Tetrahedron Lett.* **1995**, *36*, 4179.
83. Fernando, S.R.L.; Kozlov, G.V.; Ogawa, M.Y. *Inorg. Chem.* **1998**, *37*, 1900.
84. Ogawa, M.Y.; Gretchikhine, A.B.; Soni, S.D.; Davis, S.M. *Inorg. Chem.* **1995**, *34*, 6423.
85. Gretchikhine, A.B.; Ogawa, M.Y. *J. Am. Chem. Soc.* **1996**, *118*, 1543.
86. Gretchikhine, A.B. Ph.D. thesis, Bowling Green State University, 1997.
87. Wuthrich, K. *NMR of Proteins and Nucleic Acids*; John Wiley & Sons, New York, 1986.
88. Fernando, S.R.L.; Maharoof, U.S.M.; Deshayes, K.D.; Kinstle, T.H.; Ogawa, M.Y. *J. Am. Chem. Soc.* **1996**, *118*, 5783.

89. Mecklenburg, S.L.; Peek, B.M.; Erickson, B.W.; Meyer, T.J. *J. Am. Chem. Soc.* **1991**, *113*, 8540.

90. Mecklenburg, S.L.; Peek, B.M.; Schoonover, J.R.; McCafferty, D.G.; Wall, C.G.; Erickson, B.W.; Meyer, T.J. *J. Am. Chem. Soc.* **1993**, *115*, 5479.

91. Mecklenburg, S.L.; McCafferty, D.G.; Schoonover, J. R.; Peek, B.M.; Erickson, B.W.; Meyer, T.J. *Inorg. Chem.* **1994**, *33*, 2974.

92. McCafferty, D.G.; Friesen, D.A.; Danielson, E.; Wall, C.G.; Saderholm, M.J.; Erickson, B.W.; Meyer, T.J. *Proc. Natl. Acad. Sci. USA* **1996**, *93*, 8200.

93. Slate, C.A.; Striplin, D.R.; Moss, J.A.; Chen, P.; Erickson, B.W.; Meyer, T.J. *J. Am. Chem. Soc.* **1998**, *120*, 4885.

94. Aoudia, M.; Rodgers, M.A.J. *J. Am. Chem. Soc.* **1997**, *119*, 12859.

95. Santucci, R.; Picciau, A.; Antonini, G.; Campanella, L. *Biochim. Biophys. Acta* **1995**, *1250*, 183.

96. Ippoliti, R.; Picciau, A.; Santucci, R.; Antonini, G.; Brunori, M.; Ranghino, G. *Biochem. J.* **1997**, *328*, 833.

97. Low, D.W.; Winkler, J.R.; Gray, H.B. *J. Am. Chem. Soc.* **1996**, *118*, 117.

98. Fan, B.; Fontenot, D.L.; Larsen, R.W.; Simpson, M.C.; Shelnutt, J.A.; Falcon, R.; Martinez, L.; Niu, S.; Zhang, S.; Niemczyk, T.; Ondrias, M.R. *Inorg. Chem.* **1997**, *36*, 3839.

99. Fan, B.; Simpson, M.C.; Shelnutt, J.A.; Martinez, C.; Falcon, R.; Buranda, T.; Patuszyn, A.J.; Ondras, M.R. *Inorg. Chem.* **1997**, *36*, 3847.

100. Gibney, B.R.; Mulholland, S.E.; Rabanal, F.; Dutton, P.L. *Proc. Natl. Acad. Sci. USA* **1996**, *93*, 15041.

101. Choma, C.T.; Lear, J.D.; Nelson, M.J.; Dutton, P. L.; Robertson, D.E.; DeGrado, W.F. *J. Am. Chem. Soc.* **1994**, *116*, 856.

102. Robertson, D.E.; Farid, R.S.; Moser, C.C.; Urbauer, J.L.; Mulholland, S.E.; Pidikiti, R.; Lear, J.D.; Wand, A.J.; DeGrado, W.F.; Dutton, P.L. *Nature* **1994**, *368*, 425.

103. Rabanal, F.; DeGrado, W.F.; Dutton, P.L. *J. Am. Chem. Soc.* **1996**, *118*, 473.

104. Gibney, B.R.; Rabanal, F.; Reddy, K.S.; Dutton, P.L. *Biochemistry*, **1998**, *37*, 4635.

105. Mihara, H.; Nishino, N.; Hasegawa, R.; Fujimoto, T.; Usui, S.; Ishida, H.; Ohkubo, K. *Chem. Lett.* **1992**, 1813.

106. Mihara, H.; Tomizaki, K.; Fujimoto, T.; Sakamoto, S.; Aoyagi, H.; Nishino, N. *Chem. Lett.* **1996**, *187*.

107. Mutz, M.W.; McLendon, G.L.; Wishart, J.F.; Gaillared, E.R.; Corin, A.F. *Proc. Natl. Acad. Sci. USA* **1996**, *93*, 9521.

108. Mutz, M.W.; Wishart, J.F.; McLendon, G.L. in *Photochemistry and Radiation Chemistry: Complementary Methods for the Study of Electron-Transfer*; Wishart, J.F.; Nocera, D.G., Eds.; Advances in Chemistry Series 254; American Chemical Society: Washington, DC, 1998, p. 145.

109. Lieberman, M.; Sasaki, T. *J. Am. Chem. Soc.* **1991**, *113*, 1470.

110. Ghadiri, M.R.; Soares, C.; Choi, C. *J. Am. Chem. Soc.* **1992**, *114*, 825.

111. Kozlov, G.V.; Ogawa, M.Y. *J. Am. Chem. Soc.* **1997**, *119*, 8377.

112. Kozlov, G.V.; Xiao, W.; Lasey, R.C.; Shin, Y.-K.; Ogawa, M.Y., in press.

113. O'Shea, E.K.; Klemm, J.D.; Kim, P.S.; Alber, T. *Science* **1991**, *254*, 539.

114. Junius, F.K.; Mackay, J.P.; Bubb, W.A.; Jensen, S.A.; Weiss, A.S.; King, G.F. *Biochemistry* **1995**, *34*, 6164.
115. For a review, see Hodges, R.S. *Biochem. Cell Biol.* **1996**, *74*, 133.
116. Lee, H.; Faraggi, M.; Klapper, M.H. *Biochim. Biophys. Acta* **1992**, *1159*, 286.
117. Rau, H.K.; DeJonge, N.; Haehnel, W. *Proc. Natl. Acad. Sci. USA*, **1998**, *95*, 11526.

4

Tridentate Bridging Ligands in the Construction of Stereochemically Defined Supramolecular Complexes

Sumner W. Jones, Michael R. Jordan, and Karen J. Brewer
Virginia Polytechnic Institute and State University, Blacksburg, Virginia

I. INTRODUCTION

Since the discovery of the photophysical properties of $[Ru(bpy)_3]^{2+}$ (bpy = 2,2′-bipyridine, Fig. 1), the light absorbing properties of this and related chromophores have been a research topic of great interest [1,2]. Recent attention has focused on the development of polymetallic supramolecular systems designed to absorb light and carry out a specific task such as the conversion of light energy to electrical or chemical energy. As this research has progressed, the molecules have become increasingly complicated. Bridging ligands are used to hold the metal centers together, leading to the construction of polymetallic systems. The bridging ligands' structure and energetics are vitally important to the functioning of these systems. We will examine a class of polyazine tridentate bridging ligands and the resulting supramolecular systems that utilize this class of bridging ligands. These ligands bridge metals in a linear fashion, resulting in the construction of stereochemically defined multimetallic systems.

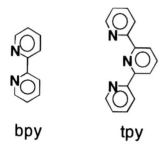

bpy   **tpy**

Figure 1   2,2'-Bipyridine and 2,2': 6',2''-terpyridine.

A. Light Absorption

When a light absorber molecule (LA) absorbs light, it is transferred to an electronically excited state (LA*), Eq. (1). This electronic excited state has properties that are unique compared to the ground state system.

$$LA + h\nu \rightarrow LA^* \tag{1}$$

The excited state of the light absorber has gained energy through the excitation. Relaxation to a ground state system can occur by many mechanisms. The excited light absorber can relax to its ground state by giving off heat in nonradiative decay, Eq. (2), or by giving off light in radiative decay, Eq. (3).

$$LA^* \rightarrow LA + \text{heat} \tag{2}$$
$$LA^* \rightarrow LA + h\nu \tag{3}$$

Bimolecular reactions can also deactivate the excited state. These include energy and electron transfer. Energy transfer occurs when the excited state of the light absorber transfers its energy to a quencher molecule (Q), Eq. (4). Electron transfer involves the transfer of an electron between the excited light absorber and an electron donor or electron acceptor.

$$LA^* + Q \rightarrow LA + Q^* \tag{4}$$

Since the excited light absorber is both a better oxidizing agent and a better reducing agent than the ground state of the light absorber, it can undergo oxidative quenching involving an electron acceptor (EA), Eq. (5), or reductive quenching involving an electron donor (ED), Eq. (6).

$$LA^* + EA \rightarrow LA^+ + EA^- \tag{5}$$
$$LA^* + ED \rightarrow LA^- + ED^+ \tag{6}$$

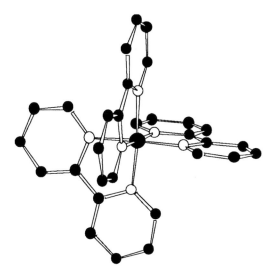

Figure 2   Representation of $[Ru(bpy)_3]^{2+}$.

One molecule that has received much attention as a light absorber with potential use in solar energy conversion schemes is $[Ru^{II}(bpy)_3]^{2+}$ (Fig. 2) [1,2]. The Jablonski diagram for $[Ru(bpy)_3]^{2+}$ is shown in Scheme 1. When this chromophore absorbs a photon of light in the visible region at 452 nm, the lowest energy singlet metal-to-ligand charge transfer (^{1}MLCT) excited state is produced. This excited state then undergoes intersystem crossing (isc) quantitatively to the ^{3}MLCT. From the ^{3}MLCT, $[Ru(bpy)_3]^{2+}$ can decay to the ground state radiatively (rad) or nonradiatively (nr). The main reasons for the interest in $[Ru^{II}(bpy)_3]^{2+}$ are its relatively high extinction coefficients in the visible spectrum, its photosta-

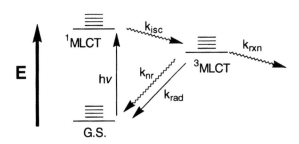

Scheme 1   Jablonski diagram for $[Ru(bpy)_3]^{2+}$.

bility, and its long-lived emissive ^{3}MLCT excited state (λ^{em}_{max} = 607, τ = 620 ns at room temperature in water) [1]. The emissive nature of the ^{3}MLCT excited state makes excited state processes easier to probe. Most importantly, the ^{3}MLCT excited state has been shown to undergo facile energy and electron transfer in the presence of suitable quenchers, k_{rxn} [1,2].

Compared to the amount of work done with bipyridine complexes, there has been relatively little interest in complexes utilizing the analogous tridentate ligand, 2,2':6',2''-terpyridine (tpy), such as [Ru(tpy)$_2$]$^{2+}$ (Fig. 3). This is due to the nonemissive nature of [Ru(tpy)$_2$]$^{2+}$ and its short excited state lifetime, 0.25 ns at room temperature (RT) [3]. Given the short excited state lifetime of [Ru(tpy)$_2$]$^{2+}$, it was thought that tridentate complexes of this type could not be effectively used in energy or electron transfer schemes. It has been shown, however, that the short lifetime of the ^{3}MLCT state of [Ru(tpy)$_2$]$^{2+}$ is due to a ligand field (LF) state that is thermally accessible at RT (Scheme 2) [4]. The ligand field state decays quickly by a nonradiative pathway to the ground state, $k_{nr'}$. The thermal accessibility of the ligand field state at RT is attributed to the nonideal bite angle of tpy for an octahedral complex that lowers the energy of the ligand field state [5,6]. The bite angle of tpy in a ruthenium complex is defined as the N(terminal pyridine)-Ru-N(terminal pyridine) angle. The ideal bite angle for a meridinal tridentate ligand in an octahedral environment would be 180°. The tpy-Ru bite angle is less than the ideal angle at approximately 158° (Fig. 4) [7–9].

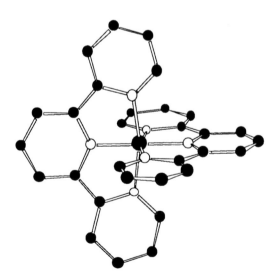

Figure 3   Representation of [Ru(tpy)$_2$]$^{2+}$.

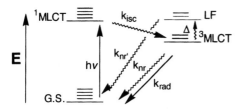

Scheme 2 Jablonski diagram for $[Ru(tpy)_2]^{2+}$.

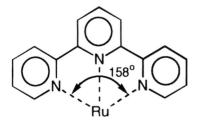

Figure 4 Bite angle in Ru–tpy complexes.

Even though energy and electron transfer reactions using complexes such as $[Ru(bpy)_3]^{2+}$ have been shown to occur with significant success, the efficiency of any bimolecular energy or electron transfer reaction is limited by the necessity for the light absorber to come into contact with a suitable quencher *during* the excited state of the light absorber. These processes are also limited by back electron transfer. These facts have sparked the interest in polymetallic supramolecular systems.

B. Supramolecular Systems

Some time ago, Balzani wrote an intriguing report on the possible use of supramolecular complexes as molecular devices for various photoinitiated processes including energy and electron transfer [10]. A molecular device is defined as an assembly of molecular components that can be either organic or inorganic moieties. The individual acts of the components add together to give the overall function of the supramolecular system. By carefully choosing the components, a supramolecular system can be designed to carry out complex functions. Since the components of a supramolecular system are chemically bonded to each other, there is no dependence on a bimolecular reaction for energy of electron transfer to occur. This has the potential of greatly increasing the efficiency of these photochemical processes.

One focus in this area has been in the development of polymetallic supra-molecular complexes designed for photoinitiated charge separation. One of the reasons for this is the desire to mimic the initial process in photosynthesis where the first step is the absorption of light and subsequent vectorial transfer of an electron to produce a separation of charge [11]. In a polymetallic supramolecular system designed for photoinitiated charge separation, a chromophore absorbs a photon of light and intramolecular electron transfer occurs involving a suitable oxidative or reductive quencher to produce a separation of charge. The energy of the charge-separated state creates a potential that is then available for use in a subsequent chemical process. A simple device for photoinitiated charge separa-tion would consist of a light absorber (LA) covalently attached to an electron donor (ED) and an electron acceptor (EA) (Scheme 3). Absorption of light by the light absorber would produce an excited state of the light absorber. The ex-

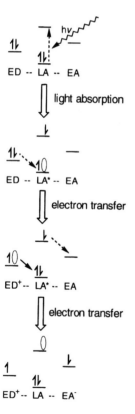

Scheme 3 Photoinitiated charge separation in an electron donor (ED)–light absorber (LA)–electron acceptor (EA) molecular device.

cited state of the light absorber would then be reductively quenched by the electron donor. This prevents decay of the light absorber back to the ground state. The light absorber would then transfer an electron to the electron acceptor, generating a separation of charge. The positively charged electron donor and the negatively charged electron acceptor are spatially separated by the light absorber, drastically retarding charge recombination.

In the construction of a polymetallic supramolecular species, bridging ligands are used to covalently connect the different metal centers together. The identity and energetics of these ligands determine many characteristics of the supramolecular complexes produced. Figure 5 shows the structure of a variety of polyazine bridging and terminal ligands that have been employed.

Bidentate polyazine bridging ligands have been used extensively in the construction of polymetallic complexes. The most widely used bidentate bridging ligand has been 2,3-bis(2-pyridyl)pyrazine (dpp) [12]. The complexes using this and related ligands display some very promising properties but they are not the subject of this chapter. One disadvantage of dpp is the difficulty in controlling the stereochemistry of dpp-bridged polymetallic complexes. The ligand dpp is

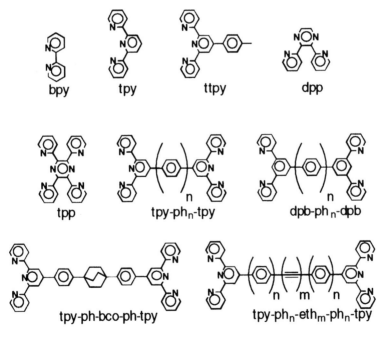

Figure 5   Polyazine bridging and terminal ligands.

an A-B type of ligand, i.e., the two nitrogens coordinated to a metal center are not chemically equivalent. Additionally, tris-bidentate complexes such as [Ru(bpy)$_3$]$^{2+}$ exist as a mixture of Δ and Λ isomers [13,14]. In polymetallic complexes using A-B bidentate bridging ligands, these two factors combine to give a large number of possible stereoisomers [13,14]. When two symmetric tridentate ligands are coordinated to a metal, there is only one possible structure. Tridentate bridging ligands therefore offer the possibility of controlling the stereochemistry of polymetallic complexes. Our focus will be on polymetallic complexes of ruthenium(II) or osmium(II) bridged by tridentate polyazine ligands highlighting the promising properties imparted by this coordination environment.

C. Intervalence Charge Transfer

Characterization of the intervalence charge transfer (IT) band is one of the most reliable and direct methods of determining the amount of intercomponent electronic communication in symmetric bimetallic systems [15,16]. For the purposes of this discussion, a symmetric bimetallic complex will be considered where both metal centers are in the 2+ oxidation state, M^{II}-M^{II}. If one metal center is oxidized, the mixed valence species M^{II}-M^{III} will result. The characteristics of the electronic transition M^{II}-M^{III} + $h\nu \to$ M^{III}-M^{II*} can give information about the degree of metal–metal coupling. For a symmetric system, Hush related the energy, intensity, and half width at maximum intensity of this IT transition to the degree of electronic coupling as in Eqs. (7)–(9), where E_{op}, ν_{max}, $\Delta\bar{\nu}_{1/2}$, and ε_{max} are the energy of the transition, the energy of maximum absorption for the band, half width at half height in cm^{-1}, and extinction coefficient at the absorption maximum of the IT band; λ is the reorganization energy of M^{III}-M^{II*}; H is the electronic coupling parameter; and r is the metal–metal distance in angstroms [16] for a symmetric system where the coordination spheres of the two metals are identical.

$$E_{op} = \lambda \tag{7}$$

$$\Delta\bar{\nu}_{1/2} = 48.06(\nu_{max})^{1/2}(\text{cm}^{-1}) \tag{8}$$

$$H = 0.0205(\varepsilon_{max}\,\Delta\bar{\nu}_{1/2}\nu_{max})^{1/2}/r \tag{9}$$

For mixed-valence bimetallic systems, Robin and Day distinguished three classes depending on the amount of metal-to-metal interaction [17]. Figure 6 is a plot of nuclear configuration vs. energy for the three classes of mixed-valence compounds. If there is essentially no interaction between the metal centers, it is called a class I system. Compounds of this type have properties that are a simple combination of the properties of the two independent metal centers. Class I compounds do not exhibit IT transitions. Mixed-valence compounds that have some limited degree of interaction between the metal centers are considered to be class II. Class II systems still have a localized valency and can be described as

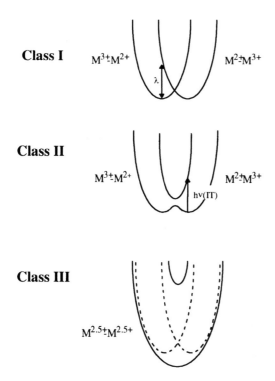

Figure 6 Classes of metal–metal interaction in ruthenium bimetallic systems bridged by polyazine bridging ligands.

M^{II}-M^{III} but have some new properties that can be attributed to the mixed-valence species. Class II compounds typically have IT transitions that can be characterized using the Hush theory. Class III systems are compounds that have such strong electronic communication between the metal centers that they are completely delocalized systems. The oxidation state of the metal centers in a class III system can best be described as $M^{2.5}/M^{2.5}$. Class III complexes typically do not have properties that can be attributed to the M^{II} or M^{III} metal center.

To differentiate between class II and class III compounds, the peak width at half height of the IT band, $\Delta \bar{v}_{1/2}$, is measured. If $\Delta \bar{v}_{1/2} \geq 48.06(v_{max})^{1/2}$, the compound can be considered class II. If $\Delta \bar{v}_{1/2} < 48.06(v_{max})^{1/2}$, the compound is considered to be class III. The solvent dependence of the IT band can also be used as an indication of the extent of delocalization. Since class III compounds do not have a change in dipole upon intervalence charge transfer, the IT bands of class III compounds do not exhibit a strong solvent dependence. A third method of determining the class of a compound is by the intensity of the IT band.

Class II compounds typically have moderately intense bands (ε < 1000 M^{-1} cm^{-1}) whereas class III compounds can have IT bands with molar extinction coefficients as high as 10,000 M^{-1} cm^{-1} [15].

The electronic coupling parameter for a class II complex is calculated from Eq. (7)

$$H = 0.0205(\varepsilon_{max} \, \Delta\bar{\nu}_{1/2}\nu_{max})^{1/2}/r \tag{10}$$

whereas the electronic coupling parameter for a class III complex is calculated as

$$H = (\nu_{max})/2 \tag{11}$$

II. TRIDENTATE BRIDGING LIGAND COMPLEXES OF RUTHENIUM AND OSMIUM

A. The tpp-Based Systems

The first polyazine tridentate ligand reported to form polymetallic Ru^{II} complexes was 2,3,5,6-(2-pyridyl)pyrazine (tpp) (Fig. 7) [18]. Petersen et al. and Thummel et al. independently showed that tpp could function as a bridging ligand in the complex [(tpy)Ru(tpp)Ru(tpy)](PF$_6$)$_4$ (Fig. 8) [19,20]. This is a linear, rigid complex with interesting electrochemical and photophysical properties. One popular technique used to study these types of polyazine complexes of Ru^{II} and Os^{II} is electrochemistry. These systems typically display metal-based oxidations indicative of the Ru- or Os-based nature of the highest occupied molecular orbital (HOMO). These systems also display ligand-based reductions indicative of the ligand-based nature of the lowest unoccupied molecular orbital (LUMO). Typically, bridging polyazine ligands can display two 1e$^-$ reductions and terminal ligands display a single 1e$^-$ reduction. The site of localization of the first electrochemical reduction is the polyazine ligand with the lowest lying π* acceptor orbital and represents the LUMO in the system. The electrochemistry of the monometallic complex [Ru(tpy)(tpp)]$^{2+}$ shows a ruthenium-based oxidation at 1.40 V vs. Ag/AgCl, a one-electron tpp-based reduction at -0.97 V, and a one-

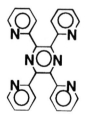

Figure 7 tpp.

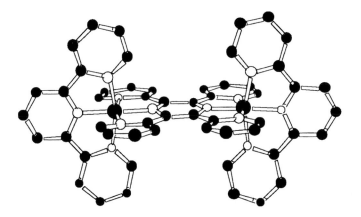

Figure 8 $[(tpy)Ru(tpp)Ru(tpy)]^{4+}$.

electron tpy-based reduction at -1.38 V (Table 1) [21,22]. The tpp-based reduction occurs at a more positive potential than the tpy-based reduction due to the lower energy π^* orbital of tpp compared to tpy.

In the electrochemistry of $[(tpy)Ru(tpp)Ru(tpy)]^{4+}$, the two ruthenium centers give rise to two oxidation couples at 1.44 and 1.76 V vs. Ag/AgCl even though the two metal centers have identical coordination environments

Table 1 Electrochemical Data for a Series of Complexes Using the Bridging Ligand tpp and Model Complexes[a]

Complex	$E_{1/2\ ox}$	$E_{1/2\ red}$	Cond.	Ref.
$[(tpy)Ru(tpp)]^{2+}$	$+1.40$ Ru$^{II/III}$	-0.97 tpp$^{0/-}$	b	21
		-1.38 tpy$^{0/-}$		
		-1.60 tpp$^{-/2-}$		
$[(tpy)Os(tpp)]^{2+}$	$+1.06$ Os$^{II/III}$	-0.97 tpp$^{0/-}$	b	21
		-1.39 tpy$^{0/-}$		
$[(tpy)\mathbf{Ru}(tpp)Ru(tpy)]^{4+}$	$+1.44$ **Ru**$^{II/III}$	-0.35 tpp$^{0/-}$	b	23
	$+1.76$ Ru$^{II/III}$	-0.84 tpp$^{-/2-}$		
		-1.30 2tpy$^{0/-}$		
$[(tpy)Os(tpp)Ru(tpy)]^{4+}$	$+1.11$ Os$^{II/III}$	-0.41 tpp$^{0/-}$	c	27
	$+1.38$ Ru$^{II/III}$	-0.85 tpp$^{-/2-}$		

[a] tpy = 2,2':6',2''-terpyridine, tpp = 2,3,5,6-tetrakis(2-pyridyl)pyrazine, and dpq = 2,3-bis(2-pyridyl)pyrazine.
[b] Potentials reported in acetonitrile solution with 0.1 M TBAH vs. Ag/AgCl (0.29 V vs. NHE).
[c] Potentials reported in acetonitrile solution with 0.1 M TBAClO$_4$ vs. SSCE.

(Table 1) [23]. The oxidation of one of the metal centers affects the oxidation potential of the second metal center and is indicative of significant metal–metal communication. The bridging tpp ligand in this bimetallic system was shown to have a more positive reduction potential than the tpp in the monometallic $[Ru(tpp)(tpy)]^{2+}$, i.e., -0.35 vs. -0.97 V, respectively. This is due to a stabilization of the tpp π^* orbital that is characteristic of bimetallic formulation of ruthenium complexes of this type. The stabilization of the tpp π^* orbital in this bimetallic system is also seen in the electronic absorption spectrum. The lowest energy spin allowed transition in $[(tpy)Ru(tpp)Ru(tpy)]^{4+}$ is a Ru $(d\pi) \rightarrow$ tpp (π^*) CT transition at 550 nm, red-shifted from the same Ru $(d\pi) \rightarrow$ tpp (π^*) CT transition for $[Ru(tpp)(tpy)]^{2+}$ at 474 nm (Table 2). Interestingly, even though $[Ru(tpy)_2]^{2+}$ is nonemissive at RT, both $[Ru(tpp)(tpy)]^{2+}$ and $[(tpy)Ru(tpp)Ru(tpy)]^{4+}$ are emissive at RT (λ_{max}^{em} = 665 and 826 nm, respectively). Our analysis of the photophysical properties of these complexes showed that $[Ru(tpp)(tpy)]^{2+}$ has an emission lifetime of 30 ns whereas $[(tpy)Ru(tpp)Ru(tpy)]^{4+}$ has a lifetime of 100 ns. We have postulated that the lower energy π^* orbital of tpp compared to that of tpy lowers the energy of the ^{3}MLCT state, reducing thermal population of the ligand field state in $[Ru(tpp)(tpy)]^{2+}$ compared to $[Ru(tpy)_2]^{2+}$ (Scheme 4) [24]. Formation of the bimetallic $[(tpy)Ru(tpp)Ru(tpy)]^{4+}$ results in a further lowering in the energy of the tpp π^* orbital and a corresponding stabilization of the ^{3}MLCT state, further reducing thermal accessibility of the ligand field state. This postulate is consistent with the RT lifetime data as well as temperature-dependent lifetime studies. At 77 K, where thermal accessibility of the ligand field state would be quite limited, $[Ru(tpy)(tpp)]^{2+}$ and $[(tpy)Ru(tpp)Ru(tpy)]^{4+}$ have emission lifetimes of 7100 and 480 ns, respectively [24]. At 77 K, the monometallic has

Table 2 Photophysical Data for a Series of Complexes Using the Bridging Ligand tpp and Model Compounds[a,b]

Compound	λ_{max}^{abs} (nm)[c]	$\varepsilon \times 10^{-4}$ (M^{-1} cm^{-1})	λ_{max}^{em} (nm)	τ (ns)	Ref.
$[(tpy)Ru(tpp)]^{2+}$	474	1.6	665	30	21
$[(tpy)Os(tpp)]^{2+}$	468	1.5	775	260	22
$[(tpy)Ru(tpp)Ru(tpy)]^{4+}$	550	3.6	826	100	23
$[(tpy)Os(tpp)Ru(tpy)]^{4+}$	546	4.4	d	d	27

[a] Measured at room temperature in deoxygenated acetonitrile.
[b] tpy = 2,2′:6′,2″-terpyridine, tpp = 2,3,5,6-tetrakis(2-pyridyl)pyrazine, CH$_3$CN, and dpq = 2,3-bis(2-pyridyl)pyrazine.
[c] The lowest energy ^{1}MLCT reported.
[d] Not detected.

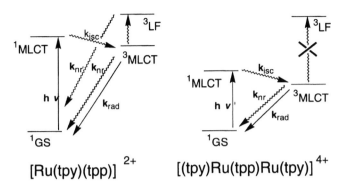

Scheme 4   Jablonski diagram for $[Ru(tpy)(tpp)]^{2+}$ and $[(tpy)Ru(tpp)Ru(tpy)]^{4+}$.

a longer excited state lifetime than the bimetallic, as expected by energy gap considerations [6,25,26].

In 1993, Abruna et al. reported some monometallic and polymetallic complexes of ruthenium and osmium incorporating the ligands tpp and tpy [27]. In the mixed-metal complex $[(tpy)Ru(tpp)Os(tpy)]^{4+}$, the osmium center was shown to oxidize at a less positive potential than the ruthenium center in the homometallic $[(tpy)Ru(tpp)Ru(tpy)]^{4+}$, 1.11 V and 1.40 V, respectively (Table 1). This is due to the higher energy $d\pi$ orbitals of osmium compared to ruthenium and illustrates the ability to tune the redox potentials of the different components of a polymetallic complex by altering the metal center.

A linear tpp-bridged trimetallic complex was also reported, $[(tpy)Ru(tpp)Os(tpy)Ru(tpy)]^{6+}$ [27]. In this complex, the osmium center oxidizes first at 1.23 V followed by sequential oxidation of the ruthenium centers at 1.59 and 1.72 V. Abruna states that this split in the oxidation potentials of the equivalent ruthenium centers implies a significant coupling between the centers across approximately 30 Å. With the metal–metal communication through the tpp bridging ligand and the ability of tpp to form linear polymetallic complexes, Abruna proposes that it should be possible to construct higher oligomer polymetallic tpp-bridged molecular "wires" with good electronic communication along the length of the wire.

B.   The tpy-ph$_n$-tpy-Based Systems

One of the main types of tridentate bridging ligands studied consists of two coordinating tpy "ends" held back-to-back by a spacer at the 4' position. The spacer that is used affects both the metal-to-metal distance and the extent of electronic coupling through the ligand. Sauvage, Constable, et al. have reported on a series of ruthenium and osmium complexes using the bridging ligand tpy-spacer-tpy where phenyl groups serve as the spacer (Fig. 9) [28–40]. The complexes have the form [(ttpy)Ru(tpy-ph$_n$-tpy)M(ttpy)]$^{4+}$ where n = 0, 1, or 2; M = Ru^{II} or Os^{II}; and ttpy = p-tolyl-2,2′,6′,2″-terpyridine. This series of stereochemically defined complexes is held in a rigid, linear fashion with well-defined metal–metal distances ranging from 11 to 20 Å.

The electrochemistry of the ruthenium–ruthenium bimetallic complex [(ttpy)Ru(tpy-tpy)Ru(ttpy)]$^{4+}$ shows that the two ruthenium centers have the same oxidation potential, 1.31 V vs. NHE (Table 3) [41]. This is in marked contrast to the tpp bridged system, [(tpy)Ru(tpp)Ru(tpy)]$^{4+}$, where the oxidation potentials of the two ruthenium centers are split by over 300 mV [28]. This seems to indicate that there is significantly less metal–metal communication in the tpy-tpy bridged system. The first reduction of all of the systems [(ttpy)Ru(tpy-ph$_n$-tpy)Ru(ttpy)]$^{4+}$, where n = 0, 1, 2, is bridging ligand–based. The reduction potential of the bridging tpy-tpy is more positive than the reduction potential of ttpy in the model monometallic [Ru(ttpy)$_2$]$^{2+}$, indicating that there is at least some interaction of the two tpy "ends" of tpy-tpy. This interaction stabilizes the bridging ligand π* orbital and makes the reduction potential more positive. The reduction potential of the bridging ligand in [(ttpy)Ru(tpy-ph$_n$-tpy)Ru(ttpy)]$^{4+}$, where n = 1 or 2, is very similar to the reduction potential of ttpy in the model monometallic, [Ru(ttpy)$_2$]$^{2+}$. This similarity of the reduction potentials, and the similarity of the energy of the bridging ligand π* orbitals, can also be seen in the emission energy and lifetimes.

In both the Ru–Ru complexes, [(ttpy)Ru(tpy-ph$_n$-tpy)Ru(ttpy)]$^{4+}$, and the Ru–Os mixed-metal complexes, [(ttpy)Ru(tpy-ph$_n$-tpy)Os(ttpy)]$^{4+}$, the lowest energy spin allowed electronic transition is an M(dπ) → BL(π*) CT transition, where BL = bridging ligand (Table 4). When two phenyl spacers are used, the M(dπ) → tpy-ph$_n$-tpy(π*) CT transition is only slightly lower in energy than the M → ttpy for the model monometallic, [M(ttpy)$_2$]$^{2+}$. As the number of phenyl spacers is decreased, the M(dπ) → tpy-ph$_n$-tpy(π*) CT transition shifts to lower

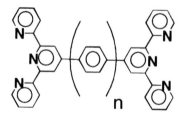

Figure 9   tpy-ph$_n$-tpy.

Table 3 Electrochemical Data for a Series of Complexes Using the Bridging Ligand tpy-ph$_n$-tpy and Model Compounds[a,b]

Complex	$E_{1/2\ ox}$	$E_{1/2\ red}$	Ref.
$[Ru(ttpy)_2]^{2+}$	$+1.25$ Ru$^{II/III}$	-1.24 ttpy$^{0/-}$	30, 31
$[Os(ttpy)_2]^{2+}$	$+0.93$ Os$^{II/III}$	-1.23 ttpy$^{0/-}$	30, 31
$[(ttpy)Ru(tpy-tpy)Ru(ttpy)]^{4+}$	$+1.31$ Ru$^{II/III}$	-0.93 tpy-tpy$^{0/-}$	41
$[(ttpy)Ru(tpy-ph-tpy)Ru(ttpy)]^{4+}$	$+1.27$ Ru$^{II/III}$	-1.18 tpy-ph-tpy$^{0/-}$	29
$[(ttpy)Ru(tpy-ph_2-tpy)Ru(ttpy)]^{4+}$	$+1.26$ Ru$^{II/III}$	-1.21 tpy-ph$_2$-tpy$^{0/-}$	29
$[(ttpy)Ru(tpy-tpy)Os(ttpy)]^{4+}$	$+0.90$ Os$^{II/III}$	-1.01 tpy-tpy$^{0/-}$	30
	$+1.27$ Ru$^{II/III}$		
$[(ttpy)Ru(tpy-ph-tpy)Os(ttpy)]^{4+}$	$+0.90$ Os$^{II/III}$	-1.20 tpy-ph-tpy$^{0/-}$	30, 34
	$+1.25$ Ru$^{II/III}$		
$[(ttpy)Ru(tpy-ph_2-tpy)Os(ttpy)]^{4+}$	$+0.90$ Os$^{II/III}$	-1.22 tpy-ph$_2$-tpy$^{0/-}$	30, 34
	$+1.24$ Ru$^{II/III}$		
$[(ttpy)Ru(tpp)Ru(ttpy)]^{4+}$	$+1.40$ Ru$^{II/III}$	-0.35 tpp$^{0/-}$	29
	$+1.70$ Ru$^{II/III}$	-0.85 tpp$^{-/2-}$	
$[(ttpy)Ru(tpy-tpy)Rh(ttpy)]^{5+}$	$+1.31$ Ru$^{II/III}$	-0.54 Rh$^{III/I}$	41
		-1.22 tpy-tpy$^{0/-}$	
$[(ttpy)Ru(tpy-ph-tpy)Rh(ttpy)]^{5+}$	$+1.29$ Ru$^{II/III}$	-0.56 Rh$^{III/I}$	41
		-1.18 tpy-tpy$^{0/-}$	
$[(ttpy)Ru(tpy-ph_2-tpy)Rh(ttpy)]^{5+}$	$+1.27$ Ru$^{II/III}$	-0.56 Rh$^{III/I}$	41
		-1.20 tpy-tpy$^{0/-}$	

[a] tpy = 2,2',2''-terpyridine, ttpy = 4'-p-tolyl-2,2',2''-terpyridine, and ph = phenyl.
[b] Potentials reported in acetonitrile solution with 0.1 M Bu$_4$NBF$_4$ vs. SCE.

energy. This is due to the increasing interaction of the two tpy ends of the bridging ligand, which lowers the energy of the bridging ligand π* orbital. The analogous Ru–Ru tpp bridged complex with terminal ttpy ligands has also been studied and displays a much lower energy MLCT transition than any of the Ru/Ru tpy-ph$_n$-tpy bridged complexes (Table 4).

Very recently, the complex $[(ttpy)Ru(tpy-tpy)Ru(ttpy)]^{4+}$ was found to have an emission centered at 720 nm with a lifetime of 570 ns at RT [42] over two magnitudes longer than the 0.95-ns emission lifetime of $[Ru(ttpy)_2]^{2+}$. These results have been explained by considering the two ^{3}MLCT excited state deactivation pathways of these complexes. The first deactivation pathway is the direct ^{3}MLCT → ^{1}GS deactivation. The second, thermally activated pathway involves population of the metal-centered ^{3}LF state, which quickly decays nonradiatively to the ^{1}GS. In $[Ru(ttpy)_2]^{2+}$, the second, thermally activated pathway is the major means of deactivation for the ^{3}MLCT excited state. In $[(ttpy)Ru(tpy-tpy)Ru(ttpy)]^{4+}$, the lower energy π* acceptor orbital of the bridging tpy-tpy compared to ttpy lowers the energy of the ^{3}MLCT state and this

Table 4 Photophysical Data for a Series of Complexes Using the Bridging Ligand tpy-ph$_n$-tpy and Model Compounds[a,b]

Compound	λ_{max}^{abs} (nm)[c]	$\varepsilon \times 10^{-4}$ (M^{-1} cm^{-1})	λ_{max}^{em} (nm)	τ (ns)	Ref.
[Ru(ttpy)$_2$]$^{2+}$	490	2.9	640[d]	0.95[d]	30, 31
[Os(ttpy)$_2$]$^{2+}$	490	2.6	734[d]	230[d]	30, 31
[(ttpy)Ru(tpy-tpy)Ru(ttpy)]$^{4+}$	520	5.8	720	570[e]	29, 42
[(ttpy)Ru(tpy-ph-tpy)Ru(ttpy)]$^{4+}$	499	6.3	656	4.0[e]	29, 42
[(ttpy)Ru(tpy-ph$_2$-tpy)Ru(ttpy)]$^{4+}$	495	7.4	g	g	29
[(ttpy)Ru(tpy-tpy)Os(ttpy)]$^{4+}$	522	6.2	800[d]	110[d]	30
[(ttpy)Ru(tpy-ph-tpy)Os(ttpy)]$^{4+}$	500	6.6	746[d]	190[d]	30
[(ttpy)Ru(tpy-ph$_2$-tpy)Os(ttpy)]$^{4+}$	496	6.6	738[d]	200[d]	30
[(ttpy)Ru(tpy-tpy)Rh(ttpy)]$^{5+}$	ca. 520	g	720[f]	<100[f]	41
[(ttpy)Ru(tpy-ph-tpy)Rh(ttpy)]$^{5+}$	ca. 500	g	655[f]	3,000[f]	41
[(ttpy)Ru(tpy-ph$_2$-tpy)Rh(ttpy)]$^{5+}$	ca. 495	g	648[f]	3,500[f]	41
[(ttpy)Ru(tpp)Ru(ttpy)]$^{4+}$	553	6.5	g	g	29

[a] Measured at room temperature in acetonitrile unless noted.
[b] tpy = 2,2′,2″-terpyridine, ttpy = 4′-p-tolyl-2,2′,2″-terpyridine, and ph = phenyl.
[c] The lowest energy ^{1}MLCT reported.
[d] Measured at room temperature in deoxygenated butyronitrile.
[e] Measured at room temperature in deoxyenated acetonitrile.
[f] Measured at 150 K in aerated 4:1 methanol/ethanol.
[g] Not reported.

decreases the amount of thermally activated deactivation through the ^{3}LF state. In the complex [(ttpy)Ru(tpy-ph-tpy)Ru(ttpy)]$^{4+}$, the phenyl spacer leads to a higher energy π^* acceptor orbital, which allows thermal activation of the ^{3}LF state resulting in an emission lifetime of 4.0 ns.

For the Ru–Os mixed-metal systems, [(ttpy)Ru(tpy-ph$_n$-tpy)Os(ttpy)]$^{4+}$, energy transfer between metal centers was examined by luminescence studies [30]. For all three of the Ru–Os complexes, it was found that there was quantitative energy transfer from the ruthenium center to the osmium center and the rate of energy transfer was extremely fast, i.e., $k_{en} > 10^{10}$ s^{-1} at room temperature [31]. No emission could be detected from the ruthenium center at 293 or 77 K. By comparison, the rate of energy transfer between Ru(bpy)$_3$$^{2+}$ and Os(bpy)$_3$$^{2+}$ separated by a rigid nonconjugated bridging ligand where the Ru–Os distance was 17 Å was 1000 times less [43]. This seems to indicate that the phenyl spacers are good bridges for efficient energy transfer even for tpy-ph$_2$-tpy where the Ru–Os distance is 20 Å.

To further characterize the degree of electronic coupling between metal centers in these (tpy-ph$_n$-tpy) bridged systems, the intervalence (IT) charge transfer transitions of the mixed valence Ru^{II}/Ru^{III} complexes were studied by Sauvage et al. (Table 5) [29].

All of the mixed-valence $[(ttpy)Ru(tpy-ph_n-tpy)Ru(ttpy)]^{5+}$ complexes, where $n = 0$, 1, or 2, showed IT bands in the 1150- to 1580-nm region [29]. From the electronic coupling parameter, H_{ab}, it can be seen that the phenyl groups only slightly attenuate the amount on intercomponent coupling, as also is indicated from the luminescence studies of the Ru–Os systems [30]. All three of the tpy-ph$_n$-tpy systems were classified by Sauvage et al. as class II mixed-valence compounds that have localized valences with some limited degree of intercomponent electronic coupling [29]. By comparison, the analogous tpp bridged system, $[(ttpy)Ru(tpp)Ru(ttpy)]^{4+}$, was shown to have an IT transition at 1520 nm with an electronic coupling parameter of 0.40 eV [29]. Sauvage et al. classified this tpp bridged system as a class III mixed-valence, delocalized compound.

The mixed metal Ru–Rh $[(ttpy)Ru(tpy-ph_n-tpy)Rh(ttpy)]^{5+}$ complexes, where $n = 0$, 1, and 2, were studied by Sauvage et al. to investigate electron transfer from the excited ruthenium center to the rhodium center. For $n = 1$, it was found that electron transfer was efficient at room temperature with a rate $\geq$ 3.0×10^9 s^{-1}. For $n = 2$, the rate of electron transfer was determined to be $\geq 5 \times 10^8$ s^{-1} [44]. The rate of electron transfer in the complex with $n = 0$ could not be determined due to the lack of a suitable model monometallic ruthenium complex.

To further investigate the ability of the phenyl spacers to electronically couple two metal centers, Sauvage et al. studied a system where an insulating, saturated hydrocarbon was used as part of the spacer between the tpy ends of the tpy-ph$_n$-tpy bridging ligand. Bicylco[2.2.2]octane (bco) was used between two phe-

Table 5 Metal-to-Metal Charge Transfer Transition Data for Mixed-Valence Compounds Using the Bridging Ligands tpy-ph$_n$-tpy and Model Compounds[a,b]

Compound	λ_{max}^{abs} (nm)	ε (M^{-1} cm^{-1})	$\Delta v_{1/2}$ (cm^{-1})	H_{ab} (eV)	Ref.
$[(ttpy)Ru(tpy-ph_2-tpy)Ru(ttpy)]^{5+}$	1150	709	4934	0.022[c]	29
$[(ttpy)Ru(tpy-ph-tpy)Ru(ttpy)]^{5+}$	1295	729	6036	0.030[c]	29
$[(ttpy)Ru(tpy-tpy)Ru(ttpy)]^{5+}$	1580	1618	4008	0.047[c]	29
$[(ttpy)Ru(tpp)Ru(ttpy)]^{5+}$	1520	e	e	0.40[d]	29

[a] tpy = 2,2′,2″-terpyridine, ttpy = 4′-p-tolyl-2,2′,2″-terpyridine, tpp = 2,3,5,6-tetra(2-pyridyl)pyrazine, and ph = phenyl.
[b] Measured at room temperature in deoxygenated acetonitrile.
[c] Calculated as a class II complex.
[d] Calculated as a class III complex.
[e] Not reported.

4+

Figure 10   [(ttpy)Ru(tpy-ph-bco-ph-tpy)Os(ttpy)]⁴⁺.

nyl rings as the spacer in the complex [(ttpy)Ru(tpy-ph-bco-ph-tpy)Os(ttpy)]⁴⁺ (Fig. 10) [32]. This complex has an Ru–Os distance of 24 Å. It was found that there was a large decrease in the rate of energy transfer from the excited state of the ruthenium component to the osmium component. At room temperature, the excited state lifetime of the ruthenium component was 1.1 ns, comparable to the 0.95-ns lifetime of the monometallic model compound, [Ru(ttpy)$_2$]$^{2+}$ [30]. There was no energy transfer from the ruthenium to the osmium component at room temperature. At 77 K, the lifetime of the excited state of the ruthenium component was 10.5 μs. This is long enough to allow energy transfer to the osmium component and the rate of energy transfer at 77 K was found to be 4.4 × 10^6 s^{-1}. This showed that the saturated bco spacer has a large insulating effect between the metal components.

Sauvage et al. also have used the (tpy)Ru^{II}(tpy) framework in the construction of a wide array of novel and promising linear supramolecular complexes [45,46]. Some very interesting work has been described incorporating porphyrin moieties in linear polymetallic molecules, but this is not the subject of this report.

C.   The typ-eth$_n$-tpy-Based Systems

Ziessel et al. have used ethyne units as spacers in the tpy-spacer-tpy framework to yield the bridging ligands tpy-eth$_n$-tpy, where eth = ethyne (Fig. 11) [47–54].

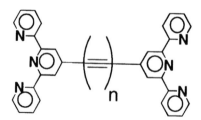

Figure 11   tpy-eth$_n$-tpy.

As in the tpy-ph$_n$-tpy ligands, the ethyne units generate linear, rigid bridging ligands.

The oxidative electrochemistry of the [(tpy)Ru(tpy-eth$_n$-tpy)Ru(tpy)]$^{4+}$ complexes, where n = 1 or 2, shows a single ruthenium oxidation (Table 6), similar to the [(ttpy)Ru(tpy-ph$_n$-tpy)Ru(ttpy)]$^{4+}$ complexes. The reductive electrochemistry of these tpy-eth$_n$-tpy bridged systems shows bridging ligand–based LUMOs. The ethyne spacer serves to lower the energy of the tpy-eth$_n$-tpy π^* orbital compared to tpy. The first tpy-based reduction in [Ru(tpy)$_2$]$^{2+}$ occurs at -1.21 V while the first reduction in [(tpy)Ru(tpy-eth-tpy)]$^{2+}$ is (tpy-eth-tpy)– based and occurs at -1.11 V (Table 6).

The lower energy (tpy-eth$_n$-tpy) π^* orbital compared to that of tpy can be seen in the electronic absorption spectroscopy (Table 7). The lowest energy ^{1}MLCT transition in [Ru(tpy)$_2$]$^{2+}$ is a Ru → tpy CT transition and occurs at 476 nm while the lowest energy ^{1}MLCT transition in [(tpy)Ru(tpy-eth-tpy)]$^{2+}$ is a Ru → (tpy-eth-tpy) CT transition and occurs at 490 nm. The formation of the bimetallic complexes further stabilizes the tpy-eth$_n$-tpy π^* orbital. [(tpy)Ru(tpy-eth-tpy)Ru(tpy)]$^{4+}$ has a lowest energy ^{1}MLCT centered at 515 nm.

The lower energy ligand π^* orbital of the (tpy-eth$_n$-tpy) containing complexes appears to limit deactivation of the ^{3}MLCT excited state through the ligand field state in polymetallic systems quite similar to the tpp bridged systems. [(tpy)Ru(tpy-eth-tpy)]$^{2+}$ has a ^{3}MLCT emission centered at 690 nm with

Table 6 Electrochemical Data for a Series of Complexes Using the Bridging Ligand tpy-eth$_n$-tpy and Model Compounds[a,b]

Complex	$E_{1/2\ ox}$	$E_{1/2\ red}$	Ref.
[Ru(tpy)$_2$]$^{2+}$	+1.30 Ru$^{II/III}$	-1.25 tpy$^{0/-}$	54
		-1.52 tpy$^{0/-}$	
[Os(tpy)$_2$]$^{2+}$	+0.96 Ru$^{II/III}$	-1.27 tpy$^{0/-}$	54
		-1.55 tpy$^{0/-}$	
[(tpy)Ru(tpy-eth-tpy)]$^{2+}$	+1.37 Ru$^{II/III}$	-1.11 tpy-eth-tpy$^{0/-}$	48
[(tpy)Ru(tpy-eth$_2$-tpy)]$^{2+}$	+1.32 Ru$^{II/III}$	-1.02 tpy-eth$_2$-tpy$^{0/-}$	54
		-1.42 tpy-eth$_2$-tpy$^{-/2-}$	
[(tpy)Ru(tpy-eth-tpy)Ru(tpy)]$^{4+}$	+1.42 Ru$^{II/III}$	-0.97 tpy-eth-tpy$^{0/-}$	53
		-1.19 tpy-eth-tpy$^{-/2-}$	
[(tpy)Ru(tpy-eth$_2$-tpy)Ru(tpy)]$^{4+}$	+1.41 Ru$^{II/III}$	-0.92 tpy-eth$_2$-tpy$^{0/-}$	53
		-1.15 tpy-eth$_2$-tpy$^{-/2-}$	
[(tpy)Ru(tpy-eth$_2$-tpy)Os(tpy)]$^{4+}$	+0.95 Os$^{II/III}$	-1.04 tpy-eth$_2$-tpy$^{0/-}$	54
	+1.39 Ru$^{II/III}$	-1.34 tpy-eth$_2$-tpy$^{-/2-}$	

[a] tpy = 2,2′,2″-terpyridine, ttpy = 4′-p-tolyl-2,2′,2″-terpyridine, and eth = ethyne.
[b] Potentials reported in acetonitrile solution with 0.1 M Bu$_4$NBF$_4$ vs. SCE.

Table 7 Photophysical Data for a Series of Complexes Using the Bridging Ligand tpy-eth$_n$-tpy and Model Compounds[a,b]

Compound	λ_{max}^{abs} (nm)[c]	$\varepsilon \times 10^{-4}$ (M^{-1} cm^{-1})	λ_{max}^{em} (nm)[d]	τ (ns)[d]	Ref.
[Ru(tpy)$_2$]$^{2+}$	475	1.1	640	0.56	30, 52
[Os(tpy)$_2$]$^{2+}$	475	1.5	718	270	30, 52
[(tpy)Ru(tpy-eth-tpy)]$^{2+}$	495	1.9	690	55	47
[(tpy)Ru(tpy-eth$_2$-tpy)]$^{2+}$	490	2.3	710	170	47
[(tpy)Ru(tpy-eth-tpy)Ru(tpy)]$^{4+}$	515	3.3	722	565	47, 53
[(tpy)Ru(tpy-eth$_2$-tpy)Ru(tpy)]$^{4+}$	512	3.3	735	720	47, 53
[(tpy)Ru(tpy-eth-tpy)Os(tpy)]$^{4+}$	536	0.7	746	225	52
[(tpy)Ru(tpy-eth$_2$-tpy)Os(tpy)]$^{4+}$	539	1.2	760	200	52

[a] Measured at room temperature in acetonitrile unless noted.
[b] tpy = 2,2′,2″-terpyridine, ttpy = 4′-p-tolyl-2,2′,2″-terpyridine, and eth = ethyne.
[c] The lowest energy ^{1}MLCT reported.
[d] Measured at room temperature in deoxygenated acetonitrile.

a lifetime of 55 ns at room temperature while [Ru(tpy)$_2$]$^{2+}$ has an emission centered at 640 nm with a lifetime of 0.6 ns. A second ethyne spacer extends the emission lifetime to 170 ns in [(tpy)Ru(tpy-eth$_2$-tpy)]$^{2+}$. When bimetallic complexes are formed, the tpy-eth$_n$-tpy π^* orbital is further stabilized resulting in even longer lifetimes. This gives [(tpy)Ru(tpy-eth-tpy)Ru(tpy)]$^{4+}$ an emission centered at 722 nm with a lifetime of 565 ns at room temperature and [(tpy)Ru(tpy-eth$_2$-tpy)Ru(tpy)]$^{4+}$ an emission centered at 735 nm with a lifetime of 720 ns [47–49].

Interesting ruthenium-, cobalt-, zinc-, and iron-containing mixed-metal trimetallic systems that use the tpy-eth$_n$-tpy bridging ligands were also studied (Fig. 12) [48]. The Fe^{II} and Co^{II} systems were found to have emission lifetimes of

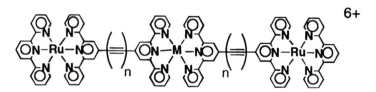

Figure 12 [(tpy)Ru(tpy-eth$_n$-tpy)M(tpy-eth$_n$-tpy)Ru(tpy)]$^{6+}$ where M = Co^{2+}, Fe^{2+}, or Zn^{2+} and n = 1 or 2.

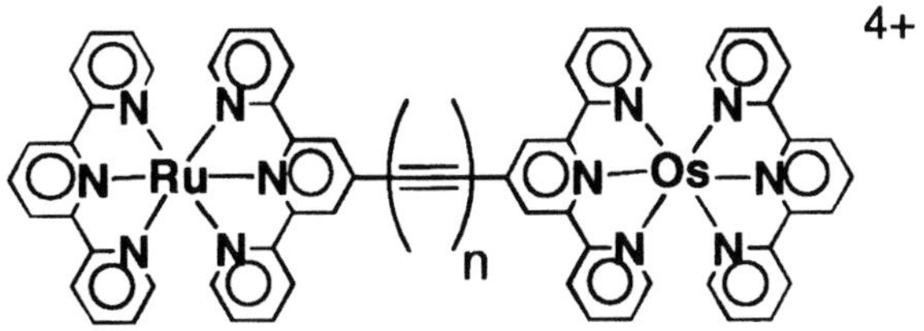

Figure 13 $[(tpy)Ru(tpy-eth_n-tpy)Os(tpy)]^{4+}$ where $n = 1, 2$.

less than 0.1 ns, significantly less than the 55-ns lifetime of the monometallic $[(tpy)Ru(tpy-eth-tpy)]^{2+}$. This is attributed to energy transfer in the case of the Ru-Fe-Ru trimetallic and electron transfer in the case of the Ru-Co-Ru trimetallic. Transient absorption studies showed that when the iron-containing complex was irradiated with a laser pulse to generate the Ru $\rightarrow$ bridging ligand CT excited state, there was energy transfer within approximately 10 ps to the $Fe^{II}(tpy)_2$ fragment. In the case of the cobalt complex, irradiation with a laser pulse to generate the Ru $\rightarrow$ bridging ligand CT excited state, there was reductive quenching by the Co^{II} metal center to produce the complex $Ru^{II}-Co^{III}-Ru^{II}$ that endured for about 130 ps before back-electron transfer occurred [48].

A set of Ru/Os bimetallics containing the bridging ligand tpy-eth$_n$-tpy were also investigated for energy transfer (Fig. 13) [52]. It was found that upon generation of the ruthenium-based MLCT excited state, there was efficient energy transfer to the osmium portion of the molecule within 20 ps [51]. For both of these complexes the rate of energy transfer at 295 K was greater than 1×10^{10} s^{-1}. Based on these and other studies, it was proposed that it should be possible to have electron transfer in ethyne bridged systems with 50% efficiency over a distance of 250 Å [51].

D. The dpb-ph$_n$-dpb^{2-}-Based Systems

A series of bridging ligands has been studied by Sauvage et al. that uses a cyclometallating benzene ring in place of the central pyridyl ring of tpy and is bridged through the 4 position of the benzene, dpb-ph$_n$-dpb, where dpb = 3,5-(dipyridyl)benzene and $n = 0, 1, 2$ (Fig. 14) [33,35–38,55–57]. The Ru–Ru, Os–Os, and Ru–Os complexes, $[(ttpy)M(dpb-ph_n-dpb)M(ttpy)]^{2+}$, where M = Ru or Os and $n = 0, 1, 2$, were synthesized and characterized (Fig. 15).

The electrochemistry of the homometallic $[(ttpy)M(dpb-dpb)M(ttpy)]^{2+}$ systems, where M = Ru and Os, shows that the metal oxidations were split by 0.15 and 0.12 V, respectively (Table 8). This indicates a large degree of metal–metal communication, in contrast to the noncyclometallated analog,

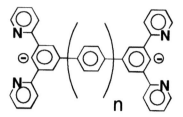

Figure 14   dpb-ph$_n$-dpb^{2-}.

[(ttpy)Ru(tpy-tpy)Ru(ttpy)]$^{4+}$, where there was no split in the oxidation potential of the ruthenium centers.

The mixed valence compounds were generated and the IT bands were characterized to aid in the determination of the amount of electronic communication between the metal centers (Table 9). The Ru–Ru complex was found to have a coupling parameter of 0.15 eV whereas the Os–Os complex was found to have a coupling parameter of 0.12 eV. This is significantly greater than the 0.047 eV coupling parameter found in the analogous Ru–Ru tpy-tpy bridged system, [(ttpy)Ru(tpy-tpy)Ru(ttpy)]$^{4+}$. In fact, characteristics of the IT band suggest that these mixed-valence dpb-dpb bridged systems are completely delocalized class III systems. The strong coupling through the dpb-dpb^{2-} ligand is attributed to the highly covalent, electron-donating nature of the bond between the cyclometallating benzene ring and the metal center. This raises the energy of the metal dπ orbitals, which can also be seen in the less positive oxidation potentials of the metals compared to those in analogous systems using noncyclometalling ligands. The higher energy metal dπ orbitals have a large degree of overlap with the dpb-dpb^{2-} π* orbital, resulting in a large electronic coupling parameter.

In luminescence studies of the mixed metal complexes, [(ttpy)Ru(dpb-ph$_n$-dpb)Os(ttpy)]$^{2+}$, it was found that the rate of energy transfer from the ruthenium center to the osmium center was orders of magnitude slower than in tpy-ph$_n$-tpy systems [55]. For [(ttpy)Ru(dpb-dpb)Os(ttpy)]$^{2+}$ the energy transfer rate was

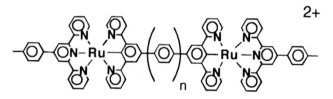

Figure 15   [(ttpy)Ru(dpb-ph$_n$-dpb)Ru(ttpy)]$^{2+}$, where $n = 0, 1, 2$.

Table 8 Electrochemical Data for a Series of Complexes Using the Bridging Ligand dpb-ph$_n$-dpb and Model Compounds[a,b]

Complex	$E_{1/2\ ox}$	$E_{1/2\ red}$	Ref.
[Ru(ttpy)$_2$]$^{2+}$	+1.25 Ru$^{II/III}$	−1.24 ttpy$^{0/-}$	30
[Os(ttpy)$_2$]$^{2+}$	+0.93 Os$^{II/III}$	−1.23 ttpy$^{0/-}$	30
[(ttpy)Ru(dpb)]$^+$	+0.48 Ru$^{II/III}$	−1.61	35
[(ttpy)Os(dpb)]$^+$	+0.33 Os$^{II/III}$	−1.60	35
[(ttpy)Ru(dpb-dpb)Ru(ttpy)]$^{2+}$	+0.34 Ru$^{II/III}$ +0.51 Ru$^{II/III}$	−1.55	38
[(ttpy)Ru(dpb-ph-dpb)Ru(ttpy)]$^{2+}$	+0.51 Ru$^{II/III}$ +1.39[c]	<−1.5	38
[(ttpy)Ru(dpb-ph$_2$-dpb)Ru(ttpy)]$^{2+}$	+0.51 Ru$^{II/III}$ +1.39[c]	<−1.5	38
[(ttpy)Os(dpb-dpb)Os(ttpy)]$^{2+}$	+0.22 Os$^{II/III}$ +0.34 Os$^{II/III}$	−1.53	35
[(ttpy)Os(dpb-dpb)Ru(ttpy)]$^{2+}$	+0.29 Os$^{II/III}$ +0.51 Ru$^{II/III}$	−1.48	37
[(ttpy)Os(dpb-ph-dpb)Ru(ttpy)]$^{2+}$	+0.33 Os$^{II/III}$ +0.53 Ru$^{II/III}$	−1.50	37
[(ttpy)Os(dpb-ph$_2$-dpb)Ru(ttpy)]$^{2+}$	+0.33 Os$^{II/III}$ +0.52 Ru$^{II/III}$	−1.50	37

[a] dpb = dipyridylbenzene, ttpy = 4′-p-tolyl-2,2′,2″-terpyridine, and ph = phenyl.
[b] Potentials reported in acetonitrile solution with 0.1 M Bu$_4$NBF$_4$ vs. SCE.
[c] Irreversible Ru^{III},Ru^{III}/Ru^{IV},Ru^{IV} two-electron oxidation.

Table 9 Metal-to-Metal Charge Transfer Transition Data for Mixed-Valence Compounds Using the Bridging Ligand dpb-ph$_n$-dpb[a,b]

Compound	λ_{max} (nm)	ε (M^{-1} cm^{-1})	$\Delta\nu_{1/2}$ (cm^{-1})	H_{ab} (eV)	Ref.
[(ttpy)Ru(dpb-ph$_2$-dpb)Ru(ttpy)]$^{3+}$	1214	2,200	5714	0.041[c]	38
[(ttpy)Ru(dpb-ph-dpb)Ru(ttpy)]$^{3+}$	1650	6,600	5112	0.074[c]	38
[(ttpy)Os(dpb-dpb)Os(ttpy)]$^{3+}$	1410	18,000	3360	0.12[c]	36
	1800	8,000			
[(ttpy)Ru(dpb-dpb)Ru(ttpy)]$^{3+}$	1820	27,000	2820	0.15[c]	36

[a] dpb = dipyridylbenzene, ttpy = 4′-p-tolyl-2,2′,2″-terpyridine, and ph = phenyl.
[b] Measured at room temperature in deoxygenated acetonitrile.
[c] Calculated as a class II complex.

found to be 2.6×10^9 s^{-1} whereas that of [(ttpy)Ru(dpb-ph$_2$-dpb)Os(ttpy)]$^{2+}$ was found to be $<2.2 \times 10^7$ s^{-1}. For the complex with two phenyl spacers, this is three orders of magnitude slower than in the analogous tpy-ph$_n$-tpy bridged system even though the cyclometallating complex has a much greater electronic coupling parameter for the IT band. These results are explained by considering the nature of the lowest energy excited state of the two series. In the tpy-ph$_n$-tpy bridged series, the lowest energy excited state is a M(dπ) $\rightarrow$ tpy-ph$_n$-tpy(π^*) CT state. This puts the excited electron on the bridging ligand and close to the other metal, facilitating energy transfer. In the dpb-ph$_n$-dpb^{2-} bridged series, it was determined that the π^* orbital of the bridging dpb-ph$_n$-dpb^{2-} is higher in energy than the π^* orbital of the terminal ttpy. This may make the lowest lying excited state a Ru(dπ) $\rightarrow$ ttpy(π^*) CT excited state, putting the excited electron a long distance from the osmium metal center and inhibiting energy transfer. While these dpb-ph$_n$-dpb^{2-} bridging ligands did not prove advantageous for energy transfer in these systems, they do offer the ability to control electron and energy transfer in polymetallic systems. By carefully choosing cyclometallating and noncyclometallating ligands, it should be possible to control the rate and direction of energy and electron transfer.

E. Other Tridentate Bridging Ligands

van Koten et al. have reported on the synthesis and characterization of the bimetallic complex [(tpy)Ru^{III}(pincer-pincer)Ru^{III}(tpy)]$^{4+}$ [58]. The Ru^{II}–Ru^{II} complex can be generated electrochemically or by reaction with hydrazine. The pincer-pincer ligand (Fig. 16) is highly conjugated and provides for strong metal–metal communication. This can be seen in the approximately 200-mV split of the ruthenium oxidation potentials and in the IT band. The Ru^{II}–Ru^{III} complex displays an intense ($\varepsilon = 33,000$ M^{-1} cm^{-1}) IT band at 1875 nm.

Recently, Balzani et al. have investigated two trimetallic systems using three 2,2′2″-terpyridine ligands linked together with -CH$_2$-CH$_2$- or -CH=CH- groups (Fig. 17) [59]. The -CH$_2$-CH$_2$- bridged system was found to behave almost identically to [Ru(tpy)$_2$]$^{2+}$ with no emission at room temperature and a lifetime of 9.4 µs at 77 K. The -CH=CH- bridged trimetallic system had a detectable

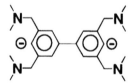

Figure 16   Pincer–pincer^{2-} ligand.

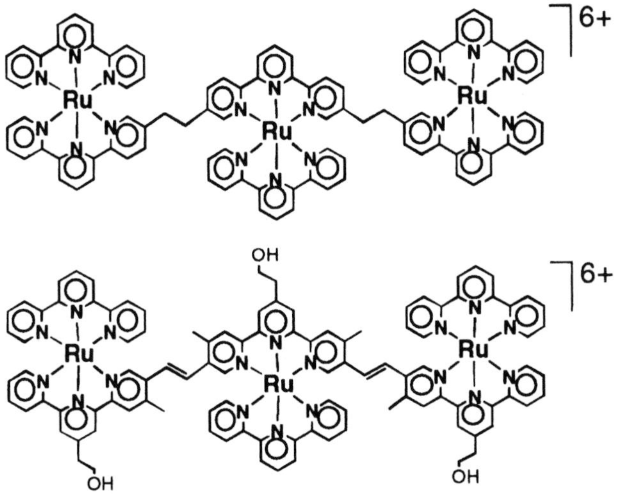

Figure 17   Trimetallic complexes bridged by -CH$_2$-CH$_2$- and -CH=CH- linked terpyridines.

emission at room temperature with a lifetime of 11 ns. At 77 K, the emission had a lifetime of 1.6 μs. These results are explained as a result of the nature of the lowest energy excited state. In both complexes the lowest energy excited state is Ru(dπ) → bridging ligand CT in character but the -CH=CH- bridging ligand has a stabilized π* orbital, which somewhat limits thermal population of the ligand field state.

Constable et al. have investigated a trinucleating tridentate ligands formed by three tpy ligands bound to a benzene ring through the 4′ position of each tpy (Fig. 18) [60]. This trinucleating ligand forms the complex {[(tpy)Ru]$_3$tris-tpy}$^{6+}$ when reacted with Ru(tpy)Cl$_3$. This compound displays only one ruthenium-based oxidation, suggesting weak metal–metal interaction. Constable et al. have proposed that this ligand could be used in starburst arrays [60].

F. Recent tpp Systems

Our recent work has focused on the development of tpp bridged mixed-metal and mixed-ligand complexes [21,23,61–64]. Some of these represent light absorber–electron acceptor, light absorber–quencher, and light absorber–electron donor systems. The complexes that we have designed and investigated have the general formula [(tpy)M(tpp)Ru(LLL)]$^{n+}$ where M = Ru^{II} or Os^{II}, and LLL = Cl$_3$, (CH$_3$CN)$_3$, tpp, or dpq and Cl$^-$ [dpq = 2,3-bis(2-pyridyl)quinoxaline]. When the ligands tpp or dpq are used as terminal ligands, they allow for the covalent attach-

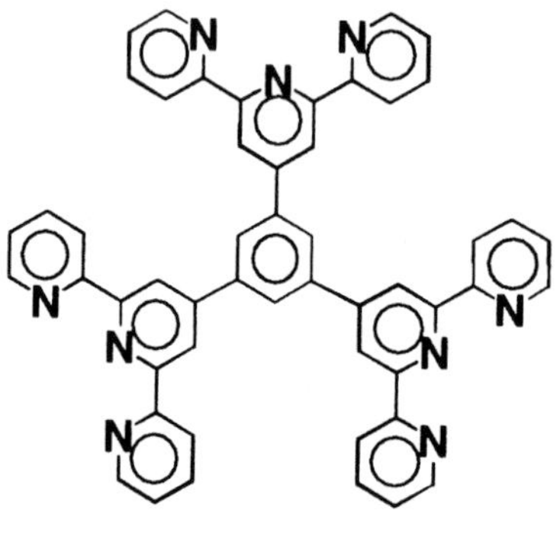

Figure 18 Tris-tpy.

ment of these versatile systems to other supramolecular devices. The utility of this approach can be seen by the use of component introduction to adjust ground and excited state properties of these systems.

The synthesis of these complexes follows a building block approach where ruthenium or osmium is coordinated to tpy followed by the stepwise attachment of the bridging tpp, ruthenium, and the terminal ligands (Scheme 5). In this way, exact control of every component in the complex is achieved.

In these tpp bridged bimetallic complexes, designed to be used in larger systems for photoinitiated electron collection, it is important to know the nature of the redox processes as they give information about the nature and energy of the HOMO and LUMO in these systems. Spectroelectrochemical studies have served to solidify the electrochemical assignments of the order of oxidation of the metal centers. In these tpp bridged bimetallic complexes, it was found that the nature of the first oxidation was dependent on the metal and the ligands used. In most cases, the first oxidation was centered on ruthenium coordinated to tpp and (LLL), (tpp)Ru^{II}(LLL), where LLL = Cl$_3$, (CH$_3$CN)$_3$, or (dpq)Cl (Table 10). However, in the case of the complexes with a terminal tpp, [(tpy)Ru(tpp)Ru(tpp)]$^{4+}$ and [(tpy)Os(tpp)Ru(tpp)]$^{4+}$, the first oxidation is centered on the metal coordinated to tpy and tpp, (tpy)M^{II}(tpp), where M = Ru or Os. In all of the complexes the first reduction is centered on the bridging tpp due to coordination of two metal centers. When these bimetallic complexes are incorporated into larger multimetallic systems, the nature of the first oxidation and/or the first reduction could change when the terminal tpp or dpq takes on the role of a bridging ligand. The electrochemical properties of these systems allow one to see the nature of the HOMO and LUMO in these complexes. This aids in the determination of the nature of the lowest lying excited state since it is normally HOMO → LUMO in nature. It can be seen that through the variation

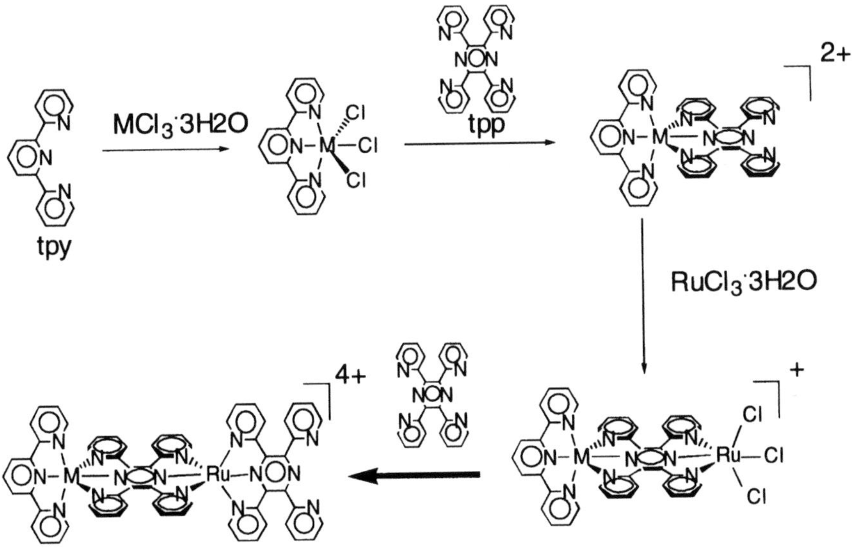

Scheme 5 Building block approach to [(tpy)M(tpp)Ru(LLL)]$^{n+}$.

of the metal centers and terminal ligands, one can change the nature of the HOMO in these systems and, therefore, the nature of the lowest lying excited state. The nature of this lowest lying excited state dictates many of the photophysical properties of these systems.

As with [(tpy)Ru(tpp)Ru(tpy)](PF$_6$)$_4$, these complexes all exhibit excited state emission lifetimes of approximately 100 ns (Table 11). This is critically important for the use of these LA-ED dyads in electron transfer since the lifetime of the dyad must be long enough to have electron transfer to an electron collector. A 100-ns lifetime is ample time to allow efficient electron transfer to an attached electron collector. The emission data on these systems also indicate that typically very efficient energy transfer occurs in these tpp bridged systems resulting in emission only from the lowest lying excited state regardless of the excitation wavelength. This is another beneficial characteristic of the tpp bridging ligand in the construction of the supramolecular systems for light harvesting and other photochemical processes.

The light absorbing properties of these complexes can be difficult to understand due to the complicated UV-Vis spectroscopy. In these complexes there are metal-to-ligand charge transfer transitions as well as ligand-based $\pi \rightarrow \pi^*$ transitions and $n \rightarrow \pi^*$ transitions. With two metals and up to three different ligands in these [(tpy)M(tpp)Ru(LLL)]$^{n+}$ complexes, the number of possible transitions is quite large and leads to very complicated UV-Vis spectra. We have used

Table 10 Electrochemical Data for tpp Bridged Metal Complexes and Model Compounds[a]

Complex	$E_{1/2\ ox}$	$E_{1/2\ red}$	Cond.	Ref.
[(tpy)Ru(tpp)]$^{2+}$	+1.40 Ru$^{II/III}$	−0.97 tpp$^{0/-}$	[b]	21
		−1.38 tpy$^{0/-}$		
		−1.60 tpp$^{-/2-}$		
[(tpy)Os(tpp)]$^{2+}$	+1.06 Os$^{II/III}$	−0.97 tpp$^{0/-}$	[b]	21
		−1.39 tpy$^{0/-}$		
[(tpy)Ru(tpp)**RuCl$_3$**]$^{4+}$	+0.73 **Ru**$^{II/III}$	−0.60 tpp$^{0/-}$	[b]	21
		−1.10 tpp$^{-/2-}$		
	+1.61 Ru$^{II/III}$	−1.50 tpy$^{0/-}$		
[(tpy)Os(tpp)RuCl$_3$]$^{4+}$	+0.66 Ru$^{II/III}$	−0.59 tpp$^{0/-}$	[b]	21
	+1.32 Os$^{II/III}$	−1.07 tpp$^{-/2-}$		
		−1.47 tpy$^{0/-}$		
[(tpy)Ru(tpp)**Ru**(tpp)]$^{4+}$	+1.51 **Ru**$^{II/III}$	−0.30 tpp$^{0/-}$	[b]	21
	+1.86 **Ru**$^{II/III}$	−0.82 tpp$^{-/2-}$		
		−1.10 tpp$^{0/-}$		
[(tpy)Os(tpp)Ru(tpp)]$^{4+}$	+1.17 Os$^{II/III}$	−0.36 tpp$^{0/-}$	[b]	21
	+1.81 Ru$^{II/III}$	−0.81 tpp$^{-/2-}$		
		−1.07 tpp$^{0/-}$		
[(tpy)Ru(tpp)**Ru**(CH$_3$CN)$_3$]$^{4+}$	+1.30 **Ru**$^{II/III}$	−0.30 tpp$^{0/-}$	[b]	61
		−0.80 tpp$^{-/2-}$		
	+1.80 Ru$^{II/III}$	−1.67 tpy$^{0/-}$		
[(tpy)Ru(tpp)**Ru**(dpq)Cl]$^{3+}$	+0.87 **Ru**$^{II/III}$	−0.72 tpp$^{0/-}$	[b]	61
		−1.19 tpp$^{-/2-}$		
	+1.36 Ru$^{II/III}$	−1.42 dpq$^{0/-}$		
[(tpy)Ru(tpp)IrCl$_3$]$^{2+}$	+1.56 Ru$^{II/III}$	−0.29 tpp$^{0/-}$	[b]	62
	+1.92 Ir$^{III/IV}$	−0.83 tpp$^{-/2-}$		
		−1.42 tpy$^{0/-}$		
[(tpy)Ru(tpp)RhCl$_3$]$^{2+}$	+1.60 Ru$^{II/III}$	−0.23 Rh$^{3+/+}$	[b]	63
		−0.60 tpp$^{0/-}$		
		−0.98 tpp$^{-/2-}$		
		−1.40 tpy$^{0/-}$		

[a] tpy = 2,2′:6′,2″-terpyridine, tpp = 2,3,5,6-tetrakis(2-pyridyl)pyrazine, and dpq = 2,3-bis(2-pyridyl)pyrazine.
[b] Potentials reported in acetonitrile solution with 0.1 M TBAH vs. Ag/AgCl (0.29 V vs. NHE).

UV-Vis spectroelectrochemistry to determine the nature of many characteristic transitions in the complicated spectra of these complexes. The lowest energy band is overlapping M $\rightarrow$ μ-tpp and Ru $\rightarrow$ μ-tpp CT transitions in the area of 540–650 nm [21–23,61,62,65]. There is a M $\rightarrow$ tpy CT transition located at approximately 470 nm and a higher energy M $\rightarrow$ tpy CT transition at around 300 nm that is lost from this region of the spectrum upon oxidation of M. This

Table 11 Photophysical Data for tpp Bridged Metal Complexes and Model Compounds[a,b]

Compound	λ_{max}^{abs} (nm)[c]	$\varepsilon \times 10^{-4}$ (M^{-1} cm^{-1})	λ_{max}^{em} (nm)	τ (ns)	Ref.
$[(tpy)Ru(tpp)]^{2+}$	474	1.6	665	30	21
$[(tpy)Os(tpp)]^{2+}$	468	1.5	775	260	22
$[(tpy)Ru(tpp)RuCl_3]^{4+}$	612	1.4	d	d	21
$[(tpy)Os(tpp)RuCl_3]^{4+}$	678	1.5	d	d	21
$[(tpy)Ru(tpp)Ru(tpp)]^{4+}$	548	4.0	833	100	21
$[(tpy)Os(tpp)Ru(tpp)]^{4+}$	546	3.6	820	120	21
$[(tpy)Ru(tpp)IrCl_3]^{2+}$	558	e	810	22	62
$[(tpy)Ru(tpp)RhCl_3]^{2+}$	516	e	665	30	63
$[(tpy)Ru(tpp)Ru(CH_3CN)_3]^{4+}$	566	2.5	790	70	23, 64
$[(tpy)Ru(tpp)Ru(dpq)Cl]^{3+}$	584	2.4	d	d	61, 64

[a] Measured at room temperature in deoxygenated acetonitrile.
[b] tpy = 2,2':6',2''-terpyridine, tpp = 2,3,5,6-tetrakis(2-pyridyl)pyrazine, and dpq = 2,3-bis(2-pyridyl)pyrazine.
[c] The lowest energy ^{1}MLCT reported.
[d] Not detected.
[e] Not reported.

drastic change in the 300-nm region upon oxidation of M is very useful in determining which metal is oxidized first in these systems. There are typically three bands at around 290, 350, and 380 nm associated with μ-tpp $\pi \rightarrow \pi^*$ transitions that are lost from this area of the spectrum upon tpp reduction and red-shifted upon metal oxidation. There are two bands at 272 and about 325 nm that are associated with the tpy $\pi \rightarrow \pi^*$ transitions. The characteristic transitions identified by spectroelectrochemistry has helped to understand the complicated light absorbing characteristics of these compounds and will also aid in understanding the excited state properties of these promising light absorber–electron donor complexes. These bimetallic complexes are useful electron donor–light absorber systems that can be incorporated into larger multimetallic complexes designed for electron transfer and electron collection.

The complexes containing terminal bridging ligands with open coordination sites, namely $[(tpy)Ru(tpp)Ru(tpp)](PF_6)_4$, $[(tpy)Os(tpp)Ru(tpp)](PF_6)_4$, and $[(tpy)Ru(tpp)Ru(dpq)Cl](PF_6)_3$, are capable of being covalently incorporated into a larger supramolecular system as light absorber–electron donor units. The first step toward this incorporation has been the synthesis of the bimetallic species $[(tpy)Ru(tpp)MCl_3](PF_6)_2$ (M = Ir, Rh) [63,64]. Despite their similarity in structure and electron count, they exhibit very different electrochemical and excited state properties.

In the iridium bimetallic, [(tpy)Ru(tpp)IrCl$_3$][PF$_6$]$_2$, the first two reductions (at −0.29 and −0.83 V) are based on the tpp ligand. The first oxidation (+1.56 V) is a Ru$^{II/III}$ process followed closely by Ir$^{III/IV}$ oxidation [63]. The proximity of the ruthenium and iridium orbitals results in the lowest energy absorption band (about 660 nm) having Ir(dπ) → tpp(π*) ^{3}MLCT character. The emission is from a triplet Ru(dπ) → tpp(π*) CT excited state. This results from energy transfer from the Ir to the Ru center.

Substitution of Rh for Ir leads to a molecule with a LUMO based on the Rh atom itself. Unlike the Ir analog, the lowest energy absorption is Ru(dπ) → tpp(π*) CT in character. This leads to rapid charge transfer to the Rh-centered LUMO. This charge transfer can be observed in the dramatically reduced room temperature emission lifetime of 22 ns compared to 100 ns for [(tpy)Ru(tpp)Ru (tpp)](PF$_6$)$_4$ and 70 ns for [(tpy)Ru(tpp)Ru(CH$_3$CN)$_3$](PF$_6$)$_4$. At 77 K, the emission lifetimes of [(tpy)Ru(tpp)Ru(tpp)](PF$_6$)$_4$ and [(tpy)Ru(tpp)RhCl$_3$](PF$_6$)$_2$ are comparable (360 vs. 330 ns) [64]. This illustrates that this complex functions as a LA-ED dyad. This Ru–Rh bimetallic is promising as a model compound for larger, stereochemically defined supramolecular compounds for photoinitiated electron collection.

III. CONCLUSIONS

Tridentate bridging ligands have the ability to form polymetallic complexes with many interesting and promising properties for photoinitiated processes. In many cases, tridentate bridging ligands eliminate the possibility of geometric isomers, which is advantageous because the distance of electron or energy transfer can be controlled in the design of complexes for photointiated processes. The ability to synthesize only one isomer also eliminates the problem of separating or studying a mixture of isomers.

Despite the limited lifetime of the excited state of [Ru(tpy)$_2$]$^{2+}$, systems with observable emission can be formed with tridentate ligands when the energy difference between the lowest ^{3}MLCT and the low-lying ligand field states are greater than in [Ru(tpy)$_2$]$^{2+}$. This has enabled the synthesis of rigid systems with fast energy transfer over 20 Å and it has been proposed that energy transfer of hundreds of angstroms could be possible.

Variation of tridentate bridging ligands has been shown to allow control of the energy and electronic coupling properties of these polymetallic assemblies. The molecules [(ttpy)Ru(tpy-ph$_n$-tpy)M(ttpy)]$^{m+}$ (M = Ru, Os; m = 4, 5) have been shown to exhibit rapid and efficient energy transfer from one metal center to another and moderate electronic coupling between the metal centers in the mixed-valence species. Substitution of dpb-ph$_n$-dpb^{2-} for tpy-ph$_n$tpy leads to a series of molecules [(ttpy)Ru(dpb-ph$_n$-dpb)M(ttpy)]$^{m+}$ (M = Ru, Os; m = 2, 3) with strong metal-metal communication in the mixed-valence species and energy transfer rates that were orders of magnitude slower than in the tpy-ph$_n$-tpy bridged case. The top ligand has provided systems with long-lived excited states with energy and electron transfer properties that hold great promise for the construction of more complex supramolecular systems.

ACKNOWLEDGMENTS

Special thanks to all of the students and research scientists who have worked in this area in my group, including Dr. Lisa Vrana, Elizabeth R. Bullock, Prof. Jae-Duck Lee, Glen Jensen, Tara Hawks, and James Harrison. Aspects of this work have been supported by generous funding from NSF CHE-9632713.

REFERENCES

1. For an excellent review, see Kalyanasundaram, K. *Coord. Chem. Rev.* **1982**, *46*, 159–244.
2. Juris, A.; Balzani, V.; Barigelletti, F.; Campagna, S.; Belser, P.; Von Zelewsky, A. *Coord. Chem. Rev.* **1988**, *84*, 85–277.
3. Winkler, J.R.; Netzel, T.L.; Creutz, C.; Sutin, N. *J. Amer. Chem. Soc.* **1987**, *109*, 2381–2392.
4. Van Houten, J.; Watts, R. *J. Am. Chem. Soc.* **1975**, *97*, 3843–3844.
5. Kirchhoff, J.R.; McMillin, D.R.; Marnot, P.A.; Sauvage, J.-P. *J. Am. Chem. Soc.* **1985**, *107*, 1138–1141.
6. Calvert, J.M.; Caspar, J.V.; Binstead, R.A.; Westmoreland, T.D.; Meyer, T.J. *J. Am. Chem. Soc.* **1982**, *104*, 6620–6627.
7. Hecker, C.R.; Fanwick, P.E.; McMillin, D.R. *Inorg. Chem.* **1991**, *30*, 659–666.
8. Leisning, R.A.; Kubow, S.A.; Churchill, M.R.; Buttrey, L.A.; Ziller, J.W.; Takeuchi, K.J. *Inorg. Chem.* **1990**, *29*, 1306–1312.
9. Constable, E.C.; Cargill Thompson, A.M.W.; Tocher, D.A.; Daniels, M.A.M. *N. J. Chem.* **1992**, *16*, 855–867.
10. Balzani, V.; Moggi, L.; Scandola, F. in *Supramolecular Photochemistry*, Balzani, V., Ed.; 1987, Reidel, Dordrecht, 1.
11. Balzani, V.; Scandola, F. *Supramolecular Photochemistry*, Ellis Horwood, Chichester, England, 1991, 355–388.
12. For a recent review of dpp and other bidentate bridging ligands see: Balzani, V.; Juris, A.; Venturi, M.; Campagna, S.; Serroni, S. *Chem. Rev.* **1996**, *96*, 759–833.
13. Kelso, L.S.; Reitsma, D.A.; Keene, F.R. *Inorg. Chem.* **1996**, *35*, 5144–5153.
14. Reitsma, D.A.; Keene, F.R. *J. Chem. Soc., Dalton Trans.* **1993**, 2859–2860.
15. Allen, G.C.; Hush, N.S. *Prog. Inorg. Chem.* **1967**, *8*, 357–383.
16. Creutz, C. *Prog. Inorg. Chem.* **1983**, *30*, 1–73.
17. Robin, M.B.; Day, P. *Adv. Inorg. Chem. Radiochem.* **1967**, *10*, 247–422.
18. Goodwin, H.A.; Lions, F. *J. Am. Chem. Soc.* **1959**, *81*, 6415–6422.
19. Petersen, J.D. in *Supramolecular Photochemistry*, Balzani, V. Ed.; Reidel, Dordrecht, **1987**, 135–152.

20. Thummel, R.P.; Chirayil, S. *Inorg. Chim. Acta* **1988**, *154*, 77–81.
21. Vogler, L.M.; Brewer, K.J. *Inorg. Chem.* **1996**, *35*, 818–824.
22. Brewer, R.G.; Jensen, G.E.; Brewer, K.J. *Inorg. Chem.* **1994**, *33*, 124–129.
23. Vogler, L.M.; Jones, S.W.; Jensen, G.E.; Brewer, R.G.; Brewer, K.J. *Inorg. Chim. Acta* **1996**, *250*, 155–162.
24. Harrison, J.; Brewer, K.J. unpublished work.
25. Caspar, J.V.; Meyer, T.J. *J. Am. Chem. Soc.* **1983**, *105*, 5583–5590.
26. Caspar, J.V.; Meyer, T.J. *Inorg. Chem.* **1983**, *22*, 2444–2453.
27. Arana, C.R.; Abruna, H.D. *Inorg. Chem.* **1993**, *32*, 194–203.
28. Constable, E.C.; Ward, M.D. *J. Chem. Soc., Dalton Trans.* **1990**, 1405–1409.
29. Collin, J.-P.; Laine, P.; Launay, J.-P.; Sauvage, J.-P.; Sour, A. *J. Chem. Soc., Chem. Commun.* **1993**, 434–435.
30. Barigelletti, F.; Flamigni, L.; Balzani, V.; Collin, J.-P.; Sauvage, J.-P.; Sour, A.; Constable, E.C.; Cargill Thompson, A.M.W. *J. Amer. Chem. Soc.* **1994**, *116*, 7692–7699.
31. Barigelletti, F.; Flamigni, L.; Balzani, V.; Collin, J.-P.; Sauvage, J.-P.; Sour, A.; Constable, E.C.; Cargill Thompson, A.M.W. *Coord. Chem. Rev.* **1994**, *132*, 209–214.
32. Barigelletti, F.; Flamigni, L.; Balzani, V.; Collin, J.-P.; Sauvage, J.-P.; Sour, A. *N. J. Chem.* **1995**, *19*, 793–798.
33. Chodorowski-Kimmes, S.; Beley, M.; Collin, J.-P.; Sauvage, J.-P.; *Tetrahedron Lett.* **1996**, *37*, 2933–2936.
34. Beley, M.; Collin, J.-P.; Sauvage, J.-P.; Sugihara, H.; Heisel, F.; Miehe, A. *J. Chem. Soc., Dalton Trans.* **1991**, 3157–3159.
35. Beley, M.; Chodorowski, S.; Collin, J.-P.; Sauvage, J.-P.; Flamigni, L.; Barigelletti, F. *Inorg. Chem.* **1994**, *33*, 2543–2547.
36. Beley, M.; Collin, J.-P.; Sauvage, J.-P.; *Inorg. Chem.* **1993**, *32*, 4539–4543.
37. Barigelletti, F.; Flamigni, L.; Guardigli, M.; Juris, A.; Beley, M.; Chodorowski-Kimmes, S.; Collin, J.-P.; Sauvage, J.-P. *Inorg. Chem.* **1996**, *35*, 136–142.
38. Beley, M.; Chodorowski-Kimmes, S.; Collin, J.-P.; Laine, P.; Launay, J.-P.; Sauvage, J.-P. *Angew. Chem. Int. Ed., Engl.* **1994**, *33*, 1775–1778.
39. Constable, E.C.; Cargill Thompson, A.M.W. *J. Chem. Soc., Dalton Trans.* **1995**, 1615–1627.
40. Liang, Y.; Schmehl, R. *J. Chem. Soc., Chem. Commun.* **1995**, 1007–1008.
41. Indelli, M.T.; Scandola, F.; Collin, J.-P.; Sauvage, J.-P.; Sour, A. *Inorg. Chem.* **1996**, *35*, 303–312.
42. Hammarstrom, L.; Barigelletti, F.; Flamigni, L.; Indelli, M.T.; Armaroli, N.; Calogero, G.; Guardigli, M.; Sour, A.; Collin, J.-P.; Sauvage, J.-P. *J. Phys. Chem.* **1999**, in press.
43. De Cola, L.; Balzani, V.; Barigelletti, F.; Flamigni, L.; Belser, P.; von Zelewsky, A.; Frank, M.; Vogtle, F. *Inorg. Chem.* **1993**, *32*, 5228–5238.
44. Indelli, M.T.; Scandola, F.; Flamigni, L.; Collin, J.-P.; Sauvage, J.-P.; Sour, A. *Inorg. Chem.* **1997**, *36*, 4247–4250.
45. Odobel, F.; Sauvage, J.-P. *N.J. Chem.* **1994**, *18*, 1139–1141.
46. Harriman, A.; Odobel, F.; Sauvage, J.-P. *J. Am. Chem. Soc.* **1994**, *116*, 5481–5482.

47. Benniston, A.C.; Grosshenny, V.; Harriman, A.; Ziessel, R. *Angew. Chem. Int. Ed., Engl.* **1994**, *33*, 1884–1888.
48. Grosshenny, V.; Harriman, A.; Ziessel, R. *Angew. Chem. Int. Ed., Engl.* **1995**, *34*, 2705–2708.
49. Grosshenny, V.; Ziessel, R. *J. Organomet. Chem.* **1993**, *453*, C19–C22.
50. Ziessel, R.; Suffert, J.; Youinou, M.-T. *J. Org. Chem.* **1996**, *61*, 6535–6546.
51. Ziessel, R. *J. Chem. Educ.* **1997**, *74*, 673–679.
52. Grosshenny, V.; Harriman, A.; Ziessel, R. *Angew. Chem. Int. Ed., Engl.* **1995**, *34*, 1100–1102.
53. Harriman, A.; Ziessel, R. *Chem. Commun.* **1996**, 1707–1716.
54. Grosshenny, V.; Harriman, A.; Hissler, M.; Ziessel, R. *J. Chem. Soc., Faraday Trans.* **1996**, *92*, 2223–2238.
55. Barigelletti, F.; Flamigni, L.; Collin, J.-P.; Sauvage, J.-P. *Chem. Commun.* **1997**, 333–338.
56. Beley, M.; Collin, J.-P.; Louis, R.; Metz, B.; Sauvage, J.-P. *J. Am. Chem. Soc.* **1991**, *113*, 8521–8522.
57. Beley, M.; Chodorowski, S.; Collin, J.-P.; Sauvage, J.-P. *Tetrahedron Lett.* **1993**, *34*, 2933–2936.
58. Sutter, J.-P.; Grove, D.M.; Beley, M.; Collin, J.-P.; Veldman, N.; Spek, A.L.; Sauvage, J.-P.; van Koten, G. *Angew. Chem. Int. Ed., Engl.* **1994**, *33*, 1282–1285.
59. Hasenknopf, B.; Hall, J.; Lehn, J.-M.; Balzani, V.; Credi, A.; Campagna, S. *N. J. Chem.* **1996**, *20*, 725–730.
60. Constable, E.C.; Cargill Thompson, A.M.W.; Tocher, D.A. *Supramolecular Chemistry*, Kluwer Academic, Netherlands, **1992**, 219–233.
61. Jones, S.W.; Vrana, L.M.; Brewer, K.J. *J. Organoment. Chem.* **1998**, *554*, 29–40.
62. Vogler, L.M.; Scott, B.; Brewer, K.J. *Inorg. Chem.* **1993**, *32*, 898–903.
63. Lee, J.-D.; Vrana, L.M.; Bullock, E.R.; Brewer, K.J. *Inorg. Chem.* **1998**, *37*, 3575–3580.
64. Jones, S.W. Ph.D. dissertation, Virginia Polytechnic Institute and State University, 1998.

5

Photophysical and Photochemical Properties of Metallo-1,2-enedithiolates

Robert S. Pilato and Kelly A. Van Houten
University of Maryland, College Park, Maryland

I. INTRODUCTION

Metallo-1,2-enedithiolates are a new important class of emissive molecules. These complexes have also been used as Q-switching laser dyes [1,2], as components in conducting [3–6] and magnetic [7,8] materials, and as models for the molybdenum and tungsten cofactors [9–15]. The emissive properties of metallo-1,2-enedithiolates were initially discovered in the late 1970s as square planar group VIII and IIB thiolate complexes became the subject of intensive study. Among these thiolates, the 1,2-enedithiolates emerged with the most interesting, diverse, and potentially useful photophysical properties [16–36]. The photophysical properties of these complexes can be associated with several different lowest lying transitions and, by inference, several excited states. These include metal-to-ligand charge transfer (MLCT), ligand-to-ligand charge transfer (LLCT), and intraligand charge transfer (ILCT) transitions.

Most 1,2-enedithiolate ligands are unstable in the absence of a metal and many of these ligands must be prepared directly on a metal center [11,30,33–43]. However, mnt (maleonitriledithiolate), bdt (1,2-benzenedithiolate), tdt (tolu-

185

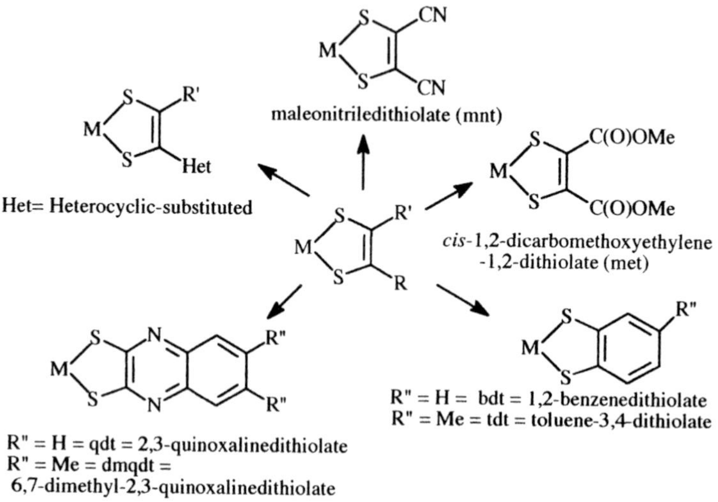

Figure 1 The metallo-1,2-enedithiolate families discussed in this chapter.

ene-3,4-dithiolate), qdt (2,3-quinoxalinedithiolate), and dmqdt (6,7-dimethyl-2,3-quinoxalinedithiolate) are either stable dianions or dithiols, and this is reflected in the propensity of studies of the corresponding metal complexes [18–28]. Recent synthetic efforts to prepare new 1,2-enedithiolate complexes have been driven by the unique ground and excited state reactivity of these complexes [30–35,44]. From these studies a new subfamily, the heterocyclic-substituted 1,2-enedithiolates has emerged. Many of these complexes are room temperature dual emitters with applications in sensing [44–46].

This chapter is intended as an overview of the photophysical and photochemical properties of emissive metallo-1,2-enedithiolates. It includes a brief description of the use of luminescent platinum 1,2-enedithiolate complexes to develop new sensors. Metal derivatives of mnt, met (*cis*-1,2-dicarbomethoxyethylene-1,2-dithiolate), bdt, tdt, dmqdt and qdt as well as complexes of the general structure type (phosphine)$_2$Pt{S$_2$C$_2$(Het)(R')} where Het = 2- or 4-pyridine and 2-quinoxaline will be discussed (Fig. 1).

II. ASSIGNING EXCITED STATES

Defining the excited state of an emissive molecule is essential to understanding its properties and potential applications. As with any assignment of electronic transitions, those of the metallo-1,2-enedithiolate complexes have relied on both theoretical and spectroscopic studies [16,19,20,26–28]. From these studies the excited states have been assigned to ILCT, LLCT, and MLCT. As will be discussed, it is often difficult to differentiate these excited states. Many complexes have electronic transitions that are best described as having mixed MLCT/ILCT or MLCT/LLCT character. While recognizing this restraint, we will attempt where possible to assign the emissive states. To do so, it is important to first define the possible orbitals involved in the lowest lying electronic transitions.

The highest occupied molecular orbital (HOMO) in most emissive metallo-1,2-enedithiolates is thought to be a 1,2-enedithiolate π orbital. In C_{2v} symmetry with the atoms placed in the xy plane and x as the primary axis of rotation, the highest occupied 1,2-enedithiolate π orbital is b_2 and comprises S_{pz}, C_{pz}, and metal d_{xz} atomic orbitals. This HOMO is C-C π-bonding/S-C π-antibonding and M(d_{xz})/p(S) antibonding (Fig. 2) [10,19,20]. The extent of metal d_{xz} contribution to the HOMO depends on the metal, the ancillary ligands, and substituents on the 1,2-enedithiolate [20,28].

The HOMO can be extended by metal-bound ancillary π-accepting ligands, such as CO. In a square planar complex the ancillary π acceptors are trans to the 1,2-enedithiolate sulfurs and there can be $d\pi$-$p\pi^*$ overlap with the metal d_{xz} orbital extending the conjugation of the metallo-1,2-enedithiolate π orbitals [20]. Unsaturated R/R′ groups, such as CN, can also extend the conjugation of the metallo-1,2-enedithiolate π orbitals [20].

The lowest unoccupied molecular orbital (LUMO) of emissive metallo-1,2-enedithiolates is generally a 1,2-enedithiolate π^* orbital unless the 1,2-enedithiolate is heterocyclic-substituted or the metal is bound by a diimine ligand. In these cases, the π^* orbital of the appended heterocycle or bound diimine is usually the LUMO [10,19,20].

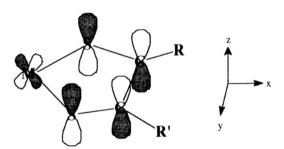

Figure 2   A representation of the typical HOMO of a metallo-1,2-enedithiolate complex. While the contribution of the metal d_{xz} orbital to the HOMO is large in this representation, it is known to vary and this accounts for variations in the metal contributions to the excited states.

Shown in Fig. 3 are crude energy level diagrams that the readers should find useful in understanding the possible transitions in the various metal complexes discussed in this chapter. The lowest lying transitions of emissive 1,2-enedithiolate complexes that do not contain an appended heterocycle or metal-bound diimine are best described as those from the 1,2-enedithiolate π orbital ($\pi_{mnt/xz}$) to the 1,2-enedithiolate π^* orbital (π^*_{S-S}) (Fig. 3A). The extent of MLCT character in this transition depends on the contribution of the metal d_{xz} atomic orbital to the HOMO. While for group VIII metals this can be a large contribution, for group IIB metals the contribution is expected to be rather small.

In most emissive (diimine)Pt(1,2-enedithiolate) and heterocyclic-substituted 1,2-enedithiolate complexes, (phosphine)$_2$Pt{S$_2$C$_2$(Het)(R')}, the lowest lying transition is from the 1,2-enedithiolate π orbital ($\pi_{mnt/xz}$ or $\pi_{C_2S_2R_2/xz}$) to the π^* orbital of the diimine (π^*_{N-N}) or heterocycle (π^*_{Het}), respectively (Fig. 3B, C). Generally, these π-to-π^* transitions are insensitive to the metal center and have been assigned to LLCT and ILCT transitions, respectively (see Sections V and VI) [10,20].

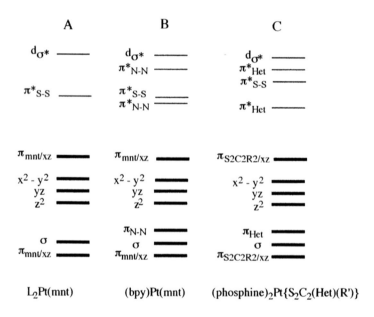

L$_2$= phosphines, amines, COD etc. (not including aromatic diimines)

Figure 3 Crude energy level diagrams of the relevant bonding and antibonding orbitals of L$_2$Pt(mnt) [20], (bpy)Pt(mnt) [20], and (phosphine)$_2$Pt{S$_2$C$_2$(Het)(R')} [47]. The filled orbitals are indicated by the dark lines. Note: The energy levels are not to scale.

The ordering of the electronic transitions in the $(phosphine)_2Pt\{S_2C_2(Het)(R')\}$ complexes depends on the appended heterocycle. For these complexes the energy of the ILCT transition tracks with the reduction potential and hence the electron affinity of the heterocycle [30]. Consistent with this observation, the energy of the transition is lowered by protonation and alkylation of the heterocycle (see Section VI) [30–35].

An important factor in determining whether a complex will be emissive is the energy of the d-to-σ* (d-to-d) transition. When a d-to-d transition is lowest lying, it generally leads to rapid nonradiative decay of emissive excited states. In most Co, Ni, and Pd 1,2-enedithiolates the d-to-d transition is generally below 20,000 cm^{-1} and is usually the lowest lying band [48]. As such, these complexes are not emissive. In the Rh, Ir, and Pt complexes, the d-to-d transition is generally greater than 24,500 cm^{-1}. Since the MLCT, LLCT, and ILCT transitions are usually lower in energy, complexes containing these metals can be emissive (see Section IV) [16,30,32,49].

Found in Tables 1a–d is a list of known emissive metallo-1,2-enedithiolate complexes. The tables are arranged by ligand type and include excitation and emission maxima, excited state assignment, and lifetimes.

III. HOMOLEPTIC METALLO-1,2-ENEDITHIOLATES

There are emissive homoleptic bis(1-2-enedithiolate) Pt, Zn, Cd, and Hg dianions. Even though these complexes are isostructural, their excited state assignments differ. The homoleptic platinum complexes, $[Pt(mnt)_2]^{2-}$ and $[Pt(qdt)_2]^{2-}$, have a lowest lying MLCT transition whereas the corresponding group IIB complexes, $[M(mnt)_2]^{2-}$ (M = Zn, Cd, and Hg), $[Zn(qdt)_2]^{2-}$, and $[Zn(tdt)_2]^{2-}$, have what is best described as a lowest lying ILCT transition [16,29,36].

$[Pt(mnt)_2]^{2-}$ is photostable in the absence of chlorinated solvents [50,51] (see Section VII) and is weakly emissive at room temperature ($\phi = 10^{-5}$) with a more intense emission at low temperature (Table 1a) [16,29]. The lowest lying transition has considerable MLCT character. The assignment is based on the metal dependence of the energy of this transition for $[M(mnt)_2]^{2-}$ (M = Ni, Pd, and Pt). A characteristic feature of MLCT transitions in the Ni triad is an energy dependence that follows Ni $\approx$ Pt $\ll$ Pd [49]. The energies of the MLCT transitions of $[M(mnt)_2]^{2-}$ are found at 19,300, 18,500, and 25,800 cm^{-1} for the Ni, Pt, and Pd complexes, respectively. Only $[Pt(mnt)_2]^{2-}$ is emissive, albeit weakly, since only this complex has a MLCT lower in energy than the corresponding metal d-to-d transition.

The homoleptic $[Pt(qdt)_2]^{2-}$ complex has a room temperature emission ($\phi = 10^{-5}$) similar to that of $[Pt(mnt)_2]^{2-}$ [16,26,29]. The emission is again attributed to a MLCT transition since, like the mnt derivative, the energy of the transition follows Ni $\approx$ Pt $\ll$ Pd (18,500, 19,600, and 22,700 cm^{-1}, respectively).

Table 1a Photophysical Properties of mnt and met Derivatives

Complex	Absorbance λ_{max} (nm)	Emission λ_{max} glass/solution (nm)	Lifetime (μs) 77 K solid	Excited state assignment[j]	Ref.
mnt					
[TBA]₂[Pt(mnt)₂]	475[e]	712[e]/775[g]	—	MLCT	28
[TBA]₂[Zn(mnt)₂]	380[l]	723[l]/—	—	LLCT	36
[TBA]₂[Cd(mnt)₂]	375[l]	720[l]/—	—	LLCT	36
[TBA]₂[Hg(mnt)₂]	395[l]	730[l]/—	—	LLCT	36
Pt(L)(L')(mnt)					
L = L' = PPh₃	585[a]	657[a]	24[a]	MLCT	16
PMe₂Ph	572[d]	584[d]	—	MLCT	24
POEt₃	553[c]	579[c]	—	MLCT	24
POPh₃	542[d]	605[d]	—	MLCT	24
(L-L') = COD	547[c]	583[c]	74[c]	MLCT	22
tdach	440[h]	610[c]	—	MLCT	20
dppe	566[c]	589[c]	76[c]	MLCT	24
dppv	579[d]	610[d]	77[d]	MLCT	24
dppb	573[d]	590[d]	119[d]	MLCT	24
chpe	570[d]	610[d]	66[d]	MLCT	24
dppm	570[c]	610[c]	—	MLCT	24
bpy	524[n]	623[n]	1[t,n]	LLCT/MLCT	18
dmbpy	498[n]	610[c]	—	LLCT/MLCT	20
dpbpy	650[b]	683[b]	(0.089)[k]	LLCT/MLCT	18,28
dbbpy	497[b]	607[e]/620[b]	(0.003)[k]	LLCT/MLCT	28
phen	562[c]	630[n]	0.67[t,n]	LLCT/MLCT	18
dpphen	492[h]	630[c]	—	LLCT/MLCT	20

[TBA][Ir(L)(L')(mnt)]					
L = L' = CO	544^a	554^a	78^a	LLCT	16
POPh₃	582^a	593^a	40^a	LLCT	16
L = CO; L' = PPh₃	591^a	605^a	30^a	LLCT	16
[PPN][Ir(L)(L')(mnt)]					
L = L' = CO	547^a	559^a	93^a	LLCT	16
POPh₃	586^a	610^a	21^a	LLCT	16
L = CO; L' = PPh₃	593^a	608^a	25^a	LLCT	16
[PPN]₂[Ir(L)(L')(mnt)]					
L = CO; L' = CN	590^a	662^a	8^a	LLCT	16
[TBA][Rh(L)(L')(mnt)]					
L = L' = CO	569^a	609^a	238^a	LLCT	16
POPh₃	594^a	602^a	281^a	LLCT	16
L = CO; L' = PPh₃	600^a	619^a	99^a	LLCT	16
[PPN][Rh(L)(L')(mnt)]					
L = L' = CO	560^a	573^a	389^a	LLCT	16
POPh₃	599^a	643^a	155^a	LLCT	16
L = CO; L' = PPh₃	603^a	617^a	303^a	LLCT	16
[AsPh₄][Rh(L)(L')(mnt)]					
L = CO; L' = PPh₃	612^a	627^a	132^a	LLCT	16
met					
Pt(dmbpy)(met)	571^m	634^c, $695^c/663^c$, 708^c	—	LLCT/MLCT	23
Pt(COD)(met)	399^m	620^c, $675^c/$—	—	LLCT	23
Pt(POMe₃)₂(met)	326^m	625^c, $680^c/$—	—	LLCT	23

^{a} EPA glass (5:5:2, ethyl ether/isopentane/ethyl alcohol). ^{b} CH₂Cl₂ solution. ^{c} DMM glass (DMF/methylene chloride/methanol, 1:1:1). ^{d} KBr Matrix. ^{e} In BuCN. ^{f} Emission from solid state. ^{g} In MeOH. ^{h} In CH₃CN. ^{i} In EtOH glass. ^{j} For a discussion of the excited state assignment, see Section II. ^{k} Lifetime measured at 298 K in CH₂Cl₂ solution. ^{l} In EtOH. ^{m} In CHCl₃. ^{n} In benzonitrile.

Table 1b Photophysical Properties of qdt, dmqdt, and tdt Complexes

Complex	Absorbance λ_{max} (nm)	Emission λ_{max} (nm) glass solution	Lifetime (ns) 298 K solution	Excited state assignment[k]	Ref.
qdt, dmqdt					
[TBA]$_2$[Zn(qdt)$_2$]	400[l]	455[m], 564[m]/440[n]	—	LLCT	36
[TBA]$_2$[Pt(qdt)$_2$]	510[f]	643[e]/606[f]	<1	MLCT	28
Pt(dbbpy)(dmqdt)	481[b]	539[e]/618[b]	80[b]	LLCT	28
Pt(COD)(qdt)	392[f]	609[g]/—	—	MLCT	26
Pt(phen)(qdt)	498[b]	607[e]/612[b]	87[b]	LLCT	26
tdt					
[TBA]$_2$[Zn(tdt)$_2$]	324[l]	345[m], 445[m]/350[n]	—	LLCT	36
Pt(tmphen)(tdt)	532[b]	638[e]/675[b]	1020[b]	LLCT	28
Pt(dbbpy)(tdt)	563[b]	641[e]/720[b]	504[b]	LLCT	28
Pt(dmbpy)(tdt)	564[b]	662[e]/720[b]	381[b]	LLCT	28
Pt(bpy)(tdt)	586[b]	666[e]/735[b]	291[b]	LLCT	28
Pt(phen)(tdt)	583[b]	674[e]/730[b]	517[b]	LLCT	28
Pt(Cl-phen)(tdt)	605[b]	683[e]/760[b]	315[b]	LLCT	28
Pt(Cl$_2$-bpy)(tdt)	630[b]	739[e]/775[b]	157[b]	LLCT	28
Pt(EC-bpy)(tdt)	679[b]	784[e]/795[b]	68[b]	LLCT	28
Pt(bpy)(bdt)	542[j]	—/750[j]	460[j]	LLCT	92
Au(η^2-C,N-ppy)(tdt)	406[b]	512[c]/—	—	LLCT	95

[a] EPA glass (5:5:2, ethyl ether/isopentane/ethyl alcohol). [b] CH$_2$Cl$_2$ solution. [c] DMM glass (DMF/methylene chloride/methanol, 1:1:1). [d] KBr Matrix. [e] In BuCN. [f] In MeOH. [g] In MeOH/EtOH glass. [h] In DMM solution. [i] In CHCl$_3$. [j] In CH$_3$CN. [k] For a discussion of the excited state assignment, see Section II. [l] In EtOH. [m] In EtOH glass.

Table 1c Photophysical Properties of Heterocyclic Substitued Platinum-1,2-Enedithiolates, $(phosphine)_2Pt\{S_2C_2(HET)(R')\}$

Complex	Absorbance λ_{max} (nm)[a]	Emission λ_{max} solution (nm)[b]	Lifetime (ns) 298 K solution[b]	Excited state assignment[c]	Ref.
$(dppe)Pt\{S_2C_2(HET)(H)\}$ HET = quinoxaline	564	660 748	0.15 4700	^1ILCT ^3ILCT	31
2-pyridinium	458	677 732	4.3 3200	^1ILCT ^3ILCT	32
4-pyridinium	474	713 742	3.1 7500	^1ILCT ^3ILCT	32
$(dppm)Pt\{S_2C_2(HET)(H)\}$ 2-pyridinium	449	610 680	0.02 1400	^1ILCT ^3ILCT	35
$(dppe)Pt\{S_2C_2(HET)(Me)\}$ HET = quinoxaline	605	701 744	0.11 720	^1ILCT ^3ILCT	31
$[(dppe)Pt\{S_2C_2(CH_2CH_2\text{-}N\text{-}2\text{-pyridinium})\}][BPh_4]$	496	677 732	0.2 8660	^1ILCT ^3ILCT	34
$[(dppm)Pt\{S_2C_2(CH_2CH_2\text{-}N\text{-}2\text{-pyridinium})\}][BPh_4]$	480	610 730	0.2 8500	^1ILCT ^3ILCT	35
$[(dppp)Pt\{S_2C_2(CH_2CH_2\text{-}N\text{-}2\text{-pyridinium})\}][BPh_4]$	499	620 700	0.2 4500	^1ILCT ^3ILCT	35

[a] In CH_2Cl_2. [b] In DMSO. [c] For a discussion of the excited state assignment, see Section II.

Table 1d Photophysical Properties of Relevant Thiolates

Complex	Absorbance λ_{max} (nm)	Emission λ_{max} (nm) glass/solution	Lifetime (ns) 298 K solution	Excited state assignment[g]	Ref.
edt					
Pt(dmbpy)(edt)	482[a]	—/624[b]	—	LLCT	20
Pt(dpphen)(edt)	488[a]	—/667[b]	—	LLCT	20
Pt(dbbpy)(edt)	480[a]	628[c]/615[a]	10[a]	LLCT	28
Pt(dbbpy)(edt)	560[d]	—/785[d]	7[d]	LLCT	91
Pt(dbbpy)(dpdt)	560[d]	—/770[d]	19[d]	LLCT	91
EtS					
Pt(dmbpy)(EtS)$_2$	482[a]	632[b]/—	—	LLCT	20
Pt(dpphen)(EtS)$_2$	488[a]	678[b]/—	—	LLCT	20
PhS					
Pt(dmbpy)(PhS)$_2$	498[a]	615[b]/—	—	LLCT	20
PhSe					
Pt(dmbpy)(PhSe)$_2$	514[a]	646[b]/—	—	LLCT	20
meo					
Pt(dmbpy)(meo)	458[a]	610[b]/—	—	LLCT	20
Heterocyclic-substituted					
(dppe)Pt{SOC$_{10}$H$_4$N$_2$S}	501[c]	690[e]	6[e]	ILCT	33
C$_3$S$_5$					
[TEA]$_2$[Zn(S$_5$C$_3$)$_2$]	487[f]	670[f]/—	—	LLCT	36

[a] CH_2Cl_2 solution. [b] DMM glass (DMF/methylene chloride/methanol, 1:1:1). [c] In BuCN. [d] In CH_3CN. [e] In DMSO. [f] In EtOH. [g] For a discussion of the excited state assignment, see Section II.

$[Pt(qdt)_2]^{2-}$ is readily protonated in organic solvents to generate $[Pt(qdt)(qdtH)]^-$ which leads to an absorbance red shift of ≈ 2600 cm^{-1}. The emission from $[Pt(qdt)_2]^{2-}$ is at 16,500 cm^{-1}, whereas the emission from $[Pt(qdt)(qdtH)]^-$ is at 13,800 cm^{-1}, consistent with the shift in the excitation maximum induced by protonation. The diprotonated complex, $Pt(qdtH)_2$ has not been studied due to its insolubility in organic solvents.

Unlike the emission from the homoleptic platinum 1,2-enedithiolates, members of the Zn triad, $[M(mnt)_2]^{2-}$ where M = Zn, Cd, Hg, as well as $[Zn(qdt)_2]^{2-}$, $[Zn(tdt)_2]^{2-}$, and $[Zn(S_5C_3)_2]^{2-}$ have emissions assigned to a lowest lying ILCT transition [36]. The ILCT assignment is supported by the lack of metal dependence on the energy of the transition [36]. This assignment is consistent with the low-lying filled d-orbital set of the group IIB metals, which does not effectively contribute to the 1,2-enedithiolate π orbitals.

$[Zn(qdt)_2]^{2-}$ and $[Zn(tdt)_2]^{2-}$ are room temperature emitters with $\phi = 0.0005$ (440 nm) and 0.005 (350 nm), respectively. In low-temperature glasses these complexes are dual emitters. In EtOH glass at 77 K, $[Zn(qdt)_2]^{2-}$ has emissions at 455 nm, $\phi = 0.002$ and 564 nm, $\phi = 0.03$ whereas $[Zn(tdt)_2]^{2-}$ has emissions at 345 nm, $\phi = 0.01$ and 445 nm, $\phi = 0.4$, respectively [36].

Interpretation of the emission spectra of $[M(mnt)_2]^{2-}$ (M = Zn, Cd, Hg) is complicated by a rapid displacement of an mnt ligand by coordinating solvent molecules to yield $L_2M(mnt)$. At 77 K in either ethanol or tetrahydrofuran glass, emissions from both $[M(mnt)_2]^{2-}$ and $L_2M(mnt)$ where L = EtOH or tetrahydrofuran (THF) are observed (see Section IV). The emission assigned to $[M(mnt)_2]^{2-}$ where M = Zn, Cd, and Hg is independent of the metal and found at 13,800 cm^{-1}. The low-temperature emission from these complexes, like those with MLCT assigned excited states, shows a pronounced vibrational progression of ≈ 1400 cm^{-1} which is assigned to the $\nu_{C=C}$ of the mnt ligand [36].

IV. L$_2$M(MNT) AND L$_2$M(MET) DERIVATIVES WHERE L = PHOSPHINES, AMINES, AND CO, OR L$_2$ = BIDENTATE PHOSPHINES AND BIDENTATE AMINES

A large number of Rh, Ir, and Pt complexes that contain either mnt or met and an ancillary 1,5-cyclooctadiene (COD), diamine, carbonyl, phosphine, or phosphite ligand have been prepared [16,20,22–24]. These complexes have excited states with mixed MLCT/ILCT character. Consistent with this assignment in $[(L)(L')M(mnt)]^-$ where M = Rh and Ir and $(L)(L')M(mnt)$ where M = Pt is the energy dependence of the lowest lying transition and emission upon the electron-donating ability of the ligand set L, L'. When the donor ability of L, L' increases, the excitation and emission energies decrease. This is seen when comparing $[TBA][(L)(L')Ir(mnt)]$, with L = L' = CO, L = CO L' = PPh$_3$, and L, L' = dppe, where the excitation energies are 18,100, 16,500, and 14,700 cm^{-1} and where the emission maxima are 18,400, 16,900, and 16,700 cm^{-1}, respectively [16]. The anionic [TBA][Rh(P(OPh)$_3$)$_2$(mnt)] and [TBA][Ir(P(OPh)$_3$)$_2$(mnt)] complexes have lowest lying transitions at 16,800 and 17,200 cm^{-1}, respectively, while the neutral (P(OPh)$_3$)$_2$Pt(mnt) complex has a transition at 18,500 cm^{-1} [16]. The difference in the energy of the lowest lying transition in the Ir and Rh complexes is rather small ($\approx$400 cm^{-1}) and should be larger for a transition with considerable MLCT character [52]. However, when compared to the neutral Pt complex, there is clearly a metal effect, supporting the MLCT assignment.

The low-temperature emissions from nearly all L$_2$M(met) and L$_2$M(mnt) complexes, where L$_2$ = COD, diamine, carbonyl, phosphine, or phosphite, have a pronounced vibrational progression of $\approx$1400 cm^{-1} that has been assigned to the $\nu_{C=C}$ of the mnt or met ligands [16,18,23,24]. This is consistent with the mnt or met π^* orbital being the acceptor in the electronic transition.

As described in Section III, the Zn, Cd, and Hg mnt complexes, [M(mnt)$_2$]$^{2-}$, undergo a photoelimination of one mnt ligand in coordinating solvents to generate L$_2$M(mnt) [36]. The lowest lying transition in these complexes is assigned to an ILCT transition since all of the L$_2$M(mnt) complexes where L = EtOH, THF, NH$_3$ and M = Zn, Cd, and Hg have an excitation within 1000 cm^{-1} (Table 1a). The featureless low-temperature emissions for the L$_2$M(mnt) complexes are independent of ligand and show little dependence on the metal center [36].

V. (DIIMINE)Pt(1,2-ENEDITHIOLATES) AND LLCT* EXCITED STATES

A large number of emissive (diimine)Pt(1,2-enedithiolate) complexes have been prepared where the diimine is a phenanthroline or bipyridyl derivative [18–20,23,26,28]. Nearly all diimine complexes possessing qdt, dmqdt, bdt, and tdt ligands have a lowest lying LLCT transition with considerable 1,2-enedithiolate π to diimine π^* character. This assignment is based on the following observations: First, the energy of the transition is insensitive to metal substitution [53]. Second, the energy of the transition is sensitive to both the accepting ability of the diimine and the donating ability of the dithiolate. This is seen when comparing (dbbpy)Pt(dmqdt), (dbbpy)Pt(tdt), and (EC-bpy)Pt(tdt) where the LLCT transitions are found at 20,800, 17,600, and 14,700 cm^{-1}, respectively [28]. Finally, the energy of the transition is sensitive to solvent polarity, suggesting some charge transfer character [28]. The corresponding mnt and met complexes appear to have more metal character in their lowest energy transition even though the transition is still primarily of the LLCT type. The LLCT transition is not limited to 1,2-enedithiolate complexes and several 1,1-dithiolate and bis(thiolate) complexes with metal-bound diimines have a similarly assigned transition (see Table 1d) [17,19,21,23–25,29].

Most of the diimine complexes containing qdt, dmqdt, tdt, and bdt are room temperature emitters with lifetimes in the 50- to 1000-ns regime (Table 1b). The lifetimes and Stokes shifts suggest the excited states have considerable LLCT triplet character. Given the ^{3}LLCT* lifetimes, these complexes have been the subject of quenching studies with both electron donors and acceptors. As expected, the quenching rate constants, k_q, range from 10^{10} to 10^6 M^{-1} s^{-1} and track with the thermodynamic driving force for the excited state electron transfer [28].

VI. HETEROCYCLIC-SUBSTITUTED METALLO-1,2-ENEDITHIOLATES

In 1997, a new synthetic route [30–34,43] made available a range of heterocyclic-substituted metallo-1,2-enedithiolate complexes (Scheme 1). The heterocycle (Het) and R′ group are readily varied using this method. Metallo-1,2-enedithiolate complexes with appended pyridines, 2-pyrazine, and 2-quinoxaline are now available. While the R′ group is limited only by the availability of the α-bromo- or tosyl ketone required for this synthesis, to date only complexes where R′ = H, Me, and CH$_2$CH$_2$OAc (and several derivatives of CH$_2$CH$_2$OH) [34] have been prepared.

The bis(triphenylphosphine)-substituted complexes readily undergo phosphine substitution, allowing a variety of phosphine complexes to be prepared (Scheme 2) [35]. By varying the phosphine, heterocycle, and R′ a large number of new complexes are available and those shown in Schemes 1 and 2 are only representative examples of the range of possible complexes.

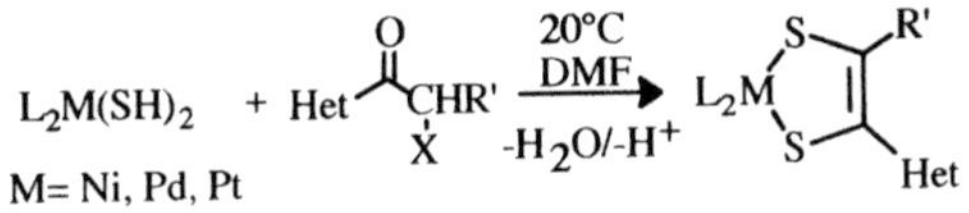

L_2= dppm, dppe, dppp, and *bis*(triphenylphosphine)

X= Br or tosyl

R′= H for Het= 2-quinoxaline, 2-, 3-, or 4-pyridine, 2-pyrazine,

  2-bromo-6-pyridine, 2-acetyl-6-pyridine, and 4-nitrobenzene[30,43]

R′= H or Me for Het= 2-quinoxaline[31]

R′= CH$_2$CH$_2$OH or CH$_2$CH$_2$OAc for Het= 2-pyridine[34,35]

Scheme 1

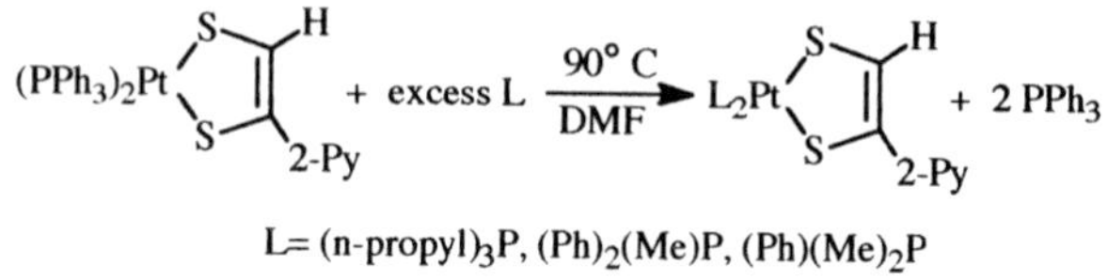

L= (n-propyl)$_3$P, (Ph)$_2$(Me)P, (Ph)(Me)$_2$P

Scheme 2

A. The ILCT (1,2-enedithiolate π to heterocycle π*) Transition of the Heterocyclic-Substituted Metallo-1,2-enedithiolates

All of the heterocyclic-substituted complexes (Schemes 1 and 2) have a band assignable to an ILCT. This transition has considerable 1,2-enedithiolate π to heterocycle π* character [30–34,43]. This assignment is made based on the following observations. First, the energy of this band is solvent-sensitive, supporting the charge transfer assignment. Second, it is nearly identical for the corresponding Ni, Pd, and Pt 1,2-enedithiolate complexes (Fig. 4) and is unaffected by variation of the phosphine ligand. This supports an ILCT assignment and rules out assignment to a MLCT and LMCT [30,43,49,54,55]. Third, the band is red-shifted by methyl substitution of the 1,2-enedithiolate and protonation of the heterocycle. The N-methylation of the heterocycle increases the energy of the transition relative to protonation. This is seen when comparing the energy of the transition for (dppe)Pt{S$_2$C$_2$(2-pyridine)}, [(dppe)Pt{S$_2$C$_2$(2-pyridinium)(H)}]$^+$, and [(dppe)Pt{S$_2$C$_2$(N-methyl-2-pyridinium)(H)}]$^+$, which are found at 24,100, 21,800, and 23,800 cm^{-1}, respectively [32,56]. Collectively, these observations strongly support the ILCT assignment where the π* orbital of the heterocycle is the acceptor. Furthermore, a plot of ILCT energy vs. the reduction potential of the appended heterocycle (or aromatic) is linear, suggesting that the accepting orbital resides on the heterocycle (Fig. 5) [30,57,58].

The d-to-d transition in these complexes is relatively insensitive to the choice of the 1,2-enedithiolate-appended groups and in most complexes it is either the ILCT or the d-to-d transition that is lowest lying. When the d-to-d is the lowest lying transition it leads to rapid nonradiative decay of emissive ILCT excited states. As such, the reduction potential of the heterocycle serves to predict not only the energy of the ILCT transition but when the ILCT transition will be the lowest lying and hence when the molecule will be emissive [30–35].

B. Quinoxaline-Substituted 1,2-Enedithiolates

Of the neutral heterocyclic-substituted dppe complexes prepared to date, only (dppe)Pt{S$_2$C$_2$(2-quinoxaline)(R')} (R' = H or Me) has an ILCT transition lower

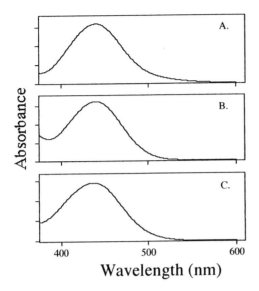

Figure 4   UV-Vis absorption spectra in CH$_2$Cl$_2$ showing the ILCT transition of (A) [(dppe)Ni{S$_2$C$_2$(2-pyridinium)(H)}]$^+$;   (B)   [(dppe)Pd{S$_2$C$_2$(2-pyridinium)(H)}]$^+$;   (C) [(dppe)Pt{S$_2$C$_2$(2-pyridinium)(H)}]$^+$.

in energy than the d-to-d transition and is emissive in solution. While (dppe)Pt{S$_2$C$_2$(2-nitrobenzene)(H)} has an ILCT lower in energy than the d-to-d transition, this complex is not emissive in room temperature solution and is currently under study in low-temperature glasses [56]. Excitation of room temperature–deaerated solutions of (dppe)Pt{S$_2$C$_2$(2-quinoxaline)(R')} (R' = H or Me) gave the characteristic dual emission of this class of molecules (Fig. 6) [30,31].

The time-resolved emission from air-free samples of (dppe)Pt{S$_2$C$_2$(2-quinoxaline)(H)} were best fit as a sum of two exponential decays with one short-lived component (τ = 0.2–0.8 ns) and one long-lived component (τ = 2.3–5.5 µs). The lifetimes and the corresponding quantum yields increase with solvent polarity. Based on the relative energies and lifetimes these excited states are thought to possess considerable ^{1}ILCT* and ^{3}ILCT* character, respectively. The ^{3}ILCT* emission of (dppe)Pt{S$_2$C$_2$(2-quinoxaline)(H)} in solution is quenched by O$_2$ (with $k_q > 3 \times 10^9$ M^{-1} s^{-1} in DMSO) at concentrations that are not sufficient to quench the shorter lived ^{1}ILCT* state. Like molecular oxygen, electron acceptors and donors of the appropriate potential selectively quench the ^{3}ILCT* emission. As expected, the quenching rate constants k_q track with the thermodynamic driving force for the excited state electron transfer reactions [31].

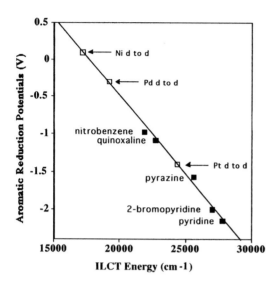

Figure 5 A plot of aromatic reduction potential vs. the energy of the ILCT transition for the family of neutral dppeM{S₂C₂(2-Het)(H)} complexes (■). Overlayed are the energies of the d-to-d transitions in the corresponding Ni, Pd, and Pt complexes (□). Note: The d-to-d transitions are relatively insensitive to 1,2-enedithiolate substitution in these complexes.

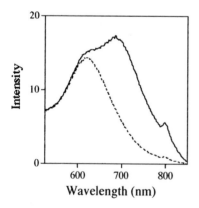

Figure 6 The uncorrected emission spectra of (dppe)Pt{S₂C₂(2-quinoxaline)(H)} in CH₃CN at 298 K: under nitrogen (solid line); in air (dashed line). The lifetimes of these excited states (τ) and the quantum yields for emission (ϕ) in CH₃CN are $^1\tau = 0.16$ ns, $^1\phi = 0.005$ and $^3\tau = 3.3$ μs, $^3\phi = 0.01$.

The ILCT* excited states have a negative charge localized on the heterocycle, which is evident from the selective quenching of the ^{3}ILCT* emission by acids that are not sufficient to protonate the heterocycles in the ground state. The quenching rate constants follow the pK_a of the acids and are in the range of 2.4×10^{10} to 3.6×10^7 M^{-1} s^{-1} (from pyridinium to MeOH in CH$_3$CN) (Fig. 7). Quenching of the ^{3}ILCT* by an acid with a $pK_a \approx 10$ units above the ground state of (dppe)Pt{S$_2$C$_2$(2-quinoxaline)(H)} [ground state pK_a(CH$_3$CN) = 11.9] has a second-order rate constant, $k_q > 1 \times 10^9$ M^{-1} s^{-1} [30,31]. The ^{3}ILCT* of (dppe)Pt{S$_2$C$_2$(2-quinoxaline)(H)} is also quenched by common hydrogen atom donors, such as p-methoxyphenol and dihydroquinone, where $k_q = 1 \times 10^9$ M^{-1} s^{-1} and 2.7×10^9 M^{-1} s^{-1}, respectively. The quenching of the ^{3}ILCT* by p-methoxyphenol and dihydroquinone are clearly hydrogen atom transfers since the thermodynamic driving forces for electron and proton donors transfers are inconsistent with the quenching rate constants [31].

While the triplet emission from (dppe)Pt{S$_2$C$_2$(2-quinoxaline)(H)} is rapidly quenched by acids insufficient to protonate the ground state, the failure to observe either luminescence or a transient absorption spectrum for [(dppe)Pt{S$_2$C$_2$(2-quinoxalinium)(H)}]$^{+*}$ strongly suggests that protonation leads to the rapid nonradiative decay of the excited state and release of the proton. Since the rapid nonradiative decay of the [(dppe)Pt{S$_2$C$_2$(2-quinoxalinium)(H)}]$^{+*}$ precludes an excited state equilibrium between (dppe)Pt{S$_2$C$_2$(2-quinoxaline)(H)}* and [(dppe)Pt{S$_2$C$_2$(2-quinoxalinium)(H)}]$^{+*}$, (dppe)Pt{S$_2$C$_2$(2-quinoxaline)(H)} should not be considered as truly photobasic but simply quenched by weak acids [31,59].

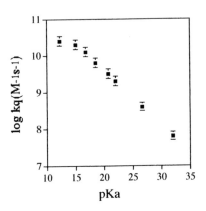

Figure 7 Dependence of k_q (log scale) vs. pK_a of the quenching agent for quenching of (dppe)Pt{S$_2$C$_2$(2-quinoxaline)(H)} in CH$_3$CN. Error bars represent the 95% confidence limits for k_q.

In all of the studies discussed, the ^{1}ILCT* emission is unaffected by quencher concentrations that result in a loss of triplet emission. At higher concentrations two of the acids studied, pyridinium (pK_a = 12.3, CH_3CN) and diammonium propane (pK_a = 15, CH_3CN), protonate the ground state of (dppe)Pt{S_2C_2(2-quinoxaline)(H)}, which results in a loss of emission. However, extensive protonation of the ground state is observed only after complete triplet quenching with these acids. As such, in all of the quenching studies, including oxygen, electron transfer, hydrogen atom transfer, and proton transfer, the ^{1}ILCT* emission acts as a reporter for the ground state and allows preassociation of quenchers to be ruled out. For applications of the dual emitters in sensing, see Section VI.E.

C. Pyridine/Pyridinium-Substituted 1,2-Enedithiolates

The luminescence and excited state electron transfer reactions of (dppe)Pt {S_2C_2(2-pyridine(ium))(H)} and (dppe)Pt{S_2C_2(4-pyridine(ium))(H)} are dependent on the protonation state of the pyridine [30–35]. The "switching on" of the luminescence in these compounds results from a change in the ordering of the electronic transitions in the pyridine and pyridinium substituted complexes. Unlike the quinoxaline-substituted complexes, the neutral pyridine complexes have a lowest lying d-to-d transition, which leads to rapid nonradiative decay of the ILCT excited states. However, upon protonation the ILCT becomes the low-lying transition. The pyridinium complexes are room temperature lumiphores with emission from ^{1}ILCT* and ^{3}ILCT* excited states (see Table 1c).

Unlike the protonated complexes the N-methylated complex [dppePt{S_2C_2 (N-methyl-2-pyridinium)(H)}]$^+$ is nonemissive in solution [56]. In the N-methylated complex the ILCT transition has the same energy as the d-to-d transition ($\approx$ 22,200 cm^{-1}) and this could account for the lack of emission. The lack of emission could also be due to steric disruption of the excited states as will be discussed in Section VI.D [34].

D. Phosphorylation Mediates an Emissive Response

Unlike [(dppe)Pt{S_2C_2(2-pyridinium)(H)}]$^+$, [(dppe)Pt{S_2C_2(2-pyridinium) (CH_2CH_2OR'')}]$^+$ (where R'' = H and Ac) [34,44,45] are nonemissive in room temperature solution ($^3\phi < 10^{-5}$). However, the reaction of (dppe)Pt{S_2C_2(2-pyridine)(CH_2CH_2OH)}, with a range of organophosphates (Scheme 3) and organosulfonates, leads to the generation of [(dppe)Pt{S_2C_2(CH_2CH_2-N-2-pyridinium)}]$^+$ which is luminescent in room temperature solution with two emissive states assigned to the ^{1}ILCT* and ^{3}ILCT* [34,44,45]. The lifetime of the ^{1}ILCT* is 0.2 ns, $^1\phi$ = 0.002 while the lifetime of the ^{3}ILCT is 8.3 μs, $^3\phi$ = 0.01 (DMSO) [34,44,45]. The difference in the quantum yields of [(dppe)Pt{S_2C_2(CH_2CH_2-N-

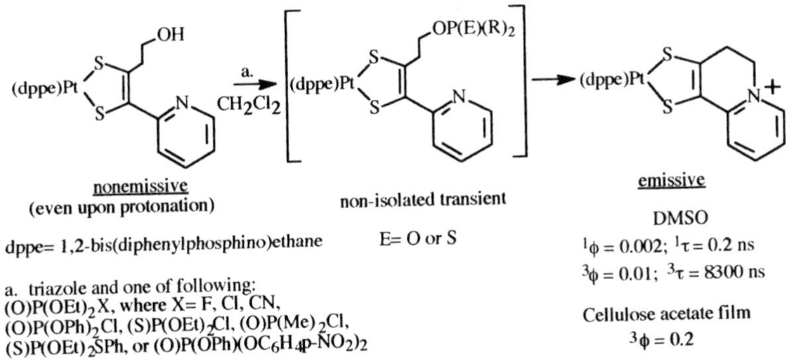

Scheme 3

2-pyridinium)}]$^+$ and [(dppe)Pt{S$_2$C$_2$(2-pyridinium)(H)}]$^+$ from that of [(dppe)Pt {S$_2$C$_2$(2-pyridinium)(CH$_2$CH$_2$OR″)}]$^+$ supports ILCT* excited states where the charge transfer formally oxidizes a 1,2-enedithiolate sulfur and reduces the pyridinium nitrogen. This leads to an increase in the double bond character between the 1,2-enedithiolate and the heterocycle in the excited state. This requires a co-planar 1,2-enedithiolate and heterocycle (Scheme 4). This criteria is forced on [(dppe)Pt {S$_2$C$_2$(CH$_2$CH$_2$-N-2-pyridinium)}]$^{+*}$ by the ground state structure (confirmed crystallographically) [34]. However, in the protonated complexes the ability of the 1,2-enedithiolate and heterocycle to be coplanar is controlled by the bulk of the R′ group. It appears that when R′ = H the heterocycle and 1,2-enedithiolate can be coplanar but that the bulk of the CH$_2$CH$_2$OR″ group impedes coplanarity [30,34].

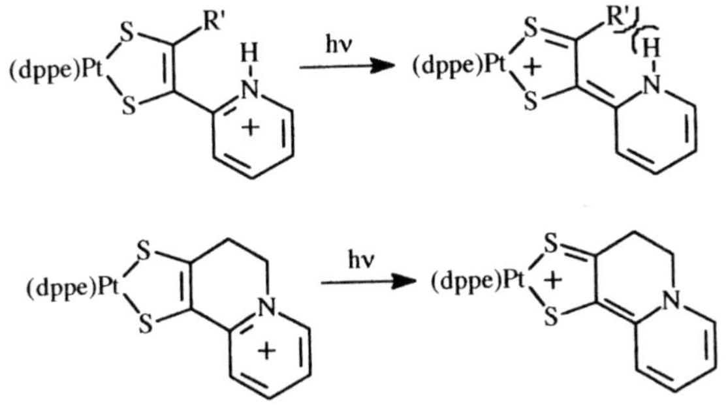

Scheme 4

It is the organophosphate-induced conversion of the nonemissive (dppe) Pt{S_2C_2(2-pyridine)(CH_2CH_2OH)} to the emissive [(dppe)Pt{S_2C_2(CH_2CH_2-N-2-pyridinium)}]$^+$ that serves as the basis for a new sensor for activated phosphate esters (Fig. 8) [44,45]. When immobilized in a polymer/plasticizer matrix, such as cellulose acetate/triethylcitrate (CA/TEC) at a 0.3% by weight loading, (dppe)Pt{S_2C_2(2-pyridine)(CH_2CH_2OH)} can be used for the rapid selective detection of volatile fluoro-, chloro-, and cyanophosphate esters (SARIN, SOMAN, and TABUN mimics). As with many film-immobilized sensor molecules, the level of detection is controlled by the immobilizing matrix [60–70]. Such an effect is seen when the sensitivity for phosphate esters increases as the plasticizer content is increased in the CA/TEC matrix (Table 2) [44,45].

E.   Dual-Emitting 1,2-Enedithiolates and Sensing

As discussed in Sections VI.A–D, the heterocyclic-substituted platinum 1,2-ene-dithiolates are dual emitters with emissions assigned to a short-lived ^{1}ILCT* and long-lived analyte-quenchable ^{3}ILCT* (Fig. 9). Since the singlet is not quenched with analyte concentrations less than molar, it serves as an internal standard making relative intensity measurements possible. In sensor technologies, relative measurements using a dual emitter would have many advantages over measuring absolute emission intensity and frequency modulation in sensor design [45,

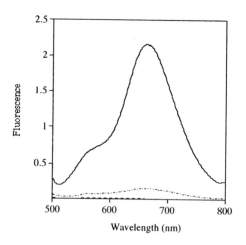

Figure 8   The luminescence spectra of (dppe)Pt{S_2C_2(2-pyridine)(CH_2CH_2OH)} (0.3%/wt) immobilized in a cellulose acetate/150% triethylcitrate film (0.5 mm thick): (- - -) Control film. (——) Film exposed to 0.1 torr OP(OEt)$_2$F in N$_2$ for 15 s. (—·—) Film exposed to HCl.

Table 2 Conversion of (dppe)Pt{S$_2$C$_2$(2-pyridine)(CH$_2$CH$_2$OH)} to
[(dppe)Pt{S$_2$C$_2$(CH$_2$CH$_2$-N-2-pyridinium)}]$^+$ in Various Polymer/Plasticizer
Combinations

Polymer/plasticizer[a]	Phosphate ester	Minimum exposure time (s)[e]	Complete exposure time (s)[f]
CA	(O)P(OEt)$_2$Cl[b]	Not observed	Not observed
CA/25% TEC	(O)P(OEt)$_2$Cl[b]	>600	Not observed
CA/50% TEC	(O)P(OEt)$_2$Cl[b]	15	600
CA/100% TEC	(O)P(OEt)$_2$Cl[b]	<15	40
CA/150% TEC	(O)P(OEt)$_2$Cl[b]	<15	30
GE-RTV 108	(O)P(OEt)$_2$Cl[b]	<15	15
GE-RTV 118	(O)P(OEt)$_2$Cl[b]	<30	180
CA/150% TEC	(O)P(OEt)$_2$F[c]	<15	
CA/150% TEC	(O)P(OEt)$_2$CN[d]	<15	

[a] Cellulose acetate (CA), triethylcitrate (TEC). Percentages listed are for TEC weight percent of CA. (dppe)Pt{S$_2$C$_2$(2-pyridine)(CH$_2$CH$_2$OH)} loading, 0.3%/weight. Silicone films impregnated in CH$_2$Cl$_2$ solution containing 0.1% NEt$_3$. [b] (O)P(OEt)$_2$Cl at 0.093 torr (0.90 g/m^3) in an N$_2$ flow of 50 ml/s. [c] (O)P(OEt)$_2$F at 0.130 torr (1.2 g/m^3) in an N$_2$ flow 50 ml/s. [d] (O)P(OEt)$_2$CN at 0.054 torr (0.88 g/m^3) in an N$_2$ flow of 50 mL/s. [e] Minimum exposure required for luminescent detection of a deareated sample, 470 nm excitation, 570 and 675 emission. [f] Minimum exposure required for maximum emission.

56,66,71,78,79]. These advantages include increased point-to-point accuracy, allowing degradation and photobleaching of the sensor to be monitored and accounted for without restandardization, allowing fluctuations in the source and detector to be tolerated, and eliminating the need to use frequency modulation allowing ease of miniaturization. While the advantages of a dual-emitting sensor molecules are well known, few systems have been studied [71]. Those systems with a similar singlet-triplet intensity are rare and those systems studied (such

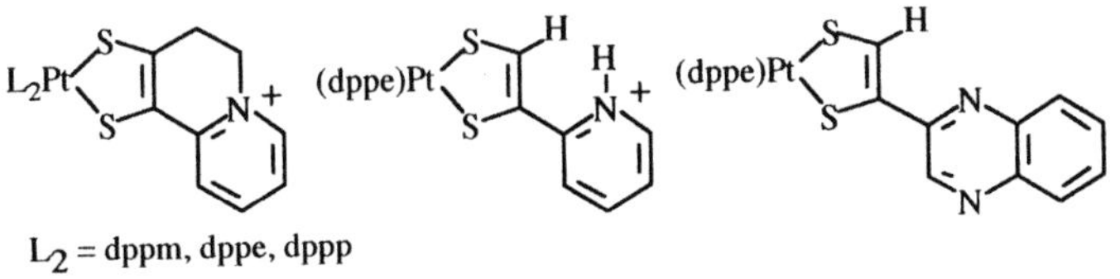

L$_2$ = dppm, dppe, dppp

Figure 9 Representative heterocyclic-substituted platinum-1,2-enedithiolates that are dual emitting.

as 4-bromo-1-naphthoyl derivatives) are not suitable for a broad spectrum of measurements.

While the advantages of the dual emitting 1,2-enedithiolates are demonstrated for dioxygen measurement, similar methods are applicable for the measurement of any ^{3}ILCT* quenching analyte [56].

Measuring Dioxygen

Given that dioxygen readily quenches long-lived emissive states, its measurement has served as an obtainable target in luminescent sensor design. Dioxygen detectors that rely on luminescence quenching are now commercially available [66,68,72–79]. Dioxygen is an important analyte in human physiology, a commodity in the field of medicine, a measure of the health of the environment, and a necessary reagent in many industrial processes. Since both fiberoptic [66,68,72–79] and polymer-encapsulated diode detectors [78,79] utilize 450 nm excitation, which is very near the λ_{max} of the heterocyclic-substituted platinum-1,2-enedithiolate complexes, these dual emitters should be "drop-in/step-out" replacements for ruthenium bipyridyl and phenanthroline complexes currently used for this application [46,56].

To take advantage of the dual emitter all that is required is measurement of both singlet and triplet intensity and analysis of data using $(^3Io/^1Io)/(^3I/^1I)$ vs. $[O_2]$ in place of the standard Stern-Volmer plot. It is the use of the singlet-triplet ratio that eliminates difficulties with fluctuation in detection and sensor molecule photobleaching. When immobilized in various polymer/plasticizer combinations, the effective measurement range for dioxygen can be controlled (Table 3) [60–70]. Representative of these ranges are the $P(O_2)_{1/2}$ values obtained in O_2/N_2 mixtures. These values are 1/slope of the Stern-Volmer plot. The $P(O_2)_{1/2}$ values for immobilized $[(dppe)Pt\{S_2C_2(CH_2CH_2\text{-}N\text{-}2\text{-pyridinium})\}]^+$ are lower than

Table 3 Oxygen Sensitivity of Lumiphores Encapsulated in GE-RTV 118 and Cellulose Acetate Butyrate

Lumiphore	Encapsulating matrix[a]	$P(O_2)_{1/2}$ (torr)	Ref.
$[Ru(dpp)_3(ClO_4)_2]$	GE-RTV 118	30	80
$[(dppe)Pt\{S_2C_2(CH_2CH_2\text{-}N\text{-}2\text{-pyridinium})\}(BPh_4)]$	GE-RTV 118	51	56
$[(dppe)Pt\{S_2C_2(CH_2CH_2\text{-}N\text{-}2\text{-pyridinium})\}(BPh_4)]$	GE-RTV 108	63	56
$[Ru(bpy)_3(ClO_4)_2]$	GE-RTV 118	377	80
Pt(ocatethylporphyrin)	CAB	13	61
$[(dppe)Pt\{S_2C_2(CH_2CH_2\text{-}N\text{-}2\text{-pyridinium})\}(BPh_4)]$	CAB	70	56
$[Ru(dpp)_3(BPh_4)_2]$	CAB	102	62

[a] GE-RTV 118 and GE-RTV 108 are commercial silicones. CAB; cellulose acetate butyrate.

[(bpy)₃Ru]²⁺ and similar to [(dpphen)₃Ru]²⁺, yet somewhat higher than Pt(ocatethylporphyrin) in the same polymer matrix [56,60,61].

VII. PHOTOCHEMISTRY

A. Oxygen Reactivity

The studies discussed earlier in this chapter focused on the photophysics of metallo-1,2-enedithiolates and possible applications for these complexes. In addition to being emissive, several (diimine)Pt(dithiolate) complexes photochemically activate molecular oxygen. Recent reports by Srivastava and co-workers showed that (bpy)Pt(tdt) and related compounds are sensitizers for the formation of singlet oxygen [81–83], the generation of which leads to decomposition of the parent metal complex. The end-products from these reactions are similar to those obtained from the thermal and photochemical oxidation of Ni(II) and Pd(II) dithiolates [84–87]. These reactions give stable metal-sulfinate (M[SO₂R]) and metal-sulfenate (M[SOR]) derivatives. Collectively, this work parallels the photo- and thermal oxidation of thioethers (RSR') [88–90].

The photooxidations of (dbbpy)Pt(edt) and (dbbpy)Pt(dpdt) were studied by Schanze and co-workers [91]. These (diimine)Pt(dithiolate) complexes are room temperature emitters with a LLCT assigned excited state (see Section V and Table 1b). (dbbpy)Pt(edt) is photooxidized upon irradiation in aerated solution to the corresponding sulfinate, disulfinate, and bis(sulfenate) complexes (Scheme 5). In contrast, (dbbpy)Pt(dpdt) is photooxidized upon irradiation in aerated solution to generate the 1,2-enedithiolate complex (dbbpy)Pt(S₂C₂Ph₂) (Scheme 5).

Gray and co-workers studied the photoreactivity of (bpy)Pt(bdt) [92]. Like (dbbpy)Pt(edt) and (dbbpy)Pt(dpdt), irradiation of an aerated solution of the 1,2-enedithiolate, (bpy)Pt(bdt), resulted in the formation of the corresponding sulfinate and disulfinate complexes (Scheme 6).

The studies by the Schanze and Gray groups suggest that these reactions proceed through a sulfoperoxide intermediate formed from the starting complex in a reaction with singlet oxygen. The formation of singlet oxygen is supported by product and mechanistic analysis including studies in deutero solvents where the lifetime of singlet oxygen is increased [91,92].

B. Electron Transfer to Halocarbons and Water

The photooxidation of the bis(mnt) complexes of Ni(II), Pd(II), and Pt(II) has been reported by Vogler and Dooley [50,51]. Photolysis of dilute solutions (10^{-5} M) of [M(mnt)₂]⁻² in halocarbon solvents leads to the formation of the corresponding one-electron oxidized species, [M(S₂C₂(CN)₂)₂]⁻¹ (Scheme 7). These species are formed by excited state electron transfer to a solvent molecule, a rapid process that competes with other modes of relaxation. The reaction rate

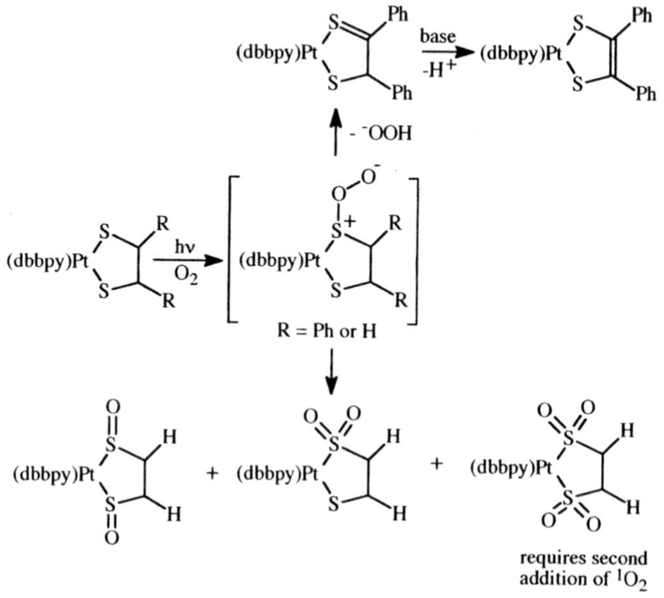

Scheme 5

depends on both the reduction potential of the halocarbon solvent and the oxidation potential of the metal complex. The quantum yield for photooxidation of $[Ni(mnt)_2]^{-2}$ increases from $\phi = 0.002$ in CH_2Cl_2 to $\phi = 0.013$ in $CHCl_3$ and to $\phi = 0.07$ in CCl_4. This also correlates with the oxidation potentials of the solvents. The photooxidation of these complexes does not occur in CH_3CN and does not

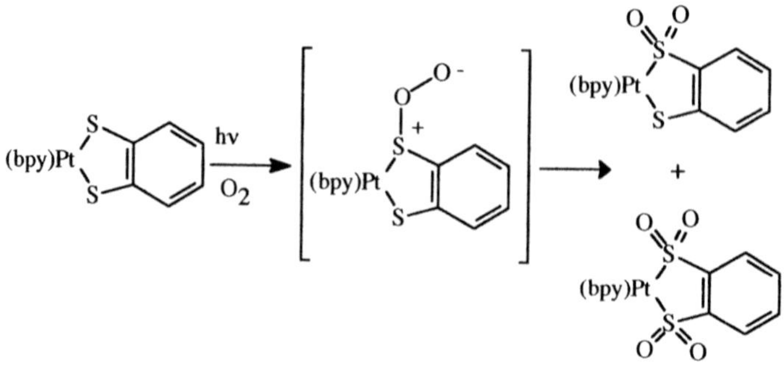

Scheme 6

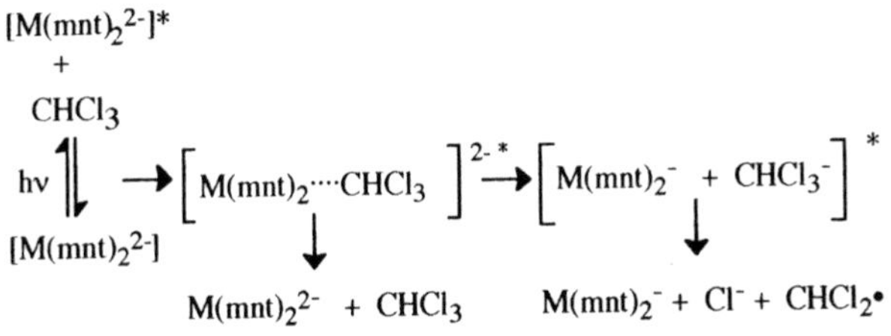

$[M(mnt)_2^{2-}]^*$
$+$
$CHCl_3$

$$ h\nu \Big\Uparrow \Big\Downarrow \longrightarrow \left[M(mnt)_2 \cdots CHCl_3 \right]^{2-*} \longrightarrow \left[M(mnt)_2^- + CHCl_3^- \right]^* $$

$[M(mnt)_2^{2-}]$

$$ M(mnt)_2^{2-} + CHCl_3 \qquad M(mnt)_2^- + Cl^- + CHCl_2^\bullet $$

Scheme 7

occur using excitation wavelengths greater than 360 nm, indicating that the photo-oxidation is not associated with the low-lying d-to-d or MLCT excited states.

A range of $[M(mnt)_2]^{2-}$ and $[M(Ph_2C_2S_2)_2]$ derivatives have been used as catalysts for the photoproduction of hydrogen from water [93,94]. Of these compounds, the $[Zn(mnt)_2]^{2-}$ and $[Ni(Ph_2C_2S_2)_2]$ are the most effective catalysts. Photolysis of the $[Zn(mnt)_2]^{2-}$ in THF/H_2O at 254 nm is reported to produce 2000 moles H_2 per mole of catalyst in 120 h. Replacing H_2O by D_2O leads to the production of D_2 and less than 15% HD/H_2, suggesting that water is the source of the hydrogen atom. While these results support an electron transfer from a photoexcited metallo-1,2-enedithiolate, given that these reactions result in the decomposition of the metal complexes and the production of a colloidal suspension, it is unclear as to whether the discrete metal complexes catalyze this reaction.

VIII. CONCLUSION

The study of emission 1,2-enedithiolates now represents a formidable body of literature, even though the chemistry, photochemistry, and photophysical properties of metallo-1,2-enedithiolate complexes are not yet as well understood or developed as those of the group metallo-VIII-diimine and metalloporphyrin complexes. However, recent developments in the synthesis of 1,2-enedithiolates have led to the discovery of room temperature emitters and complexes with useful properties. As new methods allow for the synthesis of yet unknown complexes in this family, the unique and useful properties of these complexes will become even more evident. Many of the heterocyclic-substituted 1,2-enedithiolates now available are dual emitters with a short-lived and analyte-quenchable long-lived excited states. Clearly, these dual emitters will have a unique place in the detection of quenching analytes since selective quenching of the long-lived excited state eliminates several problems encountered with luminescence-based sensing.

ACKNOWLEDGMENT

We thank Professor Neil V. Blough of the University of Maryland and Dr. Edward I. Stiefel for helpful discussions. We are indebted to the donors of the Petroleum Research Fund, administered by the American Chemical Society (Grants 32486-AC3) for support.

ABBREVIATIONS

bpy	2,2′-bipyridine
dmbpy	4,4′-dimethyl-2,2′-bipyridine
dpbpy	4,4′-diphenyl-2,2′-bipyridine
dbbpy	4,4′-di-*tert*-butyl-2,2′-bipyridine
Cl₂bpy	4,4′-dichloro-2,2′-bipyridine
EC-bpy	4,4′-bis(ethoxycarbonyl)-2,2′-bipyridine
phen	1,10-phenanthroline
dpphen	4,7-diphenyl-1,10-phenanthroline
tmphen	3,4,7,8-tetramethyl-1,10-phenanthroline
Cl-phen	5-chloro-1,10-phenanthroline
tdach	(±)-*trans*-diaminocyclohexane
mnt	maleonitriledithiolate
ecda	1-(ethoxycarbonyl)-1-cyanoethylene-2,2-dithiolate
EtS	ethanethiolate
PhS	benzenethiolate
PhSe	benzeneselenolate
meo	2-mercaptoethanolate(2-)
dppe	1,2-bis(diphenylphosphino)ethane
dppv	1,2-bis(diphenylphosphino)ethylene
dppb	1,2-bis(diphenylphosphino)benzene
chpe	1,2-bis(dicyclohexylphosphino)ethane
dppm	1,2-bis(diphenylphosphino)methane
pompom	1,2-bis(dimethoxyphosphino)ethane
PPh₃	triphenylphosphine
PMe₂Ph	dimethylphenylphosphine
PCy₃	tricyclohexyphosphine
(POPh₃)	triphenylphosphite
(PO-*i*Pr₃)	triisopropylphosphite
COD	1,5-cyclooctadiene
qdt	2,3-quinoxalinedithiolate
dmqdt	6,7-dimethyl-quinoxaline-2,3-dithiolate
tdt	toluene-3,4-dithiolate
bdt	1,2-benzenedithiolate

met	cis-1,2-dicarbomethoxyethylene-1,2-dithiolate
dpdt	meso-1,2-diphenyl-1,2-ethanedithiolate
edt	1,2-ethanedithiolate
ppy	C-deprotonated 2-phenylpyridine
THF	tetrahydrofuran
DMF	dimethylformamide
DMSO	dimethylsulfoxide
TBA	tetra-n-butyl ammonium
EPA	ethyl ether/isopentane/ethanol
DMM	DMF/methylene chloride/methanol
TEA	triethylamine

REFERENCES

1. Mueller-Westerhoff, U.T.; Vance, B.; Yoon, D.I. *Tetrahedron* **1991**, *47*, 909–932.
2. Oliver, S.; Winter, C. *Adv. Mater.* **1992**, *4*, 119–121.
3. Cassoux, P.; Valade, L.; Kobayashi, H.; Kobayashi, A.; Clark, R.A.; Underhill, A.E. *Coord. Chem. Rev.* **1991**, *110*, 115–160.
4. Veldhuizen, Y.S.J.; Veldman, N.; Spek, A.L.; Faulmann, C.; Haasnoot, J.G.; Reedijk, J. *Inorg. Chem.* **1995**, *34*, 140–147.
5. Olk, R.M.; Olk, B.; Dietzch, W.; Kirmse, R.; Hoyer, E. *Coord. Chem. Rev.* **1992**, *117*, 99–131.
6. Fourmigue, M.; Lenoir, C.; Coulon, C.; Guyon, F.; Amaudrut, J. *Inorg. Chem.* **1995**, *34*, 4979–4985.
7. Manoharan, P.T.; Noordik, J.H.; de Boer, E.; Keijzers, C.P. *J. Chem. Phys.* **1980**, *74*, 1980–1989.
8. Kuppusamy, P.; Manoharan, P.T. *Chem. Phy. Lett.* **1985**, *118*, 159–163.
9. Pilato, R.S.; Stiefel, E.I. *Molybdenum and tungsten enzymes*; in *Bioinorganic Catalysis*. Reedijk, J., Ed.; Marcel Dekker: New York, 1998, Chpt. 6.
10. Pilato, R.S.; Gea, Y.; Eriksen, K.A.; Greaney, M.A.; Stiefel, E.I.; Goswami, S.; Kilpatrick, L.; Spiro, T.G.; Taylor, E.C.; Rheingold, A.L. *Pterins, Quinoxalines, and Metallo-Ene-Dithiolates; Synthetic Approach to the Molybdenum Cofactor*; ACS Symposium Series. Stiefel, E.I.; Coucouvanis D.; Newton, W.E., Eds.; American Chemical Society: Washington, DC, 1993; Vol. 535, pp 83–97.
11. Pilato, R.S.; Eriksen, K.A.; Greaney, M.A.; Stiefel, E.I.; Goswami, S.; Kilpatrick, L.; Spiro, T.G.; Taylor, E.C.; Rheingold, A.L. *J. Am. Chem. Soc.* **1991**, *113*, 9372–9374.
12. Garner, C.D.; Armstrong, E.M.; Ashcroft, M.J.; Austerberry, M.S.; Birks, J.H.; Collision, D.; Goodwin, A.J.; Larsen, L.; Rowe, D.J.; Russell, J.R. *Strategies for the Synthesis of the Cofactor of the Oxomolybdenum*; Stiefel, E.I.; Coucouvanis D.; Newton, W.E., Eds.) American Chemical Society: Washington, DC, 1993; Vol. 535, pp 98–113.
13. Udpa, K.N.; Sarkar, S. *Polyhedron* **1987**, *6*, 627–631.
14. Udupa, K.N.; Sarkar, S. *J. Organomet. Chem.* **1985**, *284*, C36–C38.

15. Ansari, M.A.; Chandrasekaran, J.; Sarkar, S. *Inorg. Chim. Acta* **1987**, *130*, 155–156.
16. Johnson, C.E.; Eisenberg, R.; Evans, T.R.; Burberry, M.S. *J. Am. Chem. Soc.* **1983**, *105*, 1795–1802.
17. Zuelta, J.A.; Chesta, C.A.; Eisenberg, R. *J. Am. Chem. Soc.* **1989**, *111*, 8916–8917.
18. Zuleta, J.A.; Burberry, M.S.; Eisenberg, R. *Coord. Chem. Rev.* **1990**, *97*, 47–64.
19. Zuleta, J.A.; Bevilacqua, J.M.; Eisenberg, R. *Coord. Chem. Rev.* **1991**, *111*, 237–248.
20. Zuleta, J.A.; Bevilacqua, J.M.; Proserpio, D.M.; Harvey, P.D.; Eisenberg, R. *Inorg. Chem.* **1992**, *31*, 2396–2404.
21. Zuleta, J.A.; Bevilacqua, J.M.; Rehm, J.M.; Eisenberg, R. *Inorg. Chem.* **1992**, *31*, 1332–1337.
22. Bevilacqua, J.M.; Zuleta, J.A.; Eisenberg, R. *Inorg. Chem.* **1993**, *32*, 3689–3693.
23. Bevilacqua, J.M.; Eisenberg, R. *Inorg. Chem.* **1994**, *33*, 2913–2923.
24. Bevilacqua, J.M.; Zuelta, J.A.; Eisenberg, R. *Inorg. Chem.* **1994**, *33*, 258–266.
25. Bevilacqua, J.M.; Eisenberg, R. *Inorg. Chem.* **1994**, *33*, 1886–1890.
26. Cummings, S.D.; Eisenberg, R. *Inorg. Chem.* **1995**, *34*, 2007–2014.
27. Cummings, S.D.; Eisenberg, R. *Inorg. Chem.* **1995**, *34*, 3396–3403.
28. Cummings, S.D.; Eisenberg, R. *J. Am. Chem. Soc.* **1996**, *118*, 1949–1960.
29. Cummings, S.D.; Eisenberg, R. *Inorg. Chim. Acta* **1996**, *242*, 225–231.
30. Kaiwar, S.P.; Hsu, J.K.; Liable-Sands, L.M.; Rheingold, A.L.; Pilato, R.S. *Inorg. Chem.* **1997**, *36*, 4234–4240.
31. Kaiwar, S.P.; Vodacek, A.; Blough, N.V.; Pilato, R.S. *J. Am. Chem. Soc.* **1997**, *119*, 3311–3316.
32. Kaiwar, S.P.; Vodacek, A.; Blough, N.V.; Pilato, R.S. *J. Am. Chem. Soc.* **1997**, *119*, 9211–9214.
33. Kaiwar, S.P.; Hsu, J.K.; Vodacek, A.; Yap, G.; Liable-Sands, L.M.; Rheingold, A.L.; Pilato, R.S. *Inorg. Chem.* **1997**, *36*, 2406–2412.
34. Van Houten, K.A.; Heath, D.C.; Barringer, C.A.; Rheingold, A.L.; Pilato, R.S. *Inorg. Chem.* **1998**, *37*, 4647–4653.
35. Van Houten, K.A.; Pilato, R.S. Submitted *Inorg. Chem.* **1998**.
36. Fernandez, A.; Kisch, H. *Chem. Ber.* **1984**, *117*, 3102–3111.
37. Boyde, S.; Garner, C.D.; Joule, J.A.; Rowe, D.J. *J. Chem. Soc., Chem. Commun.* **1987**, 800–801.
38. Bolinger, C.M.; Rauchfuss, T.B. *Inorg. Chem.* **1982**, *21*, 3947–3954.
39. Stiefel, E.I.; Bennett, Z.D.; Crawford, T.H.; Simo, C.; Gray, H.B. *Inorg. Chem.* **1970**, *9*, 281–286.
40. Yoshinaga, N.; Ueyama, N.; Okamura, T.; Nakamura, A. *Chem. Lett.* **1990**, *9*, 1655–1663.
41. Davison, A.; Holm, R.H. *Inorg. Synth.* **1967**, *10*, 8–25.
42. Kusters, W.; de Mayo, P. *J. Am. Chem. Soc.* **1974**, *96*, 3502–3511.
43. Hsu, J.K.; Bonangelino, C.J.; Kaiwar, S.P.; Boggs, C.M.; Fettinger, J.C.; Pilato, R.S. *Inorg. Chem.* **1996**, *35*, 4743–4751.
44. Van Houten, K.A.; Pilato, R.S. *J. Am. Chem. Soc.* **1998**, *120*, 12359–12360.
45. Van Houten, K.A.; Heath, D.C.; Pilato, R.S. *Patent Pending* **1998**.
46. Van Houten, K.A.; Heath, D.C.; Blough, N.V.; Pilato, R.S. *Patent Pending* **1998**.

47. In this family of complexes the $d\sigma^*$, $\pi^*_{(Het)}$, and $\pi^*_{(S-S)}$ are closely spaced in energy and are not necessarily ordered as shown; Van Houten, K.A.; Pilato, R.S. unpublished results.
48. Gray, H.B.; Williams, R.; Bernal, I.; Billig, E. *J. Am. Chem. Soc.* **1962**, *84*, 3596–3597.
49. Shupack, S.I.; Billig, E.C.; Williams, R.; Gray, H.B. *J. Am. Chem. Soc.* **1964**, *86*, 4594–4602.
50. Vogler, A.; Kunkely, H. *Inorg. Chem.* **1982**, *21*, 1172–1175.
51. Dooley, D.M.; Patterson, B.M. *Inorg. Chem.* **1982**, *21*, 4330–4332.
52. Fordyce, W.A.; Crosby, G.A. *Inorg. Chem.* **1982**, *21*, 1023–1026.
53. Vogler, A.; Kunkely, H.; Hlavatsch, J.; Merz, A. *Inorg. Chem.* **1984**, *23*, 506–509.
54. Gray, H.B.; Ballhausen, C.J. *J. Am. Chem. Soc.* **1963**, *85*, 260–264.
55. Werden, B.G.; Billig, E.; Gray, H.B. *Inorg. Chem.* **1966**, *5*, 78.
56. Van Houten, K.A.; Blough, N.V.; Pilato, R.S. unpublished results.
57. Wiberg, K.B.; Lewis, T.P. *J. Am. Chem. Soc.* **1970**, *92*, 7154–7160.
58. Jordan, K.D.; Burrow, P.D. *Acc. Chem. Res.* **1978**, *11*, 341–348.
59. Van Houten, K.A.; Pilato, R.S.; Walters, K.A.; Schanze, K.S. Submitted *Inorg. Chem.* **1998**.
60. Mills, A.; Williams, F.C. *Thin Solid Films* **1997**, *306*, 163–170.
61. Mills, A.; Lepre, A. *Anal. Chem.* **1997**, *69*, 4653–4659.
62. Mills, A.; Thomas, M.D. *Analyst* **1998**, *123*, 1135–1140.
63. Mills, A.; Lepre, A.; Wild, L. *Anal. Chim. Acta* **1998**, *362*, 193–202.
64. Mills, A.; Thomas, M. *Analyst* **1997**, *122*, 63–68.
65. Mills, A.; Chang, Q. *Analyst* **1993**, *118*, 839–843.
66. Wolfbeis, O.S. *Oxygen sensors*; in *Fiber Optic Chemical Sensors and Biosensors*. Wolfbeis, O.S., Ed.; CRC Press: Boston, 1991; Vol. 2, pp 19–54.
67. McMurray, H.N.; Douglas, P.; Busa, C.; Garley, M.S. *J. Photochem. Photobiol.* **1994**, *80*, 283–288.
68. Li, X.-M.; Ruan, F.-C.; Wong, K.-Y. *Analyst* **1993**, *118*, 289–292.
69. Hartmann, P.; Trettnak, W. *Anal. Chem.* **1996**, *68*, 2615–2620.
70. Hartmann, P.; Leiner, M.J.P.; Lippitsch, M.E. *Anal. Chem.* **1995**, *67*, 88–93.
71. Lee, E.D.; Werner, T.C.; Seitz, W.R. *Anal. Chem.* **1987**, *59*, 279–283.
72. Wolfbeis, O.S.; Weis, L.J.; Leiner, M.J.P.; Zielgler, W.E. *Anal. Chem.* **1988**, *60*, 2028–2030.
73. Wolfbeis, O.S.; Leinger, M.J.P.; Posch, H.E. *Mikrochim. Acta* **1986**, *111*, 359–366.
74. Moreno-Bondi, M.C.; Wolfbeis, O.S.; Leiner, M.P.J.; Schaffar, B.P.H. *Anal. Chem.* **1990**, *62*, 2377–2380.
75. MacCraith, B.D.; McDonagh, C.M.; O'Keeffe, G.O.; Keyes, E.T.; Vos, J.G.; O'Kelly, B.; McGilp, J.F. *Analyst* **1993**, *118*, 385–388.
76. Velasco-Garcia, N.; Valencia-Gonzalez, M.J.; Diaz-Garcia, M.E. *Analyst* **1997**, *122*, 1405–1409.
77. Peterson, J.I.; Fitzgerald, R.V.; Buckhold, D.K. *Anal. Chem.* **1984**, *56*, 62–67.
78. Colvin, A.E. **1996** U.S. Patent #5517313, licensed exclusively to Sesnors for Medicine and Science, Germantown, MD.
79. Colvin, A.E.; Phillips, T.E.; Miragliotta, J.A.; Givens, R.B.; Bargeron, C.B. *Johns Hopkins APL Tech. Dig.* **1996**, *17*, 377–385.

80. Carraway, E.R.; Demas, J.N.; DeGraff, B.A.; Bacon, J.R. *Anal. Chem.* **1991**, *63*, 337–342.
81. Shulka, S.; Kamath, S.S.; Srivastava, T.S. *J. Photochem. Photobiol. A* **1989**, *50*, 199–207.
82. Puthraya, K.H.; Srivastava, T.S. *J. Ind. Chem. Soc.* **1985**, *62*, 843.
83. Puthraya, K.H.; Srivastava, T.S. *Polyhedron* **1985**, *4*, 1579–1584.
84. Grapperhaus, C.A.; Darensbourg, M.Y.; Summer, L.W.; Russel, D.H. *J. Am. Chem. Soc.* **1996**, *118*, 1791–1792.
85. Darnesbourg, M.Y.; Tuntulani, T.; Reibenspies, J.H. *Inorg. Chem.* **1995**, *34*, 6287–6294.
86. Mills, D.K.; Reibenspies, J.H.; Darensbourg, M.Y. *Inorg. Chem.* **1990**, *29*, 4364–4366.
87. Tuntulani, T.; Musie, G.; Reibenspies, J.H.; Darensbourg, M.Y. *Inorg. Chem.* **1995**, *34*, 6279–6286.
88. Clennan, E.I.; Zhang, H. *J. Org. Chem.* **1994**, *59*, 7952–7954.
89. Liang, J.-J.; Gu, G.-L.; Kacher, M.L.; Foote, C.S. *J. Am. Chem. Soc.* **1983**, *105*, 4717–4721.
90. Wanatabi, Y.; Kuriki, N.; Ishigureo, K.; Sawaki, Y. *J. Am. Chem. Soc.* **1991**, *113*, 2677–2682.
91. Zhang, Y.; Ley, K.D.; Schanze, K.S. *Inorg. Chem.* **1996**, *35*, 7102–7110.
92. Connick, W.B.; Gray, H.B. *J. Am. Chem. Soc.* **1997**, *119*, 11620–11627.
93. Henning, R.; Schlamann, W.; Kisch, H. *Angew. Chem. Int. Ed. Engl.* **1980**, *19*, 645–646.
94. Battaglia, R.H.; Henniny, R.; Kisch, H. *Z. Naturforsch B* **1981**, *36b*, 396–397.
95. Mansour, M.A.; Lachicotte R.J.; Gysling, H.J.; Eisenberg, R. *Inorg. Chem.* **1998**, *37*, 4625–4632. (Complexes from article were not discussed in this chapter because article appeared while this manuscript was in press.)

6

Molecular and Supramolecular Photochemistry of Porphyrins and Metalloporphyrins

Shinsuke Takagi and Haruo Inoue
Tokyo Metropolitan University, Tokyo, Japan

I. INTRODUCTION

A group of porphyrin compounds is one of the most precious groups of molecules for life. Porphyrins and their metal complexes are found in every tissue of an organism [1]. Chlorophyll has a key role in photosynthesis [2,3]. Iron porphyrin in heme is essential to the molecular oxygen transport by hemoglobin in blood and myoglobin in muscular systems [4]. Electron transport in the respiration cycle is governed by porphyrins in cytochrome. P-450 is a crucial enzyme in metabolic processes.

Porphyrins (**1**) having various kinds of metal (M) in the center of the macrocycle, substituents (R_1, R_2,), and axial ligands (L_1, L_2) are involved in all of these enzymes, proteins, and processes. Among their amazing functional activities, photosynthesis in plants and bacteria has been attracting special attention. Light harvesting systems involving energy transfer processes and the reaction center of photosynthesis with efficient irreversible multistep electron transports have long been the subject of study not only for biologists, but for chemists and physicists as well. Recent findings on the microstructure of porphyrin special

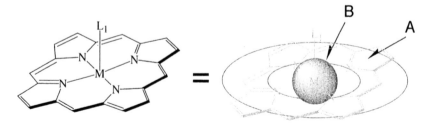

1

pair, microscopic molecular alignment of bacteriochlorophyll [428–434] and light-harvesting antenna chlorophyll [451,452], have strongly stimulated the various approaches for understanding the secrets of photosynthesis. They have simultaneously accelerated the progress of studies mimicking photosynthesis, i.e., artificial photosynthesis.

Not only in the field of photosynthesis and artificial photosynthesis but also in other fields covering the more general viewpoint of photoresponsive molecular systems, porphyrin derivatives afford one of the most promising chromophores that can absorb photons and induce various physicochemical reactions.

From the viewpoint of molecular function, porphyrins have interesting multiple features of chemistry for 1.) a large π-electron system, 2.) a metal atom in the center of the porphyrin ring, 3.) axial ligands, and 4.) an intrinsic space over the porphyrin plane provided by both axial ligands and substituents in the porphyrin ring. They are also characterized by a unique chemical reaction in the special space (A in Fig. 1) induced by an intense absorption of photon by the large π-conjugated porphyrin ring (B in Fig. 1). A combination of each part A, B provides an additional function beyond the respective intrinsic ones. Porphyrins, especially substituted metalloporphyrins, thus could form a supramolecular system by themselves.

Understanding of unimolecular properties is requisite for constructing a desired photoresponsive molecular system. Photochemical and redox properties are especially crucial for it. Intensive studies have made it possible to regulate the properties of each part A and B. Unimolecular excited states characteristics

Figure 1   Porphyrins by themselves can form a supramolecular system.

and redox properties of porphyrins will be reviewed in Section II.A and II.B. The photochemistry of porphyrins from the viewpoint of excited energy transfer, electron transfer, and unique reactions in relation to axial ligands will be described in Section III. The field of supramolecular chemistry on supramolecular systems, defined as noncovalently associated molecular assemblies having additional and unique functions [423], is expanding enormously. Porphyrins which have an inherently supramolecular nature as described above are accepted as being the most promising molecules in supramolecular systems. Recent trends in the photochemical aspects of porphyrin-related supramolecular systems will be surveyed in Sections III and IV.

II. EXCITED STATE AND REDOX PROPERTIES OF PORPHYRINS

For understanding the photochemistry of porphyrin compounds, characteristics of absorption spectrum, fluorescence quantum yield, lifetime of singlet excited state, phosphorescence quantum yield, and excited triplet state lifetime are most crucial. These data are actually helpful for a more sophisticated design of a photoresponsive molecular system. In this section, comprehensive data on excited states of porphyrins, mainly on tetraphenylporphyrin (TPP), octaethylporphyrin (OEP), and their metal complexes, are presented. Some empirical rules for estimating those data will also be described.

A. Photochemical Properties of Porphyrins

1. Absorption Spectrum

Porphine (**2**) is an aromatic molecule having 26π electrons among the molecule. The 18π electronic structure (bold line in the structural formula) has a large con-

2

tribution to the resonance among the molecule. The π electrons in both structures satisfy the Huckel rule. The bond length between each carbon is known to be within 0.135–0.144 nm and the molecule has an almost planar structure. Porphine, its substituted derivatives (free-base porphyrins), and their metalloporphyrins generally have an intense π-π^* absorption band in the visible region. A typical absorption spectrum for 5,10,15,20-tetramesitylporphyrin (H_2TMP) is shown in Fig. 2.

The absorption spectra of porphyrins and metalloporphyrins have been derived by a four-orbital model [14–18]. Molecular orbitals (MOs) and molecular states obtained by MO calculation on the D_{4h} porphyrin ring are shown in Fig. 3. The absorption bands in the visible region are explained by transitions among the two highest occupied π orbitals (a_{1u}, a_{2u} symmetry) and the two degenerate lowest unoccupied π^* ones (e_g, e_g symmetry). Transitions of $a_{1u}\pi \rightarrow e_g\pi^*$ and $a_{2u}\pi \rightarrow e_g\pi^*$ give rise to doubly degenerate four excited configurations. Configuration interaction among the four excited states affords two sets of degenerate B state in the higher energy and Q states in the lower energy. Transition to the B state is symmetry allowed and corresponds to the intense absorption band around 400 nm. (Soret band: B band). The relatively weak absorption band around 480–650 nm is explained by the symmetry-forbidden transition to the Q state (Q band). The Q band shows four vibrational peaks I–IV. The relative intensities among the four peaks are generally known to be IV > III > II > I, while substituents sometimes affect the order [6]. Electron-withdrawing substituents induce an inversion of III > IV [6]. In metalloporphyrins only two peaks in the Q band are evident due to their symmetry. The relative intensity is empirically correlated with Eq. (1) under conditions without strong interaction between the central metal and porphyrin ring [6].

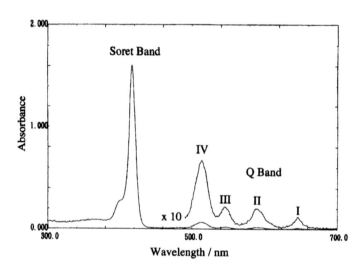

Figure 2 Absorption spectrum of 5,10,15,20-tetramesitylporphyrin free base (H_2TMP) in dichloromethane.

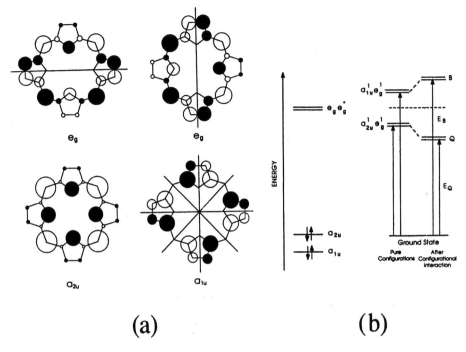

Figure 3 Molecular orbitals (a) and molecular states (b) obtained by MO calculation on the D_{4h} porphyrin ring [5].

$$\varepsilon_{0,0}/\varepsilon_{1,0} = \text{const.} \times [E^1 (a_{2u}e_g) - E^1 (a_{1u}e_g)]^2 \tag{1}$$

where $\varepsilon_{n,0}$ denotes the extinction coefficient of the absorption band $Q_{n,0}$. Equation (1) is based on the results that $Q_{1,0}$ is quite insensitive to substituents, central metal, and axial ligand. In some cases a good correlation between the 0-0 transition energy of Q band [$E(Q_{0,0})$] and oscillator strength [$f(Q)$] was observed [7]. The correlation was reported to reflect the degree of interaction between the central metal and porphyrin ring; an increase of the interaction induced an increase of $E(Q_{0,0})$ with a decrease of $f(Q)$ [13] (Fig. 4).

In addition to the central atom, substituents in the porphyrin ring also affect the absorption spectrum. The number of bromine atoms in the pyrrole position in CoTPP was correlated with the red shift of the Soret band [9], while an effect of distortion of the porphyrin ring should be considered [10] (Fig. 5). As anticipated from the molecular orbital coefficient (see Fig. 3) the a_{1u} orbital would be more dependent on the substitution on the pyrrole position than the a_{2u} orbital. A strong electron-donating substituent was reported to induce an inversion from $a_{2u} > a_{1u}$ to $a_{2u} < a_{1u}$, i.e., the a_{1u} of 2-aminoporphyrin was reported to be more destabilized by 0.71 eV than that of 2-nitroporphyrin, while a_{2u} suffers destabili-

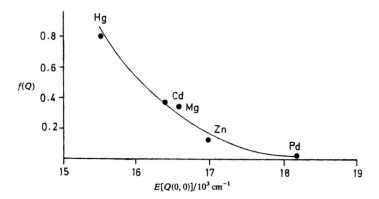

Figure 4 Relationship between the energy (E) and the oscillator strength [f(Q)] of the Q(0, 0) transition [13].

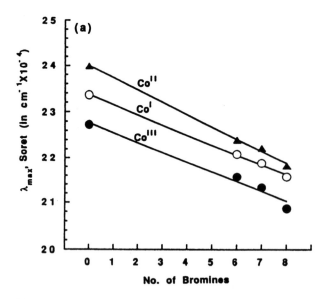

Figure 5 Soret band position of Co(I), Co(II), and Co(III) TPP derivatives with bromine atoms as substituents [9].

zation of only 0.14 eV [12]. The relative positioning between a_{1u} and a_{2u} should largely affect the chemical reactivity of the excited porphyrins in which the difference would be more evident in the lowest excited state [325]. Axial ligands induce a pretty large shift in many cases [11]. Electron-donating axial ligands can interact with the a_{2u} π orbital having a large contribution of the nitrogen orbital of porphyrin and stabilize the $a_{2u}e_g$ excited state [11].

The metal atom in the porphyrin ring has the largest effect on the absorption spectrum. According to the spectral shape, the absorption band of porphyrins has been classified into three groups of 1.) normal, 2.) hyper, and 3.) hypso type, respectively [7]. The normal-type spectrum is shown in Fig. 2 and its transitions are explained by the $a_{1u}\pi \rightarrow e_g\pi^*$ and $a_{2u}\pi \rightarrow e_g\pi^*$ as described above. Porphyrins without metal and metalloporphyrins with typical metals in normal valence state such as Mg^{II}, Al^{III}, and Sn^{IV} belong to the normal type. Even metalloporphyrins with transition metals of d^0 such as Ti^{IV} show the normal type of spectrum due to a small interaction with the porphyrin ring.

Different aspects are observed in cases where the central metals interact electronically with the porphyrin π electrons. Porphyrins with low-valent atoms such as Sn^{II}, Pb^{II}, P^{III}, As^{III}, Sb^{III}, and Bi^{III} form a group of "p-type hyper" which show a charge transfer band in the shorter wavelength region owing to a transition from p orbitals of central atoms to the porphyrin π system [8]. Low-valent transition metal ions such as Cr^{III}, Mn^{III} with half-filled d_{xy}, d_{xz}, d_{yz} orbitals form d-type hyper metalloporphyrins in which porphyrin $\pi \rightarrow d$ transition is involved. The fully occupied d_{xy}, d_{xz}, d_{yz} orbitals in the cases of Fe^{II}, Co^{III}, Ni^{II}, Cu^{II} can mix with the $e_g\pi^*$ of the porphyrin ring to destabilize the $e_g\pi^*$ and induce a blue shift of the absorption band. They are classified as hypso.

The actual peak position and absorption intensity are complicatedly dependent on substituents on the porphyrin ring, central metal, and axial ligands. The λ_{max} of various porphyrins and metalloporphyrins are listed in Table 1.

2. Excited State

Excited state parameters such as the lowest excited singlet and triplet state energy (E_S, E_T), fluorescence quantum yield (ϕ_f), intersystem crossing efficiency (triplet yield: ϕ_{isc}), singlet state lifetime (τ_S), phosphorescence yield (ϕ_P), and triplet state lifetime (τ_T) are most requisite for understanding photochemical behavior of porphyrins and for designing photoresponsive systems. They are comprehensively summarized in Tables 1 and 2. Rate constants of energy transfer and electron transfer could be estimated by the free-energy change calculated by those parameters as described in Section III.A.

In normal-type porphyrins and metalloporphyrins with typical atoms, the excited state energy diagram is simple to draw in that the lowest singlet and triplet states are derived from the porphyrin π system. The electronic states, however, depend largely on substituents as described in Section II.A.1. For example,

Table 1 Absorption Data, Energy of Excited Singlet State and Excited Triplet State

Porphyrin	λ nm($\varepsilon/10^3$)	E_S nm	E_T nm
Free			
H₂TMPyP			
H₂TPP	424(226), 520(14.5), 558(9.2), 584,638; H₂O; 65), 411, 508, 542, 585, 642; DM; 63), 419(470), 514(18.7), 549(7.7), 591(5.4), 647(3.4); Bz, rt; 64)	702; Bz, rt; 139)	865; MC, 77K; 147)
H₂TPyP			
H₂TSPP	412(530), 515(6.8), 553(16.5), 580(6.4), 633(3.7); H₂O, rt; 64)	700; H₂O, rt; 139), 649ᵇ; DMSO; 140)	865; EtOH, 77K; 85), 861ᵇ; DMSO; 65), 885ᵇ; MeOH; 148)
¹²Mg			
MgᴵᴵEtioII			675ᶜ; EPA, 77K; 149)
MgᴵᴵOEP	404, 542, 578; MeOH; 67)	585ᵃ; E/T, rt; 66)	696ᵃ; E/T; 66)
MgTSPP			873ᵇ; MeOH; 148)
MgᴵᴵTPP	426, 564, 601ᵃ; EtOH/Tol = 1/1, rt; 66)	612ᵃ; E/T, rt; 66)	796ᵃ; E/T; 66)
¹³Al			
AlᴵᴵᴵOEPOPh	399(288), 533(11.2), 572(22.4); Bz; 69)		
AlᴵᴵᴵTPPOH	415, 546, 590; CHCl₃; 68)		
AlTSPP			810ᵇ; MeOH; 148)
¹⁴Si			
SiᴵⱽOEPCl₂	408(282), 538(12.6), 573(14.5); Bz; 69)		
SiOEP(OMe)₂			695ᶜ; 2MTHF, 77K; 149)
¹⁵P			
PⱽOEP(OH)₂Cl	424, 555, 593; MeOH, rt; 71)		
PⱽOEP(OH)₂	403(235), 541(10.4), 583(17.7); EtOH; 72)		
PⱽOEP(OH)₂ClO₄		585; EtOH, 77K; 71), 72)	737; EPA/C₂H₅I = 9/1, 77K; 71), 72)

P^III OEP^+	372, 455, ~535: C_5H_5N, rt; 71)		
P^V TTPOH_2OH	431.2(282), 577.2(16.2), 600.0(8.5); DM; 70)		
^{20}Ca			
Ca^II(Meso)			697^c; EPAF, 77K; 149)
^{21}Sc			
Sc^III OEPOAc	331, 385sh, 404, 498, 534, 571; B/MP; 73) 406(331), 538(12.3), 574(33.9); Bz; 69)	577; B/MP, 300K; 73)	700; B/MP, 77K; 73)
^{22}Ti			
Ti(OEP)O	334(16.8), 384(56), 404(345), 500(2.16), 534(15.7), 572(27.2); DM; 73)	577; DM, 300K; 73)	712; DM, 77K; 73)
Ti^IV TPPO_2	423(336), 547(18); Bz; 74)		
Ti^IV TPPCl	423(330), 547(18); Bz; 76)		
Ti^IV TPPF_2	340(55), 387(70), 498(30), 650(9), 690sh; DM; 75)		
TiOTPP	421, 482, 512, 552, 590; Bz; 77)		
Ti^IV TTPCl_2	320, 376, 420, 492; Tol; 78)		
Ti^III TTPCl	428, 552; Tol; 78)		
^{23}V			
V^III OEPCl	406, 534, 572; Tol; 78)		
V^IV OOEP	570, 532, 407, 328; DM; 79)		
V^IV OTMPyP	438.5(207); H_2O; 80)		
V^IV OTPP			
V^IV OTSPP	434(169.9), 562(13.5), 602(5.1); H_2O; 81)		814; MCH/THF = 10/1; 150) 718^c; PMA, 10K; 149)
V^III TTPCl	424, 548; Tol; 78)		
^{24}Cr			
Cr^III OEPOPhPhOH	425(158), 544(11.2), 576(9.5); Bz; 69)		
CrTPPCl	447, 396, 562, 601; DM; 82)		

Table 1 Continued

Porphyrin	λ nm($\varepsilon/10^3$)	E_S nm	E_T nm
²⁵Mn			
Mnᴵᴵᴵ(Meso)(OAc)			640[c]; EPAF, 77K; 149)
Mnᴵᴵ(Meso)(OAc)			724[c]; EPAF, 77K; 149)
MnᴵᴵᴵTMPyP	441(133); H₂O; 85)		
MnᴵᴵᴵTMPyP	463(85(92)⁴⁵), 560(11); H₂O; 26), 65)		
MnᴵⱽTMPyP	420(74); H₂O; 85)		
MnᴵᴵᴵTPP	370, 471, 578, 618; DM; 63)		840; MCH, 77K; 147)
MnᴵᴵᴵTPPBr	483; MTHF; 83)		
MnᴵᴵᴵTPPCl	475; MTHF; 83)		
MnᴵᴵᴵTPPI	498; MTHF; 83)		
MnᴵᴵᴵTPPNCS	475; MTHF; 83)		
MnᴵᴵᴵTPPOAc	467; MTHF; 83)		
MnTPP₂O	345, 376, 400, 475, 578, 615; CHCl₃; 84)		
MnᴵⱽTPyP			860; EtOH, 77K; 85)
MnᴵᴵᴵTPyP	436(342); H₂O; 85)		698; EtOH, 77K; 85)
MnᴵᴵTSPP	469(95), 562(11.7), 596(7.5); H₂O; 26), 65)		
MnᴵᴵᴵTSPP		561[b], DMSO; 140)	697[b]; EtOH, 77K; 85)
MnᴵⱽTSPP	424(71); H₂O; 85)		
²⁶Fe			
Feᴵᴵᴵ(Meso)(OAc)	400(125), 518(14.2), 548(19.9); Py; 69)		665[c]; EPA, 77K; 149)
Feᴵᴵ(Meso)			663[c]; EPA, 77K; 149)
FeOEPPy2	374(71), 400(56), 493(13), 524(12), 565(13),		
Feᴵ(OEP)	684(4); DMF; 86)		
FeᴵᴵᴵTPPCl	410, 502; DM; 63)		

27Co

Compound	Data	
Co^{III}(Meso)(OAc)		664^c, EPA, 77K; 149)
Co^{III}(Meso)		656^c, EPA, 77K; 149)
Co^I(OEP)	352(70.1), 413(56.5), 510(7.3), 546(23.3); DMF; 86)	
Co^{II}OEP	384, 520, 550; DM; 89)	558^b, o-difluoroBz; 141)
Co^{III}(OEP)NH$_3$Br	411, 525, 559; EPA; 89)	
Co^{I}TPP	364.4(53.7), 428.1(70.2), 513.6(15); PhCN; 87)	
Co^{II}TPP	406, 525; hexane; 88)	
Co^{I}TPP	406, 523; DM; 63)	
Co^{I}TPP	404, 530; DM; 89)	
Co^{I}TPP	416.8(178.5), 530.7(12.6); PhCN; 87)	
Co^{III}TPP	322.8(17.2), 440.5(210.9), 552.6(13.5), 587.9(6.1); PhCN; 87)	

28Ni

Compound	Data	
Ni^{II}(Meso)	346(43.0), 407(120.9), 506(10.1), 545(21.5), 617(2.7); DMF; 86)	664^c; EPAF, 77K; 149)
Ni^I(OEP)	391(208.1), 515(11.5), 550(33.4); DMF; 86)	
Ni^{II}(OEP)	392, 516, 551; DM; 89)	
Ni^{II}(OEP)	392, 516, 551; DM; 89)	
Ni^{II}(OEP)	407, 525, 565; EPA; 89)	
Ni^{IV}(OEP)(Br)^2	416(198), 487(s), 527(25.4), 558(sh); DM; 90)	
Ni^{II}TPP	360, 418, 523; DMF; 86)	
Ni^{I}TPP	415, 527; DM; 89)	
Ni^{II}TTP		

29Cu

Compound	Data	
Cu^{II}(Meso)		649^c; EPAF, 77K; 149)
CuOEP	327(19.3), 395(136.6), 525(14.7), 562(32.3); Bz; 91)	

Table 1 Continued

Porphyrin	λ nm($\varepsilon/10^3$)	E_S nm	E_T nm
Cu porphyrins			
CuTPP			~729[b](d-d); 152) 742; 151)
³⁰Zn			
Zn[II](Meso)			
ZnOEP	407(417), 536(22.9), 572(24.5); dioxane; 69)		688[c]; EPA, 77K; 149)
ZnTMPyP⁴⁺	419(565), 585(4.9), 547(23.4); DM; 92)		762[c]; H₂O; 153)
ZnTPP	423(544), 548(22.8), 586(3.68); Bz, rt; 64)	660; Bz, rt; 139)	778; MCH, 77K; 147)
ZnTSPP	421(683), 555(22.1), 594(9.6); H₂O, rt; 64)	656; H₂O, rt; 139)	767[c]; H₂O; 153) 790[b]; MeOH; 148)
³¹Ga			
GaOEPCl	377(15.9), 400(626.5), 492(8.9), 530(33.2), 567(45.9); Bz; 93)		
GaTPPCl	397(18.0), 418(524.4), 548(20.7), 587(4.5), 621(1.3); Bz; 93)		
³²Ge			
GeOEP(OH)₂	387(48.2), 407(320), 496(1.9), 534(14), 570(14.4); PhCN; 94)		
Ge[IV]OEPCl₂			
GeTPP(OH)₂	404(44.7), 426(397.6), 517(3.8), 556(13.5), 596(10.2); PhCN; 94)		689[c]; 2MTHF, 77K; 149)
³³As			
As[V]OEP(OH)₂Cl	425, 554, 596; DM; rt; 71)	574; EPA, 77K; 71)	
As[V]OEP(OH)₂⁺			716; EPA, 77K; 71)
As[III]OEP⁺	378, 457, 565; DM; rt; 71)		750; DM, 77K; 71)

Species	Absorption	Emission (300 K)	Emission (77 K)
³⁸Sr			
Sr^II(Meso)	384, 404, 536, 572; B/MP; 96)		719ᶜ; EPAF, 77K; 149)
³⁹Y			
YOEP(OH)	374(126), 538(7.6), 674(2.4); CH; 97)	574; EPA, 77K; 143)	701; B/3MP, 77K; 96)
YOEP(acac)	420, 515, 554, 589; MeOH, rt; 95)	599; MeOH, rt; 142)	779; EPA, 77K; 95)
Y(OEP)₂			
Y^III TPP(acac)			
⁴⁰Zr			
ZrIV(OEP)₂	383(166), 490(12.6), 550(6.0), 592(24.5), 750(0.5); DM; 98)		
Zr(OEP)(OAc)₂	328(15.8), 378(66), 399(380), 490(2.24), 528(12), 565(31.6); DM; 73)	569; DM, 300K; 73)	703; DM, 77K; 73)
Zr(TPP)₂	396(347), 506(26.3), 554(9.3), 696(1.4); DM; 99)		
⁴¹Nb			
NbVOE-P(O)(O₂CCH₃)	410(127), 522(6), 538(10), 565(15), 583(10); PhCN; 100)		
[(OEP)Nb]₂O₃	399(210), 538(16), 570(34); THF; 101)		
NbOEPO	403(100), 455(7), 538(10), 577(14); THF; 102)		
Nb(OEP)OI₃			
Nb^V TPP(O)(O₂CCH₃)	430(297), 533(8), 555(12); PhCN; 100)	580; B/MP, 300K; 73)	718; B/MP, 77K; 73)
NbTTPO	428(195), 462(9), 557(15), 558(3.5); THF; 102)		
⁴²Mo			
Mo^VI TMP(O)(O₂)	431, 526, 562; DM; 103)		
Mo^V TPP(O)(OC₂H₅)	452(183), 580(15), 620(9.6); Bz; 104)		
Mo^IV TPP = O	430(355), 553(21); Tol/EtOH = 9/1; 104)		
Mo^IV TPPO	431(110), 514(20), 554(103); DM; 105)		
Mo^V TPPOClO₄	480(49), 616(16), 664(15); DM; 105)		
Mo^VI TPPO₂	425(214), 487(4.6), 535(15.2), 562(11); DM; 105)		
Mo^IV TTPCl₂	366, 398, 420, 496, 578; Tol; 78)		

Table 1 Continued

Porphyrin	λ nm($\varepsilon/10^3$)	E_S nm	E_T nm
^{43}Tc			
[Te(CO)$_3$]$_2$TPP	403, 475sh, 504, 670; DM; 106)		
TcMesoP(H)(CO)$_3$	388(t4.17), 473(t3.70), 580(t3.17); DM; 106)		619^c; Bz, 300K; 149)
^{44}Ru			
Ru^{II}OEP(CO)	393, 508, 542; DM; 109)		
Ru^{VI}(OEP)O$_2$	320(15.8), 408(93.3), 522(9.33), 600sh(2.66),		
Ru^{IV}(TMP)(OH)$_2$	660(1.29); DM/EtOH; 108)		
Ru^{VI}(TMP)O$_2$	420, 524, 560; DM; 109)		
Ru^{VI}(TMP)O$_2$	412(216), 527(23.9); DM; 109)		
Ru^{II}(TPP)CO	413, 532, 565sh; DMSO; 107)		729; DMSO, 22.8C $\pm$ 0.2; 107)
^{45}Rh			
Rh^{III}(OEP)Cl	339(18.2)403(132), 520(13.2), 554(25.1); CH$_3$Cl; 7), 111)	552^a; MeOH; 110)	664^a; MeOH, 298K; 110)
Rh^{II}TPP	418, 531, 568; CH$_3$Cl; 89)		
Rh^{III}TPPCl	418, 530, 564^a; MeOH; 110)		
Rh^{III}TPPCl	421(92.6), 534(11.9), 567(5.4); DM; 111)	570^a; MeOH; 110)	718^a; MeOH, 298K; 110)
Rh^{III}TSPP(H$_2$O)$_2$	418, 530cm^{-1}, 562^{a}cm^{-1}; H$_2$O; 110)		
RhTSPP		568^a; H$_2$O; 110)	718^a; H$_2$O, 298K; 110) 701^b; MeOH; 148)
^{46}Pd			
Pd^{II}(Meso)	390(170), 510(16), 544(50); isobutyronitrile; 112)		639^c; EPAF, 77K; 149)
Pd^{II}OEP			637^c; Bz, 300K; 149)
PdTMPyP^{4+}			687^d; H$_2$O; 153)
Pd^{II}TPP	418, 524, 554; C$_6$H$_6$; 89)		670^c; EPAF, 77K; 149)
Pd^{IV}TPPCl$_2$	423, 534, 564; DM; 89)		
PdTSPP^{4-}			695^d; H$_2$O; 153) 667^b; MeOH; 148)

Compound	Absorption	Emission
^{47}Ag		
Ag^{II}OEP	409(219), 526(12.5), 560(18.3); 114	
Ag^{III}OEPClO$_4$	404(131), 516(9.8), 552(25.2); 114	
AgTPP	423(510), 539(23), 569(4.4); DM; 113	714[c]; EPA, 77K; 149
^{48}Cd		
Cd^{II}(Meso)		
CdOEPPy	421(288), 551(21.9), 586(12.9); Py; 69	
CdTMPyP	445(169.8), 575(13.8), 618(6.8); H$_2$O; 115	
CdTPP	430(422), 602(8.5), 562(15.6); CHCl$_3$; 92	
CdTSPP		805[b]; MeOH; 148
^{49}In		
InOEPCl	411, 539, 578; Bz; 116 388(54), 408(436), 500(4), 539(19), 577(22); Bz; 117	
InTPPCl	427, 560, 601; Bz; 116 404(47), 426(570), 515(6), 559(21), 600(10); Bz; 117	
InTSPP		756[b]; MeOH; 148
^{50}Sn		
Sn^{IV}(Meso)(OAc)2		696[c]; EPA, 77K; 149
SnOEPOMe2	410(316), 542(20.4), 578(18.2); Bz; 69	689[c]; 2MTHF, 77K; 149 729[d]; H$_2$O; 153
Sn^{IV}OEPCl$_2$		
SnTMPyP4$^+$		
Sn^{IV}TTP(OH)$_2$	427(491), 563(20), 602(14); THF; 118	696[c]; 2MTHF, 77K; 149 705; EPA, 77K; 154
Sn^{IV}TPPCl$_2$		
SnTPP	400, 490, 696; Tol; 119	
Sn^{II}TPP		
SnTSPP4$^-$		757[d]; H$_2$O; 153 790[b]; MeOH; 148

Table 1 Continued

Porphyrin	λ nm(ε/10³)	E_S nm	E_T nm
⁵¹Sb			
Sb^V OEP(OH)₂Cl	422, 553, 596; DM; 122)	569; EPA; 144), 71)	702; EPA, 77K; 71)
SbOEP(OH)₂			702; EPA, 77K; 122)
Sb^III OEP	355(30.1), 376sh(64), 430sh(21.1), 460(46.4), 570(9), 600sh(2.9); EtOH, rt; 121)		
Sb^III OEPCl	379, 460, 570; DM, rt; 71)		
Sb^V TTP(OH)₂OH	423.6(355), 327.8(10), 553.3(13.8), 595.2(10.7); DM; 70)		740; EPA, 77K; 71)
Sb^V TPP(CH₃)₂	347(20), 439(320), 580(10), 623(23); PhCN; 120)		
Sb^III TPP	357(42.4), 378sh(26), 438sh(23.2), 465(119.5), 594(4.2), 644(9.6); EtOH, rt; 121)		
⁵⁶Ba			
Ba^II(Meso)			734^c; EPAF, 77K; 149)
⁵⁷La			
La(OEP)₂	394(141), 494(5.6), 540(7.1), 576(5.0); CH; 97)		
⁵⁸Ce			
Ce^IV(OEP)₂	378(162), 467(8.3), 530(5.9), 573(17.4), 636.6(3.5); DM; 98)		
Ce(TPP)₂	396(224), 486(14.1), 540(10.7), 630(3.1); DM; 99)		
⁵⁹Pr			
Pr(OEP)₂	391(151), 490(5.2), 540(6.3), 576(4.4), 670(1.6); CH; 97)		
⁶⁰Nd			
Nd(OEP)₂	390(64.6), 488(3.4), 542(9.3), 577(3.0) 668(1.2); CH; 97)		

	Absorption	Emission (300K)	Emission (77K)
[62] Sm			
Sm(OEP)$_2$	390(129), 489(6.2), 544(7.2), 578(6.0), 674(1.6); CH; 97		
[63] Eu			
EuOEP(OH)	385, 404, 536, 573; B/MP; 96	580; B/MP, 300K; 143	700; B/MP, 77K; 96
Eu(OEP)$_2$	376(110), 543(7.4), 676(2.3); CH; 97		
[64] Gd			
GdOEP(OH)	385, 405, 535, 573; B/MP; 97	579; B/MP, 300K; 143	702; B/MP, 77K; 96
Gd(OEP)$_2$	380(75.9), 542(5.1), 674(1.8); CH; 97		
GdTPPOH, OEPOH	see ref. 123; MeOH		
[65] Tb			
Tb(OEP)$_2$	378(100), 492(6.3), 544(6.5), 580(5.0), 680(2.0); CH; 97		
[66] Dy			
Dy(OEP)$_2$	386(123), 494(6.6), 546(6.9), 582(5.9), 679(1.9); CH; 97		
[67] Ho			
HoOEP(OH)	383, 403, 535, 571; B/MP; 96	577; B/MP, 300K; 143	
Ho(OEP)$_2$	376(126), 494(7.8), 546(8.1), 582(6.8), 684(1.7); CH; 97		
[68] Er			
ErOEP(OH)	384, 403, 534, 571; B/MP; 96	576; B/MP, 300K; 143	
Er(OEP)$_2$	374(112), 494(6.0), 544(6.2), 582(4.7), 679(1.7); CH; 97		
ErTPPOH, OEPOH	see ref. 123; MeOH		
[69] Tm			
TmOEP(OH)	384, 403, 535, 571; B/MP; 96	580; B/MP, 300K; 143	702; B/MP, 77K; 96
Tm(OEP)$_2$	388(120), 494(5.9), 546(5.8), 584(6.0), 672(2.0); CH; 97		

Table 1 Continued

Porphyrin	λ nm(ε/10³)	E_S nm	E_T nm
⁷⁰Yb			
YbOEP(OH)	384, 403, 534, 571; B/MP; 96)	580; B/MP, 300K; 143)	704; B/MP, 77K; 96)
Yb(OEP)₂	370(95.5), 498(5.6), 530(5.8), 672(1.6); CH; 97)		
⁷¹Lu			
LuOEP(OH)	383, 402, 534, 571; B/MP; 96)	576; B/MP, 300K; 143)	698; B/MP, 77K; 96)
Lu(OEP)₂	370(105), 490(5.8), 534(5.9), 674(2.3); CH; 97)		
LuᴵᴵᴵTPPOH	410, 550, 588; DM; 124)		
LuᴵᴵᴵTPP(acac)	418, 512, 551, 587; MeOH, rt; 95)	599; MeOH, rt; 142)	769; EPA, 77K; 95)
⁷²Hf			
Hf(OEP)₂	380(151), 486(11.0), 548(4.7), 592(20.4), 750(0.5); Tol; 99)		
Hf(OEP)(OAc)₂	328(12.4), 378(55), 398(333), 490(2.17), 526(11.5), 564(29.4); DM; 73)	570; DM, 300K; 73)	704; DM, 77K; 73)
Hf(TPP)₂	394(372), 506(28.2), 554(10), 700(1.5); DM; 99)		
⁷³Ta			
Ta(OEP)F₃	375(42.5), 401(239), 527(10), 565(30.8); CHCl₃; 73)	574; B/MP, 300K; 73)	712; B/MP, 77K; 73)
⁷⁴W			
Wⱽ O(OMe)TPP	450(254), 580(14), 625(9.9), 662sh(~7); MeOH; 125)		
Wⱽ O(H₂O)TSPP	451(156), 582(13), 626(7.6); H₂O; 125)		
Wᴵⱽ TTPCl₂	336, 370, 388, 420, 466, 558; Tol; 78)		
⁷⁵Re			
[Re(CO)₃Cl]₂TPP₂Sb Cl₆	405, 493, 532, 590; DM; 126)		
ReTPP(H)(CO)₃	402(1318), 473(417), 670(83.2); DM; 106)		
TPP[Re(CO)₃]₂	408, 485sh, 513; DM; 106)		

Compound	Absorption λ (ε); solvent; ref	Emission	Emission
^{76}Os			
$OsOEPCl_2$	386; DM; 127)		
$Os^{II}OEPCO$	389(415), 508(16), 538(31); DM; 130)		
$Os(OEP)(py)_2$		510; THF, 295K; 145)	
$Os(OEP)(CO)(py)$	391(206), 505(17), 537(20); DM; 130)	537; THF, 295K; 145)	720; THF, 295K; 145)
$Os^{III}OEPCOCl$	279(11.0), 340(52.5), 495(5.5), 527(5.0); DM; 128)		
$Os^{IV}OEP(OEt)_2$			
$Os^{IV}OEPCl_2$	377(110), 500(12), 540(11); DM; 130)		
$Os^{VI}OEP(O)_2$	377(123), 588(10); DM; 130)	594; THF, 295K; 145)	729; THF, 295K; 145)
$OsTPPCl_2$	396, 510, 539, 614; DM; 127)		
$OsTPP(OEt)_2$	270(22.4), 405(95.5), 507(10.7); DM; 128)		
$Os^{VI}TPP(O)_2$	339(14.1), 395(66.1), 479sh(12.0), 586(5.1); DM; 129)		
$Os^{VI}TPP(O)_2$	337(31), 394(119), 583(10); DM; 130)		775[b](dπ^*); Tol, 295K; 155)
$Os^{II}TPPCO$	406(280), 518(17); DM; 130)		
$OsTPP(CO)(py)$			
$Os^{III}TPPCOCl$	404(150), 517(13); DM; 130)		
$Os^{IV}TPPCl_2$	396(114), 480(16), 590(7); DM; 130)		
^{77}Ir			
$Ir^{III}(OEP)Cl(CO)$	402(170), 516(11.4), 548(28.1); DM; 132)	536; THF, 295K; 145)	641; THF, 295K; 145)
$Ir(TMP)(CO)Cl$	316, 422, 496, 532, 546; DM; 131)		
^{78}Pt			
$Pt(etio)$			
$Pt^{II}(OEP)$	382, 503, 536; C_6H_6; 89)		
$Pt^{II}TPP$	403, 510, 539; C_6H_6; 89)		640[c]; EPAF, 77K; 149)
$Pt^{IV}TTPCl_2$	420, 537, 575; DM; 89)		
^{79}Au			
$AuTMPyP$	406(275), 481(4.1), 523(15.1), 558(6.0); H_2O; 133)		687[d]; H_2O; 156)

Table 1 Continued

Porphyrin	λ nm($\varepsilon/10^3$)	E_S nm	E_T nm
AuTPP	412(331), 521(20.0); CHCl₃; 133		705[b]; H₂O; 133
AuTSPP	406(372), 483(4.2), 522(16.6); H₂O; 133		674[b]; MeOH; 148
			672[d]; H₂O; 156
⁸⁰Hg			
Hg^{II}(Meso)			750; EPAF, 77K; 149)
HgOEP	411(270), 534(11.8), 565(12.6); Bz; 69)		
Hg₂TMPyP	435.5(200), 564.5(13), 585.5sh; H₂O; 134		
HgTMPyP(OH)₂	446(170), 580(18); H₂O; 134		
HgTPP	429(458); CHCl₃; 92)		
⁸¹Tl			
TlOEPOH · H2O	415(327), 543(18.2), 581(14.2); DM; 69)		
Tl₂OEP	370(57), 478(132), 604(8); DM; 135)		
Tl₂TPP	358(47), 478(190), 684(15); DM; 135)		
⁸²Pb			
Pb^{II}(Meso)			687[c]; 3MP, 77K; 149)
PbOEP	462(138), 540(3.2), 582(14.5); Bz; 69)		695[c]; MeOH, 77K; 149)
⁸³Bi			
Bi^{III}OEP(NO₃)	355, 370, 460, 572; DM; 71)		745; EPA, 77K; 71)
Bi^{III}T(3-Py)P	463.0, 592.5, 637.5; MeOH; 136)		

Compound			
Bi:III TPP(NO$_3$)	336, 355, 467, 651; Bz; 71)		
BiTTP(OH)$_2$OH	469.4(151), 353.7(38.9), 601.4(8.5), 650.1(12.3); DM; 70)		
[90]Th			
Th^{IV}OEPCl$_2$ · 2PhCN	404, 534, 572; THF; 40)		
Th^{IV}(OEP)$_2$	345(sh), 383(142), 470(sh), 535(2), 577(8), 644(w); DM; 138)	~790; Tol, 295K; 146)	~960; Tol, 295K; 146)
Th^{IV}TPPCl$_2$ · 2PhCN	424, 512, 550, 586; THF; 137)		
Th^{IV}TPP(acac)$_2$	419, 510, 551, 591; MeOH, rt; 95)	~605; MeOH, rt; 142)	764; EPA, 77K; 95)
Th^{IV}(TPP)$_2$	400(539), 480(14), 546(12), 699(6); DM; 138)		
[92]U			
U^{IV}(OEP)$_2$	350(sh), 386(166), 464(22), 535(9), 582(19), 668(6); DM; 138)		
UOEPCl$_2$ · 2PhCN	405, 536, 574; THF; 137)		
U^{IV}(TPP)$_2$	402(280), 488(23), 549(15), 619(7); DM; 138)		
U(TPP)$_2$	404(490), 485(21.9), 550(15.1), 620(5.9); CHCl$_3$; 99)		
UTPPCl$_2$ · 2PhCN	428, 512, 553, 588; THF; 137)		

3-Methylpentane, 3MP; methylcyclohexane, MC; (ethanol/toluene = 1/1), E/T: (1-butanol/3-methylpentane = 3/7), B/MP; benzene, Bz; toluene, Tol; cyclohexane, CH; dichloromethane, DM; pyridine, Py; tetrahydrofuran, THF; octaethylporphyrin, OEP; tetraphenylporphyrin, TPP; tetra(4-sulfonatophenyl)porphyrin, TSPP; tetra(4-pyridyl)porphyrin, TPyP; tetra(N-methyl-4-pyridyl)porphyrin, TMPyP; Meso- IX-DME, Meso.

[a] Calculated value from the reported data in cm^{-1}.
[b] Calculated value from the reported data in eV.
[c] Calculated value from the reported data in kcal mol^{-1}.

Table 2 Fluorescence Yield, Triplet Yield, Phosphorescence Yield, Lifetime of Excited Singlet State and Excited Triplet State

Porphyrin	ϕ_f	ϕ_T	ϕ_p	τ_S ns	τ_T ms
Free					
H₂TPP	0.13; MeOH, rt; 157), 0.11; Bz; 123)	0.88 ± 0.10; EtOH; 165), 0.67, 0.82; Bz, rt; 1), 64), 0.82; MC; 164)	4 × 10⁻⁵; MC, 77K; 147)	10.7; MeOH, rt; 157), 12.3; Bz, 293K; 168)	6; MC, 77K; 147), 1.5; Bz, rt; 173)
H₂OEP	0.13; Bz; 123)			18.9; Bz; 158)	
H₄OEP²⁺	0.052; Bz; 123)			0.052; Bz; 158)	
H₄TPP²⁺	0.14(S2, 0.59 × 10⁻⁴); Bz; 123)			1.8; Bz; 158)	
H₂TPyP			~1 × 10⁻⁵; EtOH, 77K; 85)		4; EtOH, 77K; 85)
H₂TSPP	0.08; H₂O, rt; 64)	0.78; H₂O, rt; 64)		10.4; H₂O, rt; 64)	0.42; H₂O, rt; 64)
¹²**Mg**		0.74; H₂O; 148)			270µs; H₂O, N₂ purged; 148)
MgOEP					200; E/T, 77K; 66)
MgTPP	0.15; MC, rt; 159)	0.85 ± 0.08; EtOH; 165)	0.015; MC, 77K; 159)	9.2; MC, rt; 159)	45; MC, 77K; 159)
MgTPP(py)₂	(S2, 5.5 × 10⁻⁴); Bz; 123)				120; E/T, 77K; 66)
MgTSPP		0.77; H₂O; 148)			105µs; H₂O, N₂ purged; 148)
¹³**Al**					
AlOEPCl	0.13; EtOH; 123)	0.82; EtOH, 77K; 68)		9.9; EtOH; 158)	150; EtOH, 77K; 158)
AlᴵᴵᴵTPPOH	0.14 ± 0.01; CHCl₃; 68)			9.0 ± 1.0; CHCl₃; 68)	
AlTPPCl	0.11(S2, 5.5 × 10⁻⁴); EtOH; 123)			8.7; EtOH; 158)	46; EtOH, 77K; 158)
AlTPP		0.85 ± 0.05; EtOH; 165)	0.048; EtOH, 77K; 161)		30.5; EtOH, 77K; 161)
AlTSPP		0.81; H₂O; 148)	~0.001; EtOH, 77K; 158)		120µs; H₂O, N₂ purged; 148)

Compound					
^{15}p					
$P^V OEP(OH)_2ClO_4$	~0.1: EPA, 77K: 17), 72)		<0.001; EPA, 77K: 1), 71)		7; EPA(10%C_2H_5I), 77K: 71), 72)
$P^V TPPCl_2$	0.011; DM: 163)			0.82; DM: 163)	
^{21}Sc					
Sc(OEP)OAc	0.14; B/MP, 300K: 73)	0.046 ~ 0.0056; Bu/MP, 77K: 73)			400; B/MP, 77K: 73)
^{22}Ti					
Ti(OEP)O	0.02; 3MP, 300K: 73)	0.003; 3MP, 77K: 73)			175; MP, 77K: 73)
$Ti^{IV}TPPO$					35μs; EtOH: 178)
^{23}V					
$V^{IV}OTPP$					46.8ns; Bz, 303K: 150)
$V^{IV}OTPP$					5.2μs/26.3μs(at800nm); MCH/THF = 10/1, 77K: 150)
^{24}Cr					
$Cr^{III}TPP$	<10^{-4}; EtOH, 77K: 161)	1.0; r.t.; 166)	2.3 × 10^{-4}; EtOH, 77K: 161)		295ps; EtOH, 300K, 4T1; 161)
$Cr^{III}TPPCl$			1.2 × 10^{-5}; EtOH, 300K: 161)	<2ps; Tol/acetone; 173)	~20ns(TT*), 3+ ~ 1ns(dd); Tol/acetone; 173)
^{25}Mn					
MnTPP	<10^{-4}; MC: 147)				0.2; MC.77K: 147)
$Mn^{II}TPP$					8ps; Py/H_2O: 173)
$Mn^{II}TPPCl$					17ps; DM: 173)
$Mn^{III}TPPCl$					
$Mn^{II}TPyP$			3 × 10^{-4}; EtOH, 77K: 85)	<1ps; Py/H_2O: 173)	~0.2; EtOH, 77K: 85)
$Mn^{III}TPyP$			~5 × 10^{-5}; EtOH, 77K: 85)	0.1 ~ 0.5ps; DM: 173)	<0.1; EtOH, 77K: 85)
$Mn^{IV}TPyP$			<2 × 10^{-6}; EtOH, 77K: 85)		
^{26}Fe					
$Fe^{II}TPPCl$				<1ps; DM: 173)	50ps; DM: 173)
$Fe^{III}TPPOH$					1.78ns; DM: 173)
$Fe^{II}TPP(pip)_2$				400ps; Pip/Tol: 173)	400ps; Pip/Tol: 173)

Table 2 Continued

Porphyrin	ϕ_f	ϕ_T	ϕ_p	τ_S ns	τ_T ms
²⁷Co					
Co^II OEP					12ps(d-d); o-difluoroBz; 141)
Co^II TPP					<40ps; 173)
Co^III TPP					<40ps; 173)
²⁸Ni					
NimesoP					
Ni^II TPP	<10⁻⁴; MC; 147)		<10⁻⁵; MC, 77K; 147)	250 ± 50ps; Tol; 171) 0.25; MC; 164)	<15ps; 173)
²⁹Cu					
CuOEP	<10⁻⁵; 152)				120ns; Tol; 152) 100 ± 20ns; DM; 180)
Cu porphyrins					
CuTPP		1.00 ± 0.05; EtOH; 165)		<6ps; 173)	35ns(²·⁴T1); Tol; 152) 37 ± 15ns; DM; 180)
³⁰Zn					
ZnOEP	0.045; Bz; 123)			1.9; Bz; 158)	54; 2-MeTHF, 77K; 158)
ZnTMPyP⁴⁺	0.035; H₂O; 153)	0.90; H₂O; 153)		1.5; H₂O; 153)	1.22; H₂O, rt; 153)
ZnTPP	0.04; MC; 147)	0.88; MC; 164)	1.2 × 10⁻²; MC, 77K; 147)	2.7; MC; 164)	26; MC, 77K; 147)
ZnTPP	(S2, 3.5 × 10⁻⁴); Bz; 123)	0.90 ± 0.10; EtOH; 165)			1.2; Bz; 64)
ZnTSPP⁴⁻	0.042; H₂O; 153)	0.85; H₂O; 153), 148)		1.7; H₂O; 153)	1.40; H₂O, rt; 153) 210μs; H₂O, N₂ purged; 148)
³¹Ga					
GaOEPCl	0.065; EtOH; 123)		0.036; EtOH, 77K; 158)	5.2; EtOH; 158)	140; EtOH, 77K; 158)
GaTPPCl	0.069(S2, 4.6 × 10⁻⁴); EtOH; 123)			5.5; EtOH; 158)	44; EtOH, 77K; 158)
GaTPPOH	0.02; Bz; 123)			1.4; Bz; 158)	
³³As					
As^III OEP⁺					
As^V OEP(OH)₂⁺	0.076; EPA, 77K; 71)		<0.002; DM, 77K; 71) 0.003; EPA, 77K; 71)		50/330μs; DM, 77K; 71) 35/75; EPA, 77K; 71)

Compound						
39. Y						
YOEP(acac)	0.006; MeOH, rt; 95)			0.01; EPA, 77K; 95)		
Y^{III}TPP(acac)						55: EPA, 77K; 96) 29: EPA, 77K; 95)
40. Zr						
Zr(OEP)X$_2$L$_2$	0.016; Bu/MP, 300K; 73)					41; B/MP, 77K; 73)
Zr(TPP)$_2$		0.034; Bu/MP, 77K; 73)			1.9 ± 0.1; DM: 170	
41. Nb						
Nb(OEP)OI$_3$	0.015; Bu/MP, 300K; 73)	0.074 ~ 0.17; 2MeTHF, 77K; 73)				14.5; B/MP, 77K; 73)
NbO(MeC-OO)TPP	0.003; MC, rt; 123)				<0.5; MC, rt; 158	8.9; MC/3P, 77K; 158)
[NbTPP]$_2$O$_3$	0.0008; MC, rt; 123)				<0.5; MC, rt; 158	6.7; MC/3P, 77K; 158)
42. Mo						
Mo^{V}TPP(O)(OC$_2$H$_5$)		1.00 ± 0.05; EtOH; 165)				156ns; MC, 77K; 104)
MoTPP(O)OCH$_3$						4 ± 1ns; THF, rt; 179)
44. Ru						
Ru^{II}OEP(CO)						55 ± 2μs; DM, rt; 181)
Ru^{II}OEP(CO)(EtOH)						~80μs; EtOH, rt; 176)
Ru^{II}OEP(py)$_2$						15 ± 3μs; Py, rt; 181)
RuTPP						47 ± 4μs; DM, rt; 177)
RuTPP(CO)						36.0 ± 0.06μs; DMSO, 22.8C ± 0.2; 107) 15 ± 2ns; Py, rt; 177)
RuTPP(py)$_2$						
45. Rh						
Rh^{III}OEPCl	0.025; MeOH, 298K; 110)	0.24; MeOH, 77K; 110)	2 × 10^{-4}; MeOH, 298K; 110)		≪1.0; MeOH, 298K; 110)	10μs; MeOH, 298K; 110)
Rh^{III}TPPCl	0.012; MeOH, 298K; 110)	0.075; MeOH, 77K; 110)	5 × 10^{-4}; MeOH, 298K; 110)		≪1.0; MeOH, 298K; 110)	115μs; MeOH, 298K; 110) 0.31; MeOH, 77K; 110)

Table 2 Continued

Porphyrin	ϕ_f	ϕ_T	ϕ_p	τ_S ns	τ_T ms
RhTSPP					
Rh^{III}TSPP(H$_2$O)$_2$	3 × 10^{-4}; H$_2$O, 298K; 110	1.00; H$_2$O; 148)	0.0075; H$_2$O, 298K; 110	≪1.0; H$_2$O, 298K; 110	13μs; H$_2$O, N$_2$ purged; 148) / 130μs; MeOH, 298K; 110
^{46}Pd			0.07; H$_2$O, 77K; 110		0.35; MeOH, 77K; 110
PdOEP					360μs; isobutyronitrile; 112)
PdTMPyP^{4+}	ca. 10^{-4}; H$_2$O; 153)	1.00; H$_2$O; 153		<0.5; H$_2$O; 153	0.17; H$_2$O, rt; 153
PdTPP	5 × 10^{-4}; 110 / 2 × 10^{-4}; MC, rt; 159)	1.00 ± 0.05; EtOH; 165)	0.20; 77K; 110 / 0.17; MC, 77K; 159	0.020; MC, rt; 123	2.4; 77K; 110 / 358μs; 298K; 110
PdTSPP^{4-}	ca. 10^{-4}; H$_2$O; 153)	1.00; H$_2$O; 153), 148		<0.5; H$_2$O; 153	2.8; MC, 77K; 159) / 0.38; H$_2$O, rt; 153
^{47}Ag					
Ag^{II}mesoP					
Ag^{I}TPP					63μs; H$_2$O, N$_2$ purged; 148)
^{48}Cd			<3 × 10^{-4}; 77K; 167)		15 ± 8ps; Tol; 171)
CdTPP	4 × 10^{-4}; MC, rt; 159)	1.00 ± 0.05; EtOH; 165)	0.04; MC, 77K; 159	0.065; MC, rt; 123	2.4; MC, 77K; 159)
CdTPPpy	(S2, 0.8 × 10^{-4}); Bz, 1%py; 123)				
CdTSPP		1.00; H$_2$O; 148)			105μs; H$_2$O, N$_2$ purged; 148)
^{49}In					
InOEPCl	0.007; EtOH; 123)		0.042; EtOH, 77K; 158)	0.7; EtOH; 158)	28; EtOH, 77K; 158)
InTPPCl	0.011(S2,4.7 × 10^{-4}); EtOH; 123)		0.8; EtOH; 158)	9.5; EtOH, 77K; 158)	

InTSPP		0.95; H_2O; 148)			90μs; H_2O. N_2 purged; 148)
50Sn SnTMPy⁴⁺	0.027: H_2O; 153	0.95; H_2O; 153			0.9: H_2O. rt; 153
SnTPP	0.014 ~ 0.021; H_2O/MeOH = 1/1; 162			700 ± 100ps: DM; 173)	11.3; EPA, 77K; 154
SnTSPP⁴⁻	0.036; H_2O; 153	0.95; H_2O; 153 / 1.00; H_2O; 148			1.0: H_2O. rt; 153 / 100μs; H_2O. N_2 purged; 148
51Sb Sb^{III}OEPCl	~0.1: EPA. 77K; 122), 71		0.003: EPA, 77K; 71		88μs: EPA, 77K; 71
Sb^{V}OEP(OH)₂			0.026: EPA, 77K; 122), 71		26; EPA, 77K; 122
Sb^{V}OEP(OH)₂Cl					26: EPA, 77K; 71
58Ce Ce^{IV}(OEP)₂	<2 × 10⁻⁵; 160)			1.5 ± 0.2ps: DM, 295K; 174	
Ce^{IV}TPP₂				1.5 ± 0.5ps: DM, 295K; 174	
64Gd GdOEPOH			1 × 10⁻²; MeOH. rt; 123)		70μs; MeOH. rt; 123)
GdTPPOH			1 × 10⁻³; MeOH. rt; 123)		103μs; B/MP, 77K; 96) / 34μs; MeOH. rt; 123)
68Er ErOEPOH	2.0 × 10⁻⁴; MeOH; 123)			the order of 0.1ns; MeOH; 123)	
ErTPPOH	2.1 × 10⁻⁴; MeOH; 123)			the order of 0.1ns; MeOH; 123)	
71Lu LuOEP(OH)	~10⁻³; 160)				7.8: B/MP, 77K; 96) / ~2; 160)
LuPorphyrin	0.001: MeOH. rt; 95)		0.02: EPA, 77K; 95)		2.8: EPA, 77K; 95)
Lu^{III}TPP(acac)		~1; 160)		~50ps; 160)	
72Hf Hf(OEP)X₂L₂	0.020; Bu/MP, 300K; 73)	0.096: Bu/MP, 77K; 73)			65: B/MP, 77K; 73)
79Ta Ta(OEP)F₃	~0.004: Bu/MP, 300K; 73)	0.47: Bu/MP, 77K; 73)			2.5: B/MP, 77K; 73)

Table 2 Continued

Porphyrin	ϕ_f	ϕ_T	ϕ_p	τ_S ns	τ_T ms
[76]Os					
Os(OEP)(CO)(py)				~50ps; THF, 295K; 169	16ns; THF, 295K; 169
Os(OEP)O₂				<13ps; THF, 295K; 169	6ns; THF, 295K; 169
Os(OEP)(py)₂				<9ps; THF, 295K; 169	1ns; THF, 295K; 169
Os^{II}TPP(CO)					12.4ns; DM, r.t.; 130
OsTTP(CO)(py)			0.005; DM, r.t.; 130		33 ± 2ns; Tol, 295K; 155
OsVITPPO₂					9.2ns; DM, r.t.; 130
[78]Pt					
Pt(etio)				<15ps; THF, 295K; 169	63μs; THF, 295K; 169
					125μs; THF, 77K; 169
Pt^{II}TPP			0.45; 77K; 110		0.29; 77K; 110
Pt^{IV}TPPCl₂					54μs; 298K; 110
[79]Au					45 ± 10ps; Tol, rt; 175
AuTMPyP	<10⁻⁴; H₂O; 156	0.62; H₂O; 156	0.008 ± 0.003; H₂O; 156	<10ps; H₂O; 156	90ns; H₂O, rt; 156
AuTPPAuCl₄			~6 × 10⁻³; E/G, 77K; 167		63 ± 14μs; E/G, 77K; 167
					184 ± 6μs; E/G, 77K; 167
AuTSPP	<10⁻⁴; H₂O; 156	0.66; H₂O; 156, 148	0.008 ± 0.003; H₂O; 156	<10ps; H₂O; 156	120ns; H₂O, rt; 156
					0.12μs; H₂O, N₂ purged; 148

[80]Hg HgTPP	<10^{-3}; MC, rt: 159	0.01; MC, 77K: 159	0.2: MC, 77K, 159
[82]Pb PbOEP	(S2, 0.04 × 10^{-4}); Bz; 123		
PbTPP	(S2, 0.43 × 10^{-4}); Bz; 123		
[83]Bi Bi[III]OEP(NO$_3$)		0.008; EPA, 77K: 71	13.6μs; EPA, 77K: 71
[90]Th Th[IV](OEP)$_2$	~10^{-5}; Tol, 295K: 146	~10ps; Tol, 295K: 146	~50μs; Tol, 295K: 146
Th[IV]TPP(acac)$_2$	~0.0001: MeOH, rt: 95	~0.03; EPA, 77K: 95	<0.3: EPA, 77K: 95
ThIV(TPP)$_2$		~60ps; Tol, 295K: 172 13 ± 2ps; Tol, 295K: 172	39.4 ± 0.5μs; Tol, 295K: 172 8.9 ± 0.4μs; Tol, 295K: 172

3-Methylpentane, 3MP; methylcyclohexane, MC; (ethanol/toluene = 1/1), E/T; (1-butanol/3-methylpentane = 3/7), B/MP; benzene, Bz; toluene, Tol; cyclohexane, CH; dichloromethane, DM; pyridine, Py; tetrahydrofuran, THF; Meso-IX-DME, Meso; E/Gcerol = 11/1, E/G; piperidine, pip: octaethylporphyrin, OEP; tetraphenylporphyrin, TPP; tetra(4-sulfonatophenyl)porphyrin, TSPP; tetra(4-pyridyl)porphyrin, TPyP; tetra(N-methyl-4-pyridyl)porphyrin, TMPyP; Meso-IX-DME, Meso.

the lowest excited state of OEP has $a_{1u}e_g^*$ nature, while that of TPP has $a_{2u}e_g^*$ nature. The lowest singlet and triplet excited state energies (E_S, E_T) in normal-type porphyrins and metalloporphyrins are almost constant, irrespective of the central atom in the porphyrin ring. Other parameters besides E_S and E_T in Table 2 are governed by dynamic rate constants of radiative and nonradiative processes involving intersystem crossing. Radiative rate constants from the excited singlet states of metalloporphyrins with typical atoms were observed to coincide [19] with those calculated by Strickler-Berg's equation proposed for radiative transition [20]. Intersystem crossing in porphyrins generally has a large rate constant. This leads to a small fluorescence quantum yield, a short singlet lifetime, and a large triplet yield as shown in Table 2. An appreciable heavy atom effect on the intersystem crossing of metalloporphyrins with typical atoms was observed; fluorescence lifetimes were shortened with an increase of atomic number of the central atom [19,21]. Axial ligand was also observed to exhibit heavy atom effect on the fluorescence lifetime; τ_S = 2.1 ns(X = OH), 1.5 ns(X = Cl), <1.0 ns(X = Br) for GaOEPX in CH_2Cl_2 [19].

The lowest triplet states of porphyrins and metalloporphyrins with typical metal atoms have generally long lifetimes from several hundreds of μs to several ms even in fluid solution at ambient temperature (Table 2). Nonradiative deactivation from the triplet state of porphyrins was revealed to be well described [19,24,25] by the energy gap law [23]. As typical examples, OEP derivatives with E_T larger than that of TPP are shown to have generally a longer triplet lifetime than TPP derivatives (Table 2). For magnesium porphyrins, the nonradiative rate constant of the lowest triplet state was expressed by Eq. (2) [25,26]

$$\log k_{nr} \sim 7.5 - A(\Delta E/20{,}000) \tag{2}$$

where k_{nr} and ΔE denote the nonradiative rate constant and the energy gap between the lowest excited triplet state and the ground state. A is a constant and the value of 8.5 was obtained for magnesium porphyrins [25]. For porphyrins having heavy metal atom, k_{nr} was reported to be estimated by Eq. (3) [25,27]:

$$k_{nr} = k_{nr}^0 Z_T^2 \tag{3}$$

where k_{nr}^0 denotes a standard nonradiative rate constant under the condition without strong spin-orbit interaction and Z_T is spin-orbit matrix element for metal atom. Equation (3) was actually observed to be valid for Zn, Cd, Pd, and HgTPP [27].

On the other hand, the excited energy state diagrams of porphyrins with transition metals are rather complicated when the d orbitals of metal intervene between the π and π* orbitals of the porphyrin system. The d → d, π → d, and d → π* states are drawn over the normal porphyrin π → π* state. The relative positioning is complicatedly dependent on metal atoms, substituents, and axial ligands. Extended Huckel (EH) MO calculation has suggested that the lowest excited states should have a d → d nature for Ni[II] [32–34], Co[III] [36,37]; π → d for Co^{II} [36,37], Cu^{II} [41,42], Pt^{IV} [43]; and d $\rightarrow \pi^*$ for some Os^{II}, Ru^{II} porphyrins [38,39]. When the lowest excited triplet state is concerned with d orbital, the lifetime is generally very short as compared with that of the normal $\pi \rightarrow \pi^*$ state. Porphyrins with paramagnetic metal ions with odd d electrons have generally very short lifetimes for the excited state. The electronic structure and the magnetism largely affect the emission character. Metalloporphyrins with $Mo^V(d^1)$, $W^V(d^1)$, $Re^V(d^2)$, $Cr^{III}(d^3)$, $Mn^{III}(d^4)$, $Fe^{III}(d^5)$, $Fe^{II}(d^6)$, $Ni^{IV}(d^6)$, $Co^{III}(d^6)$, $Co^{II}(d^7)$, $Co^I(d^8)$, and $Ni^{II}(d^8)$ have very weak or almost no emission [22,44]. On the other hand, $Pd^{II}(d^8)$ and $Pt^{II}(d^8)$ porphyrins emit strong phosphorescence because they have large d-d energy gaps and their lowest excited triplet states are the normal porphyrin $\pi \rightarrow \pi^*$ [45]. The triplet state of $Rh^{III}(d^6)$, $Ru^{II}(d^6)$, and $Pd^{IV}(d^6)$ porphyrins are known to have an emissive nature. Axial ligands on metalloporphyrins, even with the same metal atom, sometimes play a substantial role. For example, the short lifetime of the lowest excited state of pyridine-coordinated $Ru^{II}TPP(Py)_2$ [15 $\pm$ 2 ns in pyridine (Py)] was reported to be considerably lengthened in carbonyl-coordinated $Ru^{II}TPP(CO)$ (47 $\pm$ 4 μs in CH_2Cl_2) [28]. The ligand CO stabilizes the d orbital of Ru through a strong π back bonding to induce an inversion from the d $\rightarrow \pi^*$ to $\pi \rightarrow \pi^*$ as the lowest excited state by increasing the d $\rightarrow \pi^*$ energy [21,29].

B. Redox Properties of Porphyrins

Redox properties estimated by the electrochemical method are also crucial for understanding the photochemistry of porphyrins. Redox potentials reported in the literature are listed in Table 3. Generally free-base porphyrins have two oxidation peaks and two reduction peaks in the usual cyclic voltammetry. They correspond to one- and two-electron oxidation and reduction of the porphyrin π system. In normal-type porphyrins with typical metal atoms the similar redox pattern is observed; the central metal cation simply acts as substituent for the porphyrin ring. The redox peak potentials were observed to have good correlations with electronegativity or inductive parameter of the central metal ion [46,47] (Fig. 6). Substitution on the porphyrin ring exhibits a good correlation between the redox peak potentials and Hamett's σ values [46]. Substitution on the phenyl group of AgTPP also showed the relation [48]. In many cases the difference between the first oxidation and reduction peaks has been observed to be well correlated with the Q-band energy, indicating that the central metal and substituents equally affect the HOMO and LUMO levels. The number of substituents was also reported to be correlated with the shift of redox peak potential in the cases of free-base porphyrins [52,53] Co [50,51], and Cu porphyrins [52,53]. Interesting relations were reported on [$Fe^{III}TPP$]Cl [49]. Introduction of a bromine atom in the pyrrole position induced a substantial shift of the reduction peak linearly with the number of bromine atoms, whereas the relation was different in the oxidation peak as shown in Fig. 7. The oxidation peak potential was explained to be largely affected by the ring distortion induced by the introduction of more than three bromine

Table 3 Redox Potential of Porphyrins

Porphyrin	Solvent/conditions	E	Ref.
Free			
H_2OEP	CH_2Cl_2, SCE, 0.1M TBAH	1.39, 0.83	185)
H_2OEP	BuCNorDMSO, SCE, TBAP	1.30, 0.81, −1.46, −1.86	186)
H_4OEP^{2+}	BuCNorDMSO, SCE, TBAP	1.65	186)
H_2TMPyP	H_2O, NHE	−0.10, −0.13	187)
H_2TPP	Benzonitrile, SCE, $TBAPF_6$	0.98, −1.23	182)
H_2TPP	DMF/SCE, TEAP	1.13, −1.04, −1.42	183)
H_2TPP	PhCN/SCE, NBu_4ClO_4	1.31, 1.06, −1.14, −1.53 all r	184)
H_2TSPP	H_2O, NHE, 0.1M Na_2SO_4	1.23, 1.10, −1.06, −1.40	188)
^{12}Mg			
MgOEP	BuCN or DMSO, SCE, TBAP	0.77r, 0.54r, −1.68r	186)
MgTPP	CH_2Cl_2, SCE, 0.1M TBAP	0.95, 0.66, −1.49	189)
^{13}Al			
AlOEPOH	BuCNorDMSO, SCE, TBAP	1.28r, 0.95r, −1.31r	186)
Al^{III}TMPyP	H_2O, NHE	−0.65, −0.72	187)
^{14}Si			
SiOEP(OH)$_2$	BuCN or DMSO, SCE, TBAP	1.19r, 0.92r, −1.35r	186)
^{15}p			
$PTTPCl_2$	CH_2Cl_2, SCE, 0.1M TBAP	−0.33, −0.71	190)
$PTTPCl_2^+$	CH_2Cl_2, SCE, 0.2M TBAP, 20°C	−0.33, −0.79	191)
^{20}Ca			
CaOEPOH	BuCNorDMSO, SCE, TBAP	0.86r, 0.50r, −1.68r	186)
^{21}Sc			
ScOEPOH	BuCNorDMSO, SCE, TBAP	1.03r, 0.70r, −1.54r	186)

Compound	Solvent, conditions	Potentials	Ref.
[22]Ti			
$Ti^{IV}OEPF_2$	CH_2Cl_2, SCE	−1.75r, −0.56m, 1.16r, 1.76r	(192)
$Ti^{IV}OEPO_2$	CH_2Cl_2, Ag/Ag₃₄NBu₄	−1.61, −1.14, −1.07, 1.21, 1.67	(193)
$Ti^{IV}TPPF_2$	CH_2Cl_2, SCE	−1.98r, −1.53r, −0.45m, 1.25r, 1.82r	(192)
$Ti^{IV}TPPO_2$	CH_2Cl_2, Ag/Ag₃₄NBu₄	−1.37, −0.93, −0.83, 1.27, 1.53	(193)
[23]V			
VOOEP	BuCNorDMSO, SCE, TBAP	1.25r, 0.96r, −1.25r, −1.72r	(186)
VOTPP	DMF, SCE, TBAP	1.32, 1.01, −0.98, −1.50	(194)
VOTSPP	DMF, SCE, TBAP	1.45, 1.02, −0.96, −1.48	(194)
[24]Cr			
CrOEPOH	BuCNorDMSO, SCE, TBAP	1.22r, 0.99r, 0.79m, −1.14m, −1.35r	(186)
Cr^{III}tetrakis(2,6-dimethyl-3-sulfonatophenyl)PX	H_2O, SCE, 0.2N $NaClO_4$	X = $(H_2O)2$, −1.04m, 0.61m, 1.01m; X = $(H_2O)OH$, −1.13m, 0.50m, 0.98m	(195)
$Cr^{III}TPP(H_2O)_2$	H_2O, SCE, $NaNO_3$ or $NaClO_4$	−0.96m, 0.71m, 1.07m	(196)
CrTPPCl	CH_2Cl_2, SCE, TBAPF₆	−0.88, −1.16, −1.79	(197)
$CrTPPClO_4$	CH_2Cl_2, SCE, TBAP	1.39, 1.03, −0.81, −1.19	(198)
[25]Mn			
MnOEPOH	BuCN or DMSO, SCE, TBAP	>1.4r, 1.12r, −0.42m, −1.61r	(186)
MnTPP	Pyridine/SCE	−0.24(III/II), −1.31r, −1.80r	(184)
MnTPP	PhCN/SCE, NBu₄ClO₄	1.64r, 1.28r, −0.21III/II, −1.47r, −1.79r	(184)
MnTPPCl	CH_2Cl_2, SCE, 0.1M TBAP	1.55, 1.16, −0.29, −1.68	(189)
[26]Fe			
FeOEPOH	BuCNorDMSO, SCE, TBAP	1.24r, 1.00r, −0.24m, −1.33r	(186)
FeTPP	Pyridine, SCE	0.16(III/II), −1.51(II/I), −1.71r	(184)
FeTPP	PhCN, SCE, NBu₄ClO₄	1.52r, 1.20r, −0.29III/II, −1.06II/I	(184)
FeTPPCl	CH_2Cl_2, SCE, 0.1M TBAP	1.49, 1.19, −0.29, −1.04	(189)
$Fe^{III}TPP(OH)_2$	H_2O, SCE, $NaNO_3$ or $NaClO_4$	1.06m, 1.34m	(196)

Table 3 Continued

Porphyrin	Solvent/conditions	E	Ref.
^{27}Co			
CoOEP	BuCNorDMSO, SCE, TBAP	1.00r, −1.05m	186)
Co^{II}TPP	CH$_2$Cl$_2$, SCE	0.71m, −0.93m	199)
	DMF, SCE	0.57, 1.20, −0.80, −1.90	200)
	Pyridine, SCE	−0.21(III/II), −1.03(II/I), −1.93r	184)
CoTPP	PhCN, SCE, NBu$_4$ClO$_4$	1.39, 1.20r, 0.52III/II, −0.85II/I, −1.97r	184, 201)
	CH$_2$Cl$_2$, SCE, 0.1M TBAP	1.15, 0.96, 0.80, −0.92, −1.43	189)
Co^{III}(TPP)(H$_2$O)$_2$BF$_4$	CH$_2$Cl$_2$, SCE	0.72m, −1.14m	199)
Co^{III}(TPP)(H$_2$O)Cl	CH$_2$Cl$_2$, SCE	0.27m, −1.12m	199)
^{28}Ni			
NiOEP	BuCNorDMSO, SCE, TBAP	0.73r, −1.5r	186)
NiTPP	CH$_2$Cl$_2$, SCE, 0.1M TBAP	1.17, 1.04, −1.26	189)
^{29}Cu			
CuOEP	CH$_2$Cl$_2$, SCE	1.25, 0.75, −1.49	203)
	BuCNorDMSO, SCE, TBAP	1.19, 0.79r, −1.46r	186)
Cu^{II}TPP	CH$_2$Cl$_2$, SCE	1.045, 0.670, −1.20	202)
CuTPP	CH$_2$Cl$_2$, SCE, 0.1M TBAP	1.36, 1.07, −1.49	189)
^{30}Zn			
ZnOEP	BuCNorDMSO, SCE, TBAP	1.02r, 0.63r, −1.61r	186)
ZnTMPyP^{4+}	H$_2$O,Na$_2$SO$_4$, NHE	1.18	204)
ZnTPP	CH$_2$Cl$_2$, SCE, 0.1M TBAP	1.14, 0.82, −1.33, −1.70	189)
ZnTSPP^{4-}	H$_2$O,Na$_2$SO$_4$, NHE	0.90	204)
ZnTSPP	H$_2$O, NHE, 0.1M Na$_2$SO$_4$	1.14, 0.87, −1.16	188)
Zn^{II}TMPyP	H$_2$O, NHE	−0.69, −0.73	187)

Compound	Conditions	Potentials	Ref.
[31]Ga			
GaOEPCl	CH_2Cl_2, SCE, 0.1M TBAPF$_6$	1.01, 1.45, −1.41, −1.85	205)
GaOEPOH	BuCNorDMSO, SCE, TBAP	1.32r, 1.01r, −1.34r, −1.80r	186)
Ga^{III}TMPyP	H_2O, NHE	−0.59, −0.72	187)
GaTPPCl	CH_2Cl_2, SCE, 0.1M TBAPF$_6$	1.19, 1.42, −1.16, −1.56	205)
[32]Ge			
GeOEP(OH)$_2$	SCE, PhCN, 0.1M TBAP	1.42, 1.01, −1.37, −1.84	206)
Ge^{IV}TMPyP	H2O, NHE	−0.56, −0.70	187)
GeTPP(OH)$_2$	SCE, PhCN, 0.1M TBAP	1.46, 1.15, −1.09, −1.56	206)
[39]Y			
Y(OEP)$_2$	DMF, SCE, 0.2M NBu$_4$PF$_6$	0.14, −0.21	207)
[40]Zr			
Zr(OEP)$_2$	CH_2Cl_2, SCE, NBu$_4$PF$_6$	0.576, −0.026, −1.663	209)
Zr(TTP)$_2$	CH_2Cl_2, Ag/AgCl, Bu$_4$NPF$_6$	0.940, 0.515, −1.322, −1.674	208)
[41]Nb			
Nb^{V}OEP(O)(O$_2$CCH$_3$)	CH_2Cl_2, SCE, 22°C, TBAP	−1.18m, 0.05m, −1.24r, −1.71r, 0.98r	210)
[Nb^V(OEP)]$_2$O$_3$	CH_2Cl_2, SCE, 22°C, TBAP	−1.40, −1.86, 0.63, 1.09	211)
Nb^{IV}TPPO	CH_2Cl_2, SCE, TBAP	0.18m, 0.66m, −0.62m, −1.01r, −1.39r	210)
Nb^{V}TPP(O)(O$_2$CCH$_3$)	CH_2Cl_2, SCE, 22°C, TBAP	−0.94m, 0.17m, −1.00r, −1.38r, 1.21r	210)
[Nb^V(TTP)]$_2$O$_3$	CH_2Cl_2, SCE, 22°C, TBAP	−1.28, −1.53, −1.75, 0.89, 1.20	211)
[42]Mo			
Mo^{VI}OOEPOH	BuCN or DMSO, SCE, TBAP	1.43r, −0.21m, −1.30r, −1.72r	186)
Mo^{IV}TPPO	CH_2Cl_2, SCE, TBAP	−1.13, −1.48, 0.02, 1.49	212)
Mo^{V}TPPOClO$_4$	CH_2Cl_2, SCE, TBAP	0.02, −1.13, −1.48, 1.48	212)
Mo^{V}TPPO$_2$	CH_2Cl_2, SCE, TBAP	−1.13, −1.48, −0.92, 1.49, 1.22	212)
[44]Ru			
RuTFPPCl$_8$(CO)	CH_2Cl_2, AgCl/Ag, 0.1M TBAPF$_6$	1.71, −0.64	216)
RuTPP(CO)	CH_2Cl_2, SCE, 0.1M TBAP	0.87, −1.59	213)
RuTPP(CO)	DMSO, 22.8°C ± 0.2, SSCE	1.20m, 0.79r, −1.32	214)
RuTPP(CO)	CH_2Cl_2, SCE, 0.1M TBAP	1.43, 0.87, −1.59, −2.08	215)

Table 3 Continued

Porphyrin	Solvent/conditions	E	Ref.
^{45}Rh			
RhTPPCl(Me$_2$N)	PhCN, SCE, TBAP0.1M, 24°C	1.43, 1.00, −0.24, −1.28, −1.96	217
Rh^{III}TPPCl	CH$_2$Cl$_2$, SCE, TBAP0.1M	1.36, 0.98	218
^{46}Pd			
PdOEP	CH$_2$Cl$_2$, SCE	1.54, 0.82, −1.52	203
PdTMPyP^{4+}	H$_2$O, NHE, Na$_2$SO$_4$	1.41	204
PdTPP	CH$_2$Cl$_2$, SCE, 0.1M TBAP	1.62, 1.20, −1.24	189
PdTSPP^{4-}	H$_2$O, NHE, Na$_2$SO$_4$	1.16	204
^{47}Ag			
Ag^{II}OEP	DMF, SCE, 0.1M THAP	−0.95	220
AgOEP	BuCNorDMSO, SCE, TBAP	0.44m, −1.29r	186
Ag^{II}TPP	CH$_2$Cl$_2$, SCE, TBAP	1.62, 0.59m, −1.01	219
	DMF, SCE, 0.1M THAP	−0.88, −1.65, −2.01	220
^{48}Cd			
CdOEP	BuCNorDMSO, SCE, TBAP	1.04r, 0.55r, −1.52r	186
CdTPP	CH$_2$Cl$_2$, SCE, 0.1M TBAP	0.64	221
	SCE, (see ref.52),53)	0.63, 0.93	222, 223
^{49}In			
InOEPOH	BuCNorDMSO, SCE, TBAP	1.36r, 1.08r, −1.19r, −1.59r	186
In^{III}TMPyP	H$_2$O, NHE	−0.51, −0.65	187
InTPPCl	CH$_2$Cl$_2$, SCE, 0.1M TBAClO$_4$	1.45, 1.16, −1.09, −1.48	224, 225
^{50}Sn			
SnOEP(OH)$_2$	BuCNorDMSO, SCE, TBAP	>1.4r, −0.90r, −1.30r	186
Sn^{II}OEP	THF, SCE, TBAClO$_4$	−1.28, −1.73	227
SnTMPyP^{4+}	H$_2$O, NHE, Na$_2$SO$_4$	>1.50V	204
Sn^{IV}TMPyP	H$_2$O, NHE	−0.28, −0.45	187
SnTSPP^{4-}	H$_2$O, NHE, Na$_2$SO$_4$	1.14V	204
SnTTP(OH)$_2$	THF, SCE, TBAClO$_4$	−0.86, −1.32	226
Sn^{II}TTP	THF, SCE, TBAClO$_4$	−1.08, −1.50	227

^{51}Sb			
Sb^{III}OEPOH	BuCNorDMSO, SCE, TBAP	>1.4r, 0.75r, −1.07r	186)
Sb^{V}OEP(OH)$_2$		>1.6, −0.55	229)
SbTPP(CH$_3$)$_2$	PhCN, SCE, 0.1M TBAP	1.51, −0.50, −0.96	228)
Sb^{V}TPyP	H$_2$O, NHE	−0.20, −0.40	187)
^{56}Ba			
BaTPP	SCE, see ref.52),53)	0.46, 0.75	222), 223)
^{57}La			
La(OEP)$_2$	DMF, SCE, 0.2M NBu$_4$PF$_6$	−0.01, −1.935	207)
^{58}Ce			
Ce(OEP)$_2$	DMF, SCE, 0.2M NBu$_4$PF$_6$	0.33, −0.47, −1.935	207)
	CH$_2$Cl$_2$, SCE, NBu$_4$PF$_6$	0.745, 0.17, −0.58	209)
Ce(TPP)$_2$	CH$_2$Cl$_2$, SCE, NBu$_4$PF$_6$	1.075, 0.675, −0.27	209)
^{59}Pr			
Pr(OEP)$_2$	DMF, SCE, 0.2M NBu$_4$PF$_6$	0.24, −0.08, −1.96	207)
^{60}Nd			
Nd(OEP)$_2$	DMF, SCE, 0.2M NBu$_4$PF$_6$	0.225, −0.095, −1.96	207)
^{63}Eu			
Eu(OEP)$_2$	DMF, SCE, 0.2M NBu$_4$PF$_6$	0.185, −0.15, −1.97	207)
^{64}Gd			
Gd(OEP)$_2$	DMF, SCE, 0.2M NBu$_4$PF$_6$	0.18, −0.165, −1.97	207)
^{65}Tb			
Tb(OEP)$_2$	DMF, SCE, 0.2M NBu$_4$PF$_6$	0.17, −0.18, −1.98	207)
^{66}Dy			
Dy(OEP)$_2$	DMF, SCE, 0.2M NBu$_4$PF$_6$	0.155, −0.20, −1.985	207)
^{67}Ho			
Ho(OEP)$_2$	DMF, SCE, 0.2M NBu$_4$PF$_6$	0.15, −0.215, −1.985	207)
^{68}Er			
Er(OEP)$_2$	DMF, SCE, 0.2M NBu$_4$PF$_6$	0.135, −0.225, −1.995	207)

Table 3 Continued

Porphyrin	Solvent/conditions	E	Ref.
[69]Tm			
Tm(OEP)$_2$	DMF, SCE, 0.2M NBu$_4$PF$_6$	0.12, −0.235, −1.995	207)
[70]Yb			
Yb(OEP)$_2$	DMF, SCE, 0.2M NBu$_4$PF$_6$	0.115, −0.25, −1.995	207)
[71]Lu			
Lu^{III}OEPOH	CH$_2$Cl$_2$, Ag/AgCl, 0.1M TBAClO$_4$	0.69, −1.68	230)
Lu(OEP)$_2$	DMF, SCE, 0.2M NBu$_4$PF$_6$	0.10, −0.265, −2.00	207)
Lu^{III}TPPOH	CH$_2$Cl$_2$, Ag/AgCl, 0.1M TBAClO$_4$	0.86, −1.26	230)
[72]Hf			
Hf(OEP)$_2$	CH$_2$Cl$_2$, SCE, NBu$_4$PF$_6$	0.573, −0.035, −1.661	209)
Hf(TPP)$_2$	CH$_2$Cl$_2$, SCE, NBu$_4$PF$_6$	0.986, 0.542, −1.286, −1.647	209)
[76]Os			
OsOEPCl$_2$	CH$_2$Cl$_2$, ferrocene/nium, NBu$_4$BF$_4$	0.7, −0.54, −1.69	231)
OsTPPCl$_2$	CH$_2$Cl$_2$, ferrocene/nium, NBu$_4$BF$_4$	0.77, −0.33, −1.42	231)
Os^{Ir}TPPCO	SCE, CH$_2$Cl$_2$	−1.17r, −0.72r, 0.51m, 1.31m	232)
OsTPP(OEt)$_2$	CH$_2$Cl$_2$, ferrocene/nium, NBu$_4$BF$_4$	0.29, −1.15	231)
[77]Ir			
Ir^{III}OEP(CO)Cl	CH$_2$Cl$_2$, SCE, TBAClO$_4$	1.45, 1.05, −1.48	233)
[78]Pt			
PtTPP	CH$_2$Cl$_2$, SCE, 0.1M TBAP	1.65, 1.30, −1.21, −1.79	189)

^{79}Au			
AuTMPyP	H$_2$O, NHE	1.70, −0.20, −0.50	235)
Au^{III}TPP(AuCl$_4$)	CH$_2$Cl$_2$, SCE	1.68, −0.59	234)
Au^{III}TPP	CH$_2$Cl$_2$, NHE)	2.02, −0.38, −1.00	235)
AuTSPP	H$_2$O, NHE	1.50, −0.36	235)
^{80}Hg			
HgTPP	CH$_2$Cl$_2$, SCE, 0.1M TBAP	0.71	221)
^{81}Tl			
Tl^{III}OEP$^+$	CH$_2$Cl$_2$, SCE, THAP	−0.47, −1.30	236)
TlOEPOH	BuCNorDMSO, SCE, TBAP	1.31r, 1.00r, −1.24r	186)
Tl^{III}TPP$^+$	DMF, SCE, 0.1M THAP	−0.35, −1.00	220)
Tl^{III}TPP$^+$	CH$_2$Cl$_2$, SCE, THAP	−0.35, −1.00	236)
^{82}Pb			
PbOEP	CH$_2$Cl$_2$, SCE, 0.1M TBAH	1.03, 0.68	185)
	BuCNorDMSO, SCE, TBAP	0.91r, 0.65r, −1.30r	186)
Pb^{II}TMPyP	H$_2$O, NHE	−0.58, −0.72	187)
PbTPP	SCE, see ref.52),53)	0.63, 0.96	222), 223)
	SCE, see ref.52),54)	−1.10, −1.52	222), 237)
^{90}Th			
Th(OEP)$_2$	CH$_2$Cl$_2$, SCE, TBAClO$_4$	1.91, 0.85, 0.32, −1.53	238)
Th(TPP)$_2$	CH$_2$Cl$_2$, SCE, NBu$_4$PF$_6$	0.990, 0.580	209)
	CH$_2$Cl$_2$, SCE, TBAClO$_4$	1.79, 1.13, 0.79, −1.27, −1.55	238)
^{92}U			
U(OEP)$_2$	CH$_2$Cl$_2$, SCE, TBAClO$_4$	1.50, 0.79, 0.23, −1.55	238)
U(TPP)$_2$	CH$_2$Cl$_2$, SCE, NBu$_4$PF$_6$	0.990, 0.580	209)
U(TPP)$_2$	CH$_2$Cl$_2$, SCE, TBAClO$_4$	1.78, 1.10, 0.71, −1.27, −1.57	238)

TBA, (tetrabutylammonium salt; m, redox of metal; r, redox of porphyrin ring; octaethylporphyrin, OEP; tetraphenylporphyrin, TPP; tetra(4-sulfonatophenyl) porphyrin, TSPP; tetra(4-pyridyl)porphyrin, TPyP; tetra(N-methyl-4-pyridyl)porphyrin, TMPyP; Meso- IX-DME, Meso.

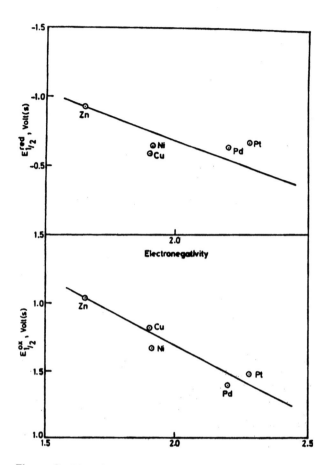

Figure 6 Plot of the first ring oxidation and reduction potentials of the various (metal-looctabromoporphyrins) [(M(OBO)s)] versus Pauling's electronegativity of the central metal ions [46].

atoms [49]. Theoretical studies also suggested that the oxidation potential should largely depend on a porphyrin ring distortion, while the reduction potential did not [54,55].

In metalloporphyrins with transition metal ions, the redox peaks of central metal ions intervene between the redox peaks of porphyrin ring in many cases as shown in Table 3. Delicate positioning among them are often observed [726]. Comprehensive interpretation of redox potentials of porphyrins has been attempted with the use of ligand parameter E_L [60]. Axial ligands also affect the redox properties of metalloporphyrins. Detailed studies were reported on Fe, Mn, and Cr porphyrins [56,57] (Fig. 8).

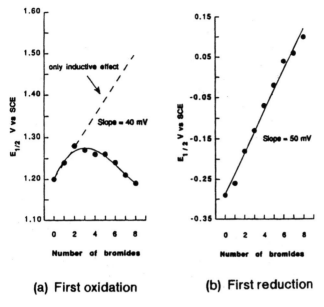

Figure 7   Dependence of $E_{1/2}$ on the number of Br groups of $(Br_xTPP)FeCl$ for (a) the first oxidation and (b) the first reduction in benzonitrile containing 0.1 M TBAP [49].

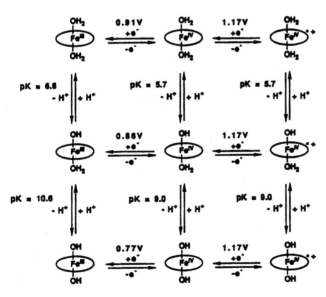

Figure 8   Equilibrium of the axial hydroxyl group of Fe porphyrin and each redox potential [56,57].

In some cases, redox properties were observed to be affected by solvent [58]. Local solvation to the basket-handle porphyrins was reported to shift the redox peaks [59]. Donor number of solvent was observed to be well correlated with the first oxidation peak of Cd-, Hg-, and ZnTPP [61]. Even the anion of supporting electrolyte can affect the redox peak[62].

III. MOLECULAR PHOTOCHEMISTRY OF PORPHYRINS

A. Excited Energy Transfer [281–284]

Excited energy transfer among chromophores is one of the most fundamental photophysical processes. According to the mechanism the excited energy transfer is classified into 1.) radiative trivial type, 2.) nonradiative Forster type [286], and 3.) nonradiative Dexter type [285].

In the radiative trivial mechanism, an energy-accepting molecule absorbs the photon emitted from an energy-donating molecule in solution. The energy transfer efficiency is directly dependent on the emission efficiency of donor molecule and the degree of overlap between the emission spectrum of the donor molecule and the absorption spectrum of the acceptor molecule. Since the fluorescence quantum yield of porphyrins is generally small and the emission spectrum is in the long-wavelength region of 600–700 nm, this radiative trivial mechanism is almost negligible for porphyrins.

For nonradiative energy transfer, a simple quantum chemical treatment on the process in Eq. (4) affords the transition probability expressed by Eq. (5) [727]:

$$D^*(1 \text{ or } 2) + A(1 \text{ or } 2) \rightarrow D(1 \text{ or } 2) + A^*(1 \text{ or } 2) \tag{4}$$

$$\begin{aligned} k \propto [&\int \psi_{D^*}(1)\psi_A(2)H'\psi_D(1)\psi_{A^*}(2)d\tau \\ &- \int \psi_{D^*}(1)\psi_A(2)H'\psi_D(2)\psi_{A^*}(1)d\tau]^2 \end{aligned} \tag{5}$$

where $\psi(D^*(i \text{ or } j))$, $\psi(D(i \text{ or } j))$, $\psi(A^*(i \text{ or } j))$, $\psi(A(i \text{ or } j))$ denote the wave function of excited donor (D^*), ground state donor (D), acceptor (A^*), and acceptor (A) having electron i or j, respectively. The equation affords a clear understanding of the nonradiative energy transfer. The energy transfer through the first term in parentheses, which is expressed in coulombs without an electron exchange, is called a Forster or a coulombic mechanism. The perturbation of Hamiltonian H' was postulated to be a dipole–dipole interaction by Forster [286,451]. As easily understood from the spin rule in the integral, a singlet-singlet energy transfer [Eq. (6)] is allowed, whereas a triplet-triplet transfer [Eq. (7)] is forbidden by the Forster mechanism.

$$^1D^* + {}^1A \rightarrow {}^1D + {}^1A^* \tag{6}$$

$$^3D^* + {}^1A \rightarrow {}^1D + {}^3A^* \tag{7}$$

The rate constant of energy transfer by the Forster mechanism is experimentally expressed as Eq. (8):

$$k_{ET} = \frac{9000c^4(\ln 10)}{128\pi^5 N_A n^4 \tau_D^0 r^6}\kappa^2 \int f_D(\nu)\varepsilon_A(\nu)\frac{d\nu}{\nu^4}$$ (8)

where c, n, τ_D, r denote velocity of photon, refractive index of solvent, fluorescence lifetime of the energy donor, distance between the donor and acceptor. The κ, $f_D(\nu)$, $\varepsilon(\nu)$, ν are the molecular orientation parameter in dipole–dipole interaction, relative fluorescence intensity distribution function of the energy donor, extinction coefficient of absorption by the acceptor, and frequency, respectively. Since the Forster mechanism does not require a collisional interaction, the excited energy can be transferred even in a distance of 100 Å.

The orientation parameter κ in Eq. (8) is expressed as Eq. (9), where a geometrical relation between the two dipoles of chromophore is indicated in Fig. 9. The validity was beautifully confirmed by a series of molecules prepared by Osuka et al. [287,288].

$$\kappa^2 = (\cos\alpha + 3\cos\theta_1\cos\theta_2)^2$$ (9)

The energy transfer through the second term in Eq. (5) postulated by Dexter [285] involves electron exchange between the energy donor and acceptor molecules and

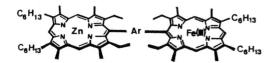

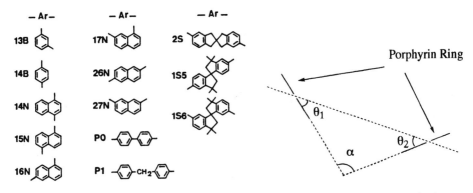

Figure 9 Osuka's molecules and geometrical orientation between the two porphyrin planes [287,288].

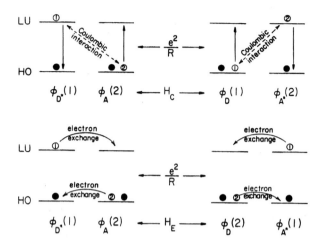

Figure 10 Orbital comparison of the coulombic (Forster) and exchange (Dexter) mechanism of electronic energy transfer [292].

is called the Dexter type of mechanism or exchange mechanism. In Dexter-type energy transfer, the triplet-triplet energy transfer is spin-allowed [Eq. (7)], while the singlet-singlet one [Eq. (6)] is forbidden contrary to Forster mechanism. Since Dexter-type mechanism involves electron exchange, a collision between the donor and acceptors and overlapping of molecular orbitals are required. An experimental expression of the rate constant of exchange mechanism is shown as Eq. (10):

$$k_{eT} \propto e^{-2r/L} \int_0^\infty f_D(\nu)\varepsilon_A(\nu)d\nu \tag{10}$$

where r and L denote a distance between the energy donor and acceptor, and a constant, respectively. An orbital explanation of the two energy transfer mechanisms is given in Fig. 10. In many cases of chromophoric assemblies as described in Section IV.B, both the coulombic and exchange mechanisms are supposed to operate [284,293].

B. Electron Transfer

Among the fundamental photophysical and photochemical processes, electron transfer has a high priority. Porphyrins and metalloporphyrins form one of the most promising series of chromophoric compounds for designing suitable redox systems as described above, since they have intense visible absorption bands and the λ_{max} can be varied by the substituents and the central metal ions. The redox properties are also easily regulated according to purpose. In recent intensive studies on electron transfer by various approaches from the viewpoints of theory, experimental verification, photosynthesis, artificial photosynthesis, and redox catalytic reactions, porphyrin chromophores have played very crucial roles. They can be classified into the following types of studies: 1.) an intermolecular electron transfer, 2.) a covalently linked molecular system, and 3.) a supramolecular system linked by a noncovalent bond such as hydrogen bond and coordination interaction [239]. Since an intermolecular electron transfer has already been reviewed comprehensively [243] characteristics of an intermolecular electron transfer are briefly summarized in this section. The current topics on 2) and 3) will be reviewed in Sections IV.C and IV.D. An intermolecular electron transfer is further classified into (1, 2) a photoinduced charge separation, (3) a charge recombination, and (4, 5) a charge shift reaction, respectively.

$$(1) \quad A^* \ldots B \rightarrow A^{+\cdot} \ldots B^{-\cdot}$$
$$(2) \quad A^* \ldots B \rightarrow A^{-\cdot} \ldots B^{+\cdot}$$
$$(3) \quad A^{+\cdot} \ldots B^{-\cdot} \rightarrow A \ldots B$$
$$(4) \quad A^{+\cdot} \ldots B \rightarrow A \ldots B^{+\cdot}$$
$$(5) \quad A^{-\cdot} \ldots B \rightarrow A \ldots B^{-\cdot}$$

In the general reaction scheme for the net electron transfer (Fig. 11), the rate constant of intrinsic electron transfer process k_{eT} can be classically expressed by Marcus's equation [Eq. (11)] according to a simplified potential energy curve in Fig. 12a [240–242,244,728–730].

$$k_{eT} = ve^{-\Delta G^{\neq}/RT} = v \exp\left[-\frac{(\Delta G + \lambda)^2}{4\lambda RT}\right]$$

$$\lambda = \lambda_{in} + \lambda_{out}$$

$$\lambda_{out} = \frac{(\Delta e)^2}{2}\left[\frac{1}{n^2} - \frac{1}{\varepsilon}\right]\left[\frac{1}{r_D} + \frac{1}{r_A} - \frac{2}{r_{DA}}\right]$$

$$\Delta G = 23.06[E(D^+/D) - E(A/A^-)] - \Delta G_{00} - wp$$

$$wp = (Z_D + Z_A) e^2/r_{DA}\varepsilon_s$$

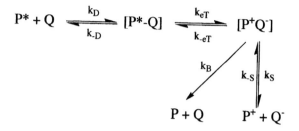

Figure 11 Reaction scheme of electron transfer.

where ν is nuclear frequency; $\Delta G^{\neq}$ and ΔG are activation free energy and net free energy difference for electron transfer; λ, λ_{in}, and λ_{out}: net, inner-sphere, and outer-sphere nuclear reorganization energy; ε: dielectric constant of solvent; r_{DA}, r_{D}, and r_{A}: intermolecular distance, molecular radius of donor and acceptor; $E(D^{+}/D)$, $E(A/A^{-})$, and ΔG_{00}: oxidation potential of donor, reduction potential of acceptor, and 0–0 transition energy of the excited state; $Z_{D^{+}}$ and $Z_{A^{-}}$, formal charge of D^{+} and A^{-}, respectively. The Marcus theory predicted an "inverted region" where a largely exoergonic free energy difference for electron transfer should induce a substantial decrease of rate constant, while the so-called bell-shaped curve for electron transfer (Fig. 12b) had not long been found in solution. Experimental observation generally shows a saturation of rate constant in exoergonic region partly due to a bimolecular process being diffusion limited in solution

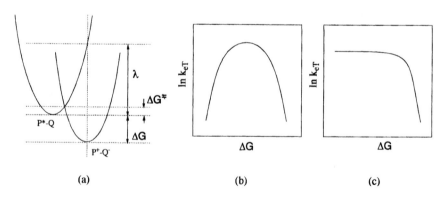

Figure 12 (a) Potential energy curve for electron transfer from P* to Q. (b) Bell-shaped curve of rate constant of electron transfer vs. free energy change (ΔG) predicted by Marcus theory. (c) Saturation curve of rate constant of bimolecular electron transfer observed in solution.

[731] (Fig. 12c). Empirical equation [Eq. (12)] was postulated by Weller et al. [654,732] to fit the data in solution.

$$\Delta G^{\neq} = \frac{\Delta G}{2} + \left[\left(\frac{\Delta G}{2}\right) + \Delta G^{\neq}(0)^2\right]^{1/2} \tag{12}$$

where $\Delta G^{\neq}(0)$ denotes a constant to fit the observed data.

In 1983 Closs and Miller found the bell-shaped curve for charge shift reaction in a linked molecular system to verify the validity of Marcus theory [479,480,488,490,530,540,733–735]. A porphyrin-linked molecular system also exhibited the bell-shaped curve as described in Section IV.C.

Quantum mechanical consideration with Fermi's golden rule for electron transfer led to Eq. (13):

$$k_{eT} = \frac{2\pi}{h}|H_{el}|^2(FC) \tag{13}$$

where H_{el} and FC denote an electronic coupling matrix element and the Franck-Condon factor, respectively.

At ambient temperature, Marcus's semiclassical expression [245] in Eq. (14) is used:

$$k_{eT} = \left(\frac{2\pi}{h}\frac{H_{el}^2}{(4\pi\lambda RT)^{1/2}}\right)\exp\left[-\frac{\Delta G^{\neq}}{RT}\right] \tag{14}$$

The electronic coupling depends on intermolecular distance r and is expressed as Eq. (15):

$$H_{el} = H_{el}^0 \exp[-\beta(r - \sigma)/2] \tag{15}$$

where H_{el}^0 and β denote electronic coupling at the distance $\sigma(= r_D + r_A)$ and a constant, respectively. The electronic coupling term suffers an exponential decay with an increase of distance r. The outer-sphere reorganization energy λ_{out}, mainly ascribable to solvent motion [248–252], monotonically increases with the reaction coordinate. In a "normal" endoergic region, an increase of λ induces an increase of the activation free energy $\Delta G^{\neq}$ to reduce the rate constant k_{eT}, while the relation becomes opposite in the inverted region [246,247].

In solution even a long-range electron transfer sometimes proceeds as described in Section IV.D; while such a phenomenon is not observed in vacuum where the β value is very large. The environmental media surrounding the corresponding molecules should have crucial roles in an electron transfer process. Those effects have been interpreted in terms of super exchange mechanism of electron transfer. The third molecule such as solvent and protein molecule surrounding the two chromophores or covalent bonds intervening between them could conceivably mediate the long-range electron transfer through virtual couplings of their HOMO or LUMO with the two chromophores. The superexchange mechanism has been intensively discussed on biological systems and linked molecular systems described in Section IV.D. From the viewpoint of electron transfer, porphyrins and metalloporphyrins have served as one of a group of typical chromophores of which redox properties could easily be regulated.

C. Photoinduced Catalytic Redox Reaction

Photoinduced electron transfer has attracted much attention especially in relation to a charge separation which leads to a net redox reaction and a catalytic reaction. One of the crucial factors that govern the actual charge separation is the back electron transfer process k_B in the electron transfer scheme in Fig. 13. Very efficient back electron transfers competing with the charge separation k_S have been observed in many cases to reduce the charge separation efficiency. Net redox reactions are not observed in such cases, while excited state quenchings are only observed. Interesting effects of molecular size on the back electron transfer were observed [264–266]. The lifetime of ion pair ($P^{+\cdot}$, $Q^{-\cdot}$) was the longer for the electron acceptor with the larger π system in the electron transfer from the excited ZnTSPP to the viologen acceptors in Fig. 14. The electron delocalization in the larger π-electron system was considered to reduce the actual overlapping between the molecular orbitals and electronic coupling for the back electron transfer [264]. Electrostatic repulsion between the ion pair was revealed to have a remarkable accelerating effect on the charge separation [736–738]. Another suggestive example of maintaining the net charge separation was reported on a microheterogeneous system [270–280]. Free ion obtained by charge separation was effectively stored on the viologen linked bilayer membrane [280] (Fig. 15).

The charge separation has been revealed to be affected by many factors of 1.) the free energy difference for back electron transfer, 2.) the molecular environment such as solvent viscosity, solvent polarity, and ionic strength, 3.) the electrostatic nature of ion pair, i.e., the formal charge of respective donor

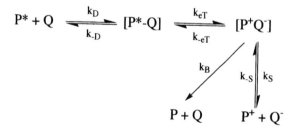

Figure 13   Reaction scheme of electron transfer.

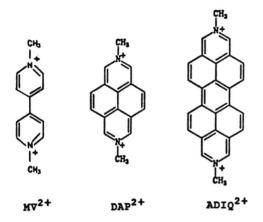

Figure 14 Viologen derivatives with differently sized π systems [264].

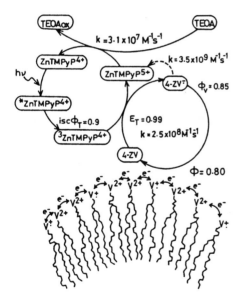

Figure 15 Schematic diagram of an efficient harvesting of photoliberated electrons from ZnTMPyP⁴⁺ as a sensitizer by the use of PSV as an electron mediator, TEOA as a sacrificial electron donor, and a bilayer membrane of LEV²⁺ as an electron pool. ZnTMPyP⁴⁺ tetrakis(*N*-methylpyridinium)porphyrinatozinc; TEOA, triethanolamine; PSV, sulfonato-propyl)viologen; LEV²⁺, *N*-ethyl-*N*′-[5-(*N,N*-dihexadecylcarbamoyl)pentyl]-4,4′-bipyri-dinium dibromide [280].

and acceptor, 4.) the molecular size [269], 5.) the spin multiplicity of ion pair [268], 6.) the addition of salt [267,739,740,784], 7.) the microheterogeneous interface [270–280]; and 8.) others.

Among various photoinduced catalytic redox reactions, tremendous efforts have been made in relation to artificial photosynthesis [373,736–738,741–744]. The target reactions in artificial photosynthesis should not necessarily be restricted to those of natural photosynthesis. One of the most crucial points should be how the uphill reaction can be induced with the use of water molecule as an ideal electron donor. Special attention has been paid to a splitting of water upon visible light irradiation [373,736–738,741–744]:

$$H_2O(l) \rightarrow H_2(g) + O_2(g) \tag{16}$$

The water splitting is a highly endoergic process with $\Delta G = 237.2 \text{ kJ mol}^{-1}$ and has not yet been established completely upon visible light irradiation. Heterogeneous photocatalytic systems such as semiconductors [373] and homogeneous systems with appropriate sensitizers such as porphyrins and cocatalyst such as colloidal platinum [374–380] are the typical approaches for the water splitting. Tremendous efforts have been surveyed elsewhere [373,741–743] and recent topics are reviewed here. For hydrogen evolution in the reduction terminal end of photoredox cycles, sacrificial electron donors such as amines have often been used to complete the redox cycles as indicated in Fig. 16. With such sacrificial electron donors, the hydrogen evolution reactions are of course interesting as model reactions; however, the net chemistry of reaction still has an exoergic downhill character in most cases. Usefully recyclable electron donors such as NADPH have been used in place of sacrificial donors [381–384]. In the catalytic redox cycles involving a bimolecular process with excited porphyrins, the excited state responsible for the reaction is generally the excited triplet state, which has a long lifetime. In a linked molecular system with NADPH as an electron donor, however, an electron transfer via the excited singlet state of Zn porphyrins and hydrogen evolution was reported [388–390] (Fig. 17).

Persaud et al. tried to construct the redox system in Fig. 18, whereby the electron transfer also proceeded through the excited singlet state of Zn porphyrin adsorbed outside of zeolite L to methylviologen in the zeolite [385]. Hydrogen

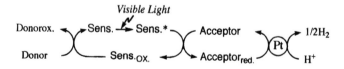

Figure 16 Photoredox cycles coupled with hydrogen evolution in the reduction terminal end.

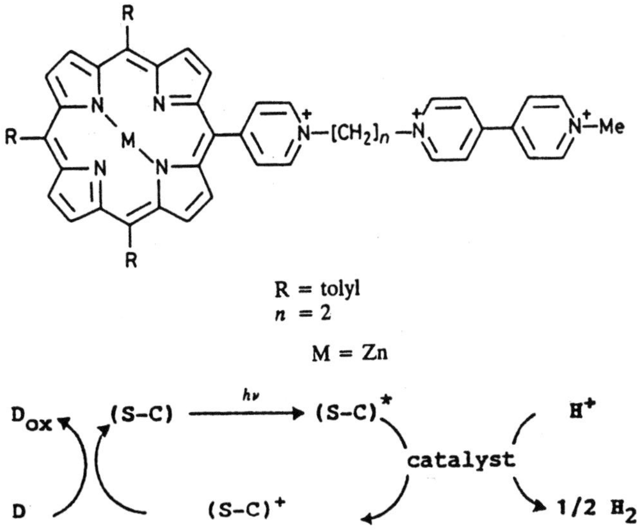

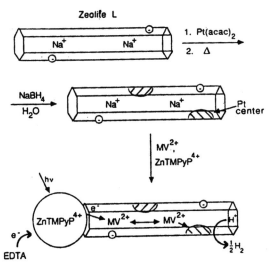

R = tolyl
n = 2

M = Zn

Figure 17 Hydrogen evolution in a linked molecular system of Zn porphyrins with NADPH as an electron donor [388–390].

Figure 18 Electron transfer from the excited singlet state of Zn porphyrin adsorbed outside of zeolite L to methylviologen in the zeolite [385].

evolution mediated by platinum was observed, while the efficiency was modest. Interesting photoinduced redox reactions in zeolite have also been reviewed [386].

Photoinduced electron transport from EDTA to viologen through the excited Zn porphyrin and TiO_2 particle entrapped within lithium aluminum–layered double hydroxide was attempted [387] (Fig. 19). Since the fluorescence of Zn porphyrin was quenched by TiO_2, the excited singlet state of porphyrin was supposed to transfer an electron to TiO_2 particle.

Another example of a photoredox system in an organized molecular assembly was reported on hydrogen evolution in LB membrane where porphyrin-viologen-linked molecules were assembled on a glass plate with platinum particles [391] (Fig. 20). Photoinduced hydrogen evolution upon visible light irradiation of Zn porphyrin mediated by natural cytochrome c_3 and H_2ase have also been reported [392].

Another target reaction in the reduction terminal end of photoredox cycles has been the reduction of carbon dioxide. Pioneering works have been reported [393–403]. Interesting examples of photochemical fixation of carbon dioxide have also been reported for Al porphyrins [404–406]. The insertion reaction of CO_2 into Al-X bond (X = R, OR, SR, NR_2) of Al porphyrin (**3**) was observed

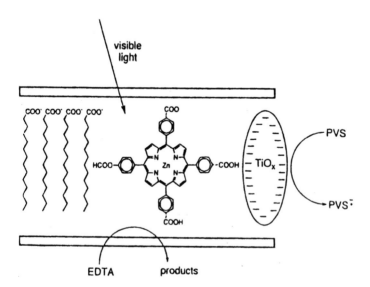

Figure 19   Photoinduced electron transport from EDTA to viologen through the excited Zn porphyrin and TiO_2 particle entrapped within lithium aluminum layered double hydroxide [387].

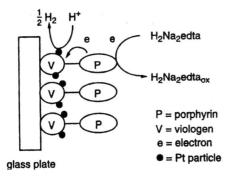

Figure 20   Hydrogen evolution in LB membrane where porphyrin-viologen linked molecules were assembled on a glass plate with platinum particles deposited [391].

to be remarkably accelerated. The reaction could be widely applicable and meaningful from the viewpoint of synthetic chemistry.

TPPAlEt

3

Recently, the reduction of carbon dioxide by Fe porphyrin upon light irradiation has been reported to form carbon monoxide [788].

Oxygen evolution in the oxidation terminal end of photoredox cycle has also attracted much attention [373,741–752]. In natural photosynthesis, a water molecule is the electron donor in a PSII system where the water-oxidizing complex, i.e., Mn complex, forms oxygen [407]. The details of water oxidation have not yet been fully understood. Among the challenging attempts aiming at artificial photosynthesis involving oxygen evolution upon light irradiation, Harriman et al. reported a highly efficient oxygen evolution (quantum yield = 0.72) in the system with Zn porphyrin as a sensitizer, $IrO(OH)_2$ as an oxygen evolving catalyst, and $S_2O_8^{2-}$ as a sacrificial electron donor [409]. Other catalysts such as PtO_2 and RuO_2 were also reported to be effective for oxygen evolution [408,410–412]. The respective hydrogen evolution with appropriate sacrificial electron donor or oxygen evolution with electron acceptors is called "a half reaction." Simultaneous evolution of oxygen and hydrogen without sacrificial electron donor and acceptor in homogeneous photocatalytic system has not yet been established except for the cases of semiconductor photocatalysts which can only utilize ultraviolet light. Since the oxygen evolution from a water molecule requires a four-electron oxidation, a simple redox system involving a one-electron transfer process cannot solve the problem and a more sophisticated multielectrons conversion device is required. After establishing the half reaction of hydrogen evolution and a "rush of oxygen evolution," attention has been shifting to how a water molecule as an ideal electron donor could be incorporated in the redox cycles and several interesting studies have appeared [413,415–418,791]. As indicated in Table 4, possible oxidation processes of water or hydroxide ion involve 1.) one-electron oxidation, 2.) two-electron oxidation, and 3.) four-electron oxidation, respectively [407]. Though the one-electron abstraction from water or hydroxide ion requires a rather high oxidation potential around 2.0 V vs. NHE, visible light irradiation of high valent metalloporphyrin such as SbTPP with the excited state reduction potential of 1.38V vs. NHE was observed to form the radical anion of SbTPP in the presence of OH^- [413] (Fig. 21). Hydrogen evolution coupled with the electron transfer was observed. The quantum yield of electron transfer from OH^- to the excited SbTPP was pretty small, around 0.01, in aqueous acetonitrile with a high content of water (15 vol%), while the reactivity drastically increased up to $\phi = 0.3$ at low water content (3%). The predicted decrease of oxidation potential of OH^- [414] in acetonitrile with low water content was supposed to render the electron transfer possible.

Table 4 Standard Redox Potentials in Aqueous Solutions

Reaction	E
Four-Electron Oxidation	
$2H_2O \rightarrow O_2 + 4H^+ + 4e^-$	1.229
$OH^- + H_2O \rightarrow O_2 + 3H^+ + 4e^-$	1.022
$2OH^- \rightarrow O_2 + 2H^+ + 4e^-$	0.815
$4OH^- \rightarrow O_2 + 2H_2O + 4e^-$	0.401
Two-Electron Oxidation	
$2H_2O \rightarrow H_2O_2 + 2H^+ + 2e^-$	1.776
$2OH^- \rightarrow H_2O_2 + 2e^-$	0.948
$H_2O_2 \rightarrow O_2 + 2H^+ + 2e^-$	0.682
$H_2O_2 + 2OH^- \rightarrow O_2 + 2H_2O + 2e^-$	−0.146
One-Electron Oxidation	
$H_2O \rightarrow OH + H^+ + e^-$	2.848
$OH^- \rightarrow OH + e^-$	2.020
$H_2O_2 \rightarrow HO_2 + H^+ + e^-$	1.495
$H_2O_2 \rightarrow HO_2 + H^+ + e^-$	0.667
$HO_2 \rightarrow O_2 + H^+ + e^-$	−0.180
$HO_2 + OH^- \rightarrow O_2 + H_2O + e^-$	−0.958

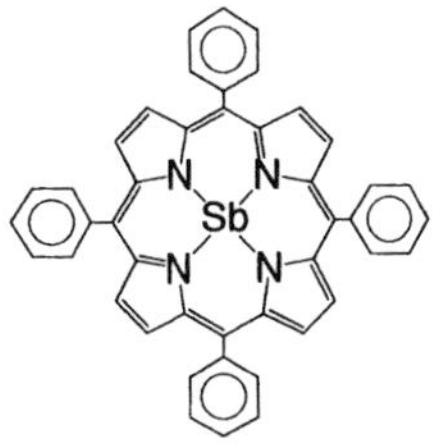

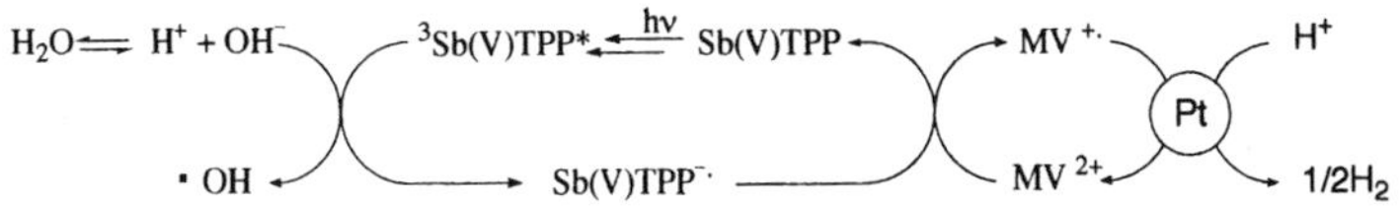

Figure 21   Hydrogen evolution coupled with the electron transfer from OH⁻ to the excited SbTPP [413].

Interesting reports on the two-electron oxidative activation of water have appeared [415–417,791]. Visible light irradiation to [SbTPP(OH)$_2$]PF$_6$ in the presence of K$_2$PtCl$_6$ as an electron acceptor induced oxygenation of alkene with a high quantum yield of 0.2. The oxygen atom of oxygenated product such as epoxide was confirmed to be derived from a water molecule by an experiment using ¹⁸O. Hydroxide ion coordinated to SbTPP was activated by one-electron oxidation and successive deprotonation to afford a metal–oxo complex that exerted two-electron oxidation of alkene (Fig. 22). The reaction mechanism and the net chemistry indicated that the water molecule served as both electron and oxygen atom donor in the reaction. The process afforded the similar oxo complex with that postulated in P-450 models (Fig. 23). Efficient oxygenation of alkene was also reported to proceed through a hole transfer mechanism from the cation radical of [SbTPP(OMe)$_2$]Br with quantum yield of 0.9 [418]. Reductive quenching of the excited triplet state of SbTPP also induced the efficient photochemical oxygenation reaction of cyclohexene in the presence of triphenylphosphine [781,782].

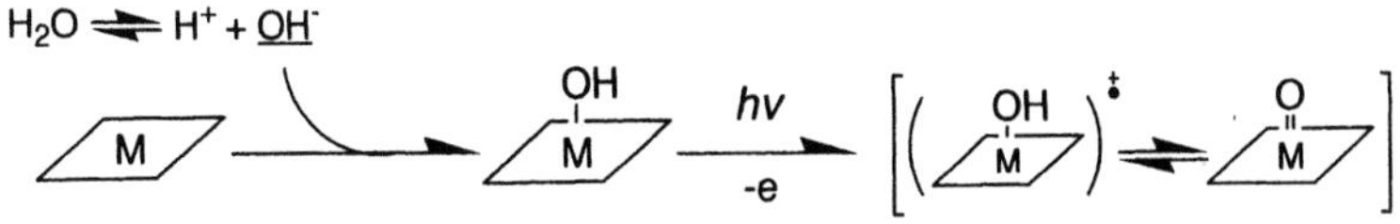

Figure 22   Activation of hydroxide ion coordinated to SbTPP by one-electron oxidation and successive deprotonation to afford a metal–oxo complex that exerts two-electron oxidation of alkene [415–417,791].

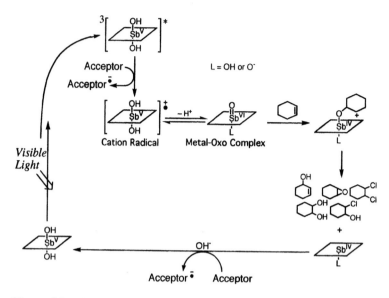

Figure 23 Photochemical P-450 model reaction sensitized by [SbTPP(OH)$_2$]PF$_6$ with water as electron and oxygen donor [415–417,791].

D. Photochemistry Involving Axial Ligand and Other Photochemistry

Metalloporphyrins have contributed to many aspects of porphyrin photochemistry. In particular, axial ligands on the central metal ions have been playing one of the most crucial roles. In this section, metalloporphyrin photochemistry based on axial ligand–related behavior such as 1.) photochemical dissociation of the metal–axial ligand bond and ligand substitution, 2.) activation of ligand through electron transfer and energy transfer, and 3.) other photochemical reactions will be surveyed.

1. Photochemical Dissociation of Metal-Axial Ligand's Bond and Ligand Substitution

In biological systems, coordination and dissociation of axial ligands in metalloporphyrins often have a key role in biological activities. Photochemical dissociation of nitric oxide, carbon monoxide, and molecular oxygen as the axial ligands has been extensively studied [294–296]. Rapid dissociation of NO from the excited Co^{II}TPPNO has been studied by picosecond laser flash photolysis [297–299]. For the dissociation of axial ligand in a model heme protein, a simple contact pair was revealed to be equilibrated with the free ligand [300–302], while a four-state dissociation mechanism (Fig. 24) involving contact pair and protein pair was postulated for myoglobin [303–307]. Several free-energy barriers were

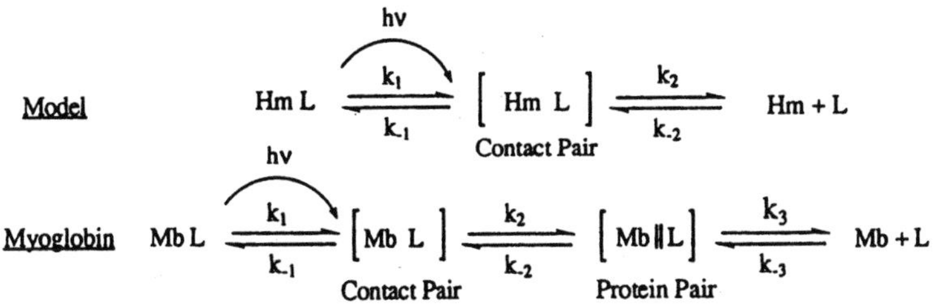

Figure 24 Four-state dissociation mechanism involving contact pair and protein pair for myoglobin [308].

reported to be involved in ligand dissociation of heme protein [306,310,313]. Hydrogen bonding interaction with solvent and solvent viscosity affected the quantum yield of the dissociation [303]. Their topics have been reviewed by Traylor [308].

Photochemical dissociation of CO from carboxyhemoglobin has been studied by time-resolved infrared (IR) measurements. The dissociation and recombination mechanism had been discussed in relation to protein structure [309,310]. The dissociated CO was supposed to be trapped in protein pocket below 200 K, while at a higher temperature the trapped CO was liberated from the protein [311]. The detailed time-resolved study indicated that the dissociation proceeded through the metastable heme protein with the lifetime of 2 ± 1 ps, while the dissociated CO bounded with heme pocket for at least 500 ps [783]. Model compounds such as Fe and Ru porphyrins have also been studied. Rather high quantum yields of dissociation of CO (0.53 and 0.9) were reported on $Fe^{II}TPP(CO)C_2H_5OH$ and $Fe^{II}TPP$ $(CO)Im$ in ethanol, respectively [312]. Basket-handle porphyrins as model compounds for heme have been studied [314–316] (**4**):

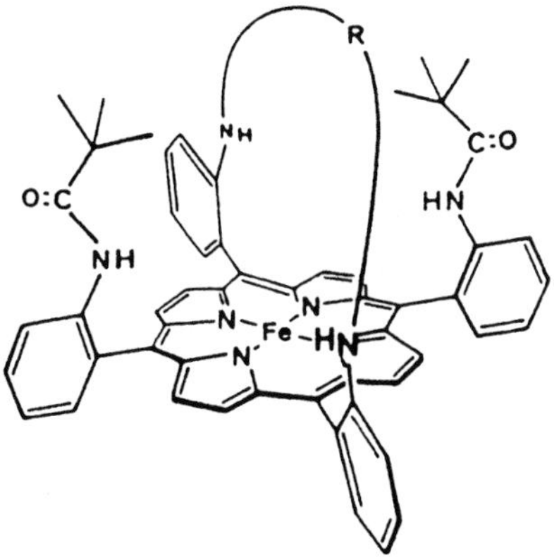

$R = CO-(CH_2)_{10}-CO$

$\quad = CO-(CH_2)_8-CO$

$\quad = CO-(CH_2)_6-CO$

4

The dissociation of CO from the excited $Ru^{II}OEP(CO)$ was observed to be less efficient as $\phi = 0.01 \pm 0.02$ by the excitation of $550 > \lambda > 350$ nm and $\phi = 0.08 \pm 0.01$ for $550 > \lambda > 350$ nm excitation [317].

Many artificial heme models mimicking reversible binding with molecular oxygen have been reported [318]. Picket-fence porphyrin (5) forms a stable complex with molecular oxygen. Upon laser photolysis at 314 nm, 43% of the dissociated molecular oxygen was observed to suffer recombination with the parent picket-fence porphyrin with methylimidazole as another axial ligand in trans position, while the recombination decreased to 20% with dimethylimidazole as axial ligand. The lifetime of the caged geminate pair was observed to be only around 30 ps [319].

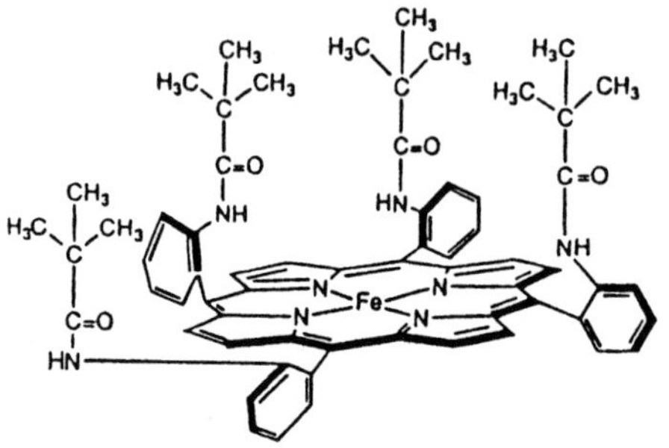

5

Co porphyrin having appropriate axial ligand can form a stable complex with molecular oxygen (6) and was observed to liberate molecular oxygen upon light irradiation [324].

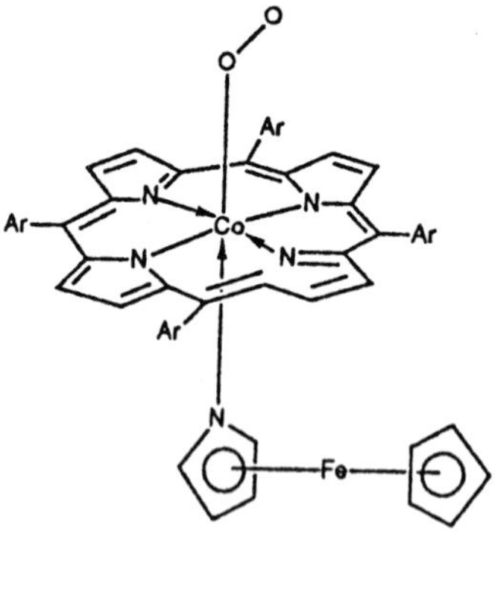

Ar = Ph
Ar = p-MeC$_6$H$_4$
Ar = p-MeOC$_6$H$_4$

6

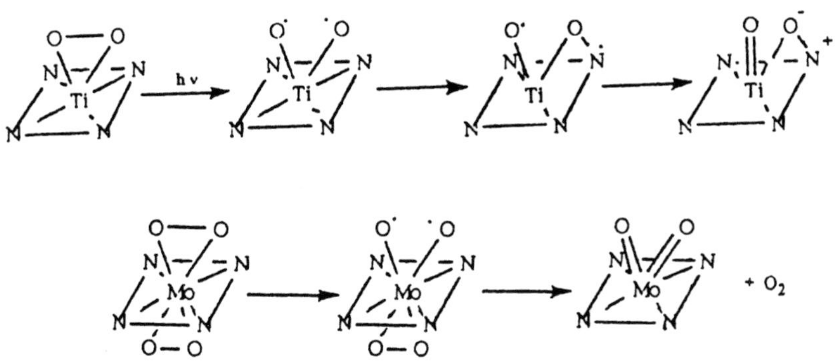

Figure 25 Photochemical transformation of oxygen complex with Ti and Mo porphyrins [320,321].

A unique photochemical transformation of oxygen complex with Ti and Mo porphyrins was reported [320] (Fig. 25). Excitation of the complexes induced oxygen–oxygen bond cleavage to form metal–oxo complexes with a quantum yield of 0.095 for $Ti^{IV}TPP(O_2)$ and 0.054 ± 0.005 for $Mo^{VI}TPP(O_2)_2$ upon excitation by the light of wavelengths shorter than their Soret bands [320–323]. Photochemical cleavage of the nitrogen–nitrogen bond in azide-coordinated Mn^{III} porphyrins was also reported. Light irradiation to $Mn^{III}TPP(N_3)$ in methyltetrahydrofuran afforded $Mn^{V}TPP(N)$ and $Mn^{II}TPP$ [329] (Fig. 26).

The acidity (pK_a) of organic acids is well known to vary in the excited states. Enhancement of deprotonation of the axial ligand hydroxy group was also found for the excited triplet state of $SbTPP(OH)_2$ as the first example for metal complexes [791]. The two-step deprotonations were studied in detail by a double-beam laser flash photolysis [792] (Fig. 27).

Different coordination characteristics in the excited state from those in the ground state were observed in Mg porphyrins [325]. The enhanced coordination of pyridine to Mg was interpreted by a different electronic structure in the excited state (Fig. 28).

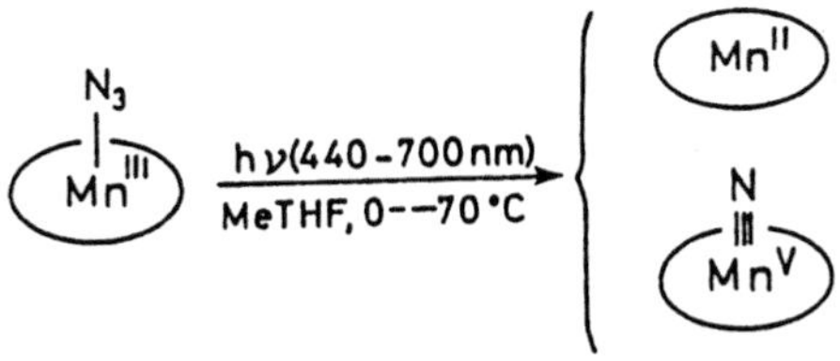

Figure 26 Photochemical cleavage of nitrogen–nitrogen bond in azide-coordinated Mn^{III} porphyrins [329].

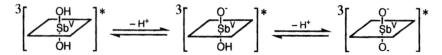

Figure 27   Two-step deprotonations of axial ligand hydroxyl groups of [SbTPP(OH)₂]Br in the excited triplet state [792].

Double substitutions of two halogen axial ligands in phosphorous porphy-rins by an alkoxy group were reported to proceed in a single step [326,327], whereas the similar disubstitution of axial ligands of [SbTPP(Br)₂]Br suffered a stepwise reaction and monosubstituted complex was actually isolated [328].

An interesting chain reaction on axial ligand sensitized by energy transfer from metalloporphyrin was reported by Whitten et al. [330]. Azastilbenes such as 4-stilbazole axially coordinated to Zn, Mg porphyrins exhibited efficient cis → trans isomerizations upon excitation of porphyrins. The quantum yields were observed to be larger than unity; 1-(1-naphthyl)-2-(4-pyridyl)ethylene exhib-ited a quantum yield of 7. A chain reaction mechanism involving energy transfer and back energy transfer among the metalloporphyrin and the axial ligand within the complex was postulated; the energy transfer from metalloporphyrin to the cis form induced the adiabatic isomerization and the resultant excited trans form transferred the energy back to the metalloporphyrin followed by ligand substitu-tion with the unreacted cis form within the lifetime of porphyrin [330]. Photore-sponsive coordination was reported on the basis of the above examples [331,332].

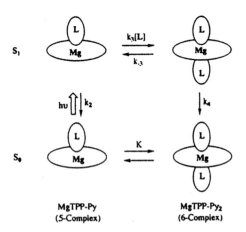

Figure 28   Enhanced coordination of pyridine to Mg porphyrin in the excited state [325].

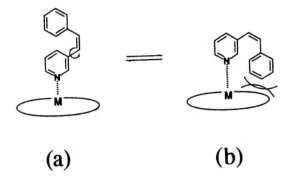

(a)   **(b)**

Figure 29 Recognition of two possible conformations of (a) "favored" and (b) "nonfavored" one for 3-stilbazole on AlTMP [331].

Two possible conformations—a "favored" (Fig. 29a) and a "nonfavored" one (Fig. 29b)—for 3-stilbazole on Al tetramesitylporphyrin (TMP) were recognized by the Al porphyrins upon visible light irradiation. Predominant coordination of the trans form over the cis form was observed. The recognition may afford one of the typical examples of "photoresponsive and photoregulative systems.

Efficient electron transfer from axial ligand to the porphyrin chromophore was observed in azaferrocene-coordinated Co porphyrin [372] (Fig. 30). Electron transfer from axial ligand carboxylate to the excited Fe^{III} porphyrin induced a Kolbe-type decarboxylation reaction [787].

Homolysis of the ligand–metal bond is often observed upon light irradiation as indicated in Eq. (17):

$$[\text{X-M}^n\text{-porphyrin}]^* \rightarrow \text{X} \cdot + \text{M}^{n-1}\text{-porphyrin} \tag{17}$$

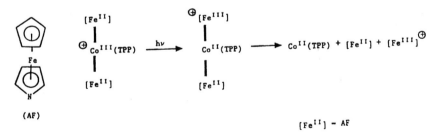

Figure 30 Efficient electron transfer from axial ligand azaferrocene to Co porphyrin [372,725].

where X is a halogen, alkoxy, hydroxy, and alkyl group, respectively [333–336]. For example, ethyl indium^{III} tetraphenylporphyrin in benzene liberated ethyl radical in the lowest excited triplet state [338,339] [Eq. (18)]:

$$^3[C_2H_5In^{III}TPP]^* \rightarrow C_2H_5\cdot + In^{II}TPP \qquad (18)$$

The quantum yield was observed to depend on the concentration of pyridine added to the reaction mixture; $\phi = 0.046$ without pyridine, while the quantum yield increased up to 0.3 at a high concentration of pyridine. The pyridine-coordinated species was supposed to be reactive. The similar homolytic dissociation of alkyl group was also observed in Rh [340] and Co [341,342] porphyrins. The quantum yield generally depends on C-H bond energy of solvent; the liberated alkyl radical recombinates with the central metal competing with hydrogen abstraction from solvent [337]. The recombination is also affected by the solvent viscosity. Apparent ligand substitution through the homolytic dissociation of alkyl group was reported in the case of TPP-Al-Et upon light irradiation [343] (Fig. 31).

Synthetic application of aluminum porphyrins afforded a series of interesting reactions. Light irradiation to TPP-Al-Et enhanced the addition reaction of ethyl group to vinyl ketone and polymerization [349] (Fig. 32). Photochemical insertion of carbon dioxide into Al-Et bond was reported in the presence of 1-methylimidazole [352]. Similar insertion of CO_2 into the Al–enolate bond was also reported [353,354]. These reactions have attracted much attention in relation to fixation of carbon dioxide in artificial photosynthesis as described in Sec. III.C (Fig. 33).

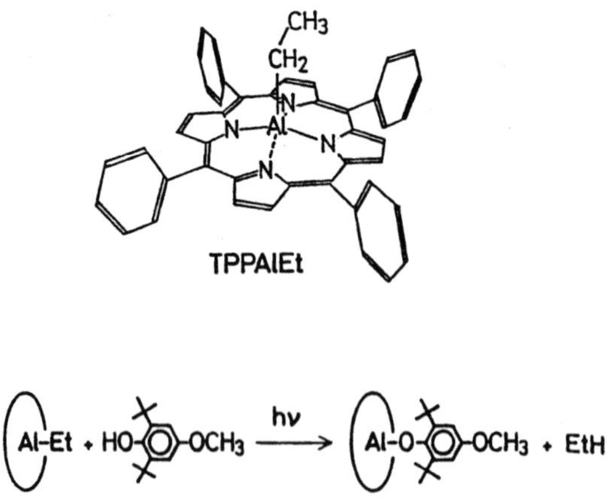

TPPAlEt

Figure 31 Ligand substitution through the homolytic dissociation of alkyl group of TPP-Al-Et upon light irradiation [343].

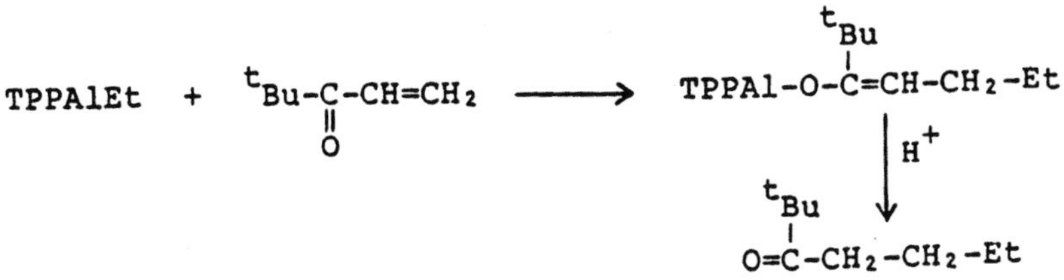

$$TPPAlEt \; + \; ^tBu\text{-}\underset{\underset{O}{\|}}{C}\text{-}CH=CH_2 \; \longrightarrow \; TPPAl\text{-}O\text{-}\underset{\underset{tBu}{|}}{C}=CH\text{-}CH_2\text{-}Et$$

$$\downarrow H^+$$

$$\underset{\underset{tBu}{|}}{O}=C\text{-}CH_2\text{-}CH_2\text{-}Et$$

Figure 32 Addition of ethyl group of TPP-Al-Et to vinyl ketone enhanced by light irradiation [349].

A living polymerization with the use of a similar reaction was reported on Zn porphyrins [350,351]. A unique reaction through the photochemical homolytic alkyl dissociation was observed in dirhodium diporphyrin [344,345]; the enhanced reactions of bimetalloradical with hydrogen and methane were interpreted by a four-centered transition state [346–348] (Fig. 34).

Photochemical homolytic dissociations of the oxygen–metal bond also induce various reactions [Eq. (19)]:

$$[RO\text{-}M^n\text{-}porphyrin]^* \rightarrow RO\cdot + M^{n-1}\text{-}porphyrin \tag{19}$$

Ito et al. reported on an oxidation of alkoxy group in alkoxy-coordinated Fe porphyrins upon light irradiation [355,356].

Oxidation of ethanol by a chain reaction initiated by the photochemical homolysis of the oxygen–metal bond was reported to proceed with a turnover

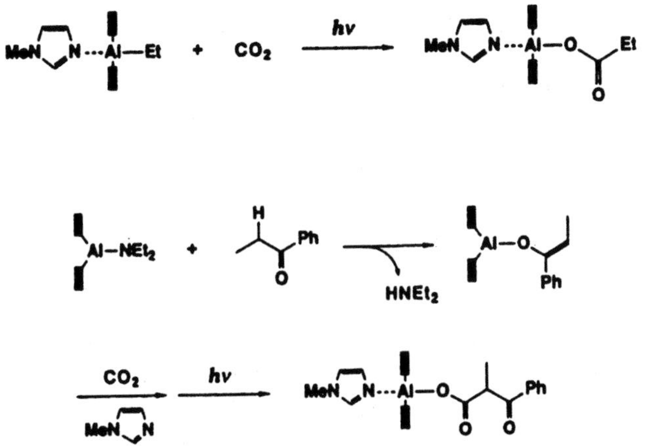

Figure 33 Photochemical insertion of carbon dioxide into Al-Et bond [354].

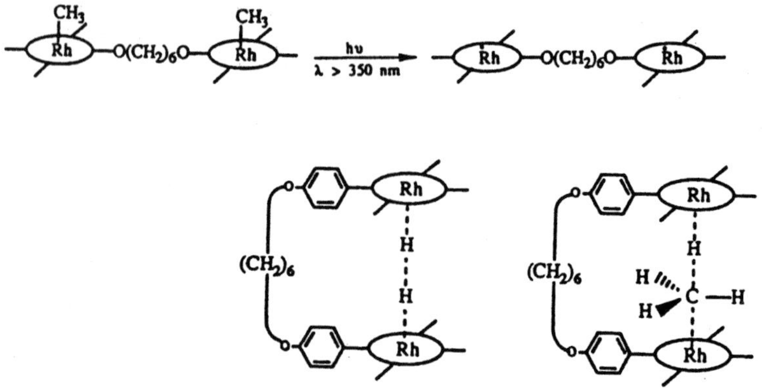

Figure 34 Photochemical homolytic alkyl dissociation in dirhodium diporphyrin and a four-centered transition state [346–348].

number up to 10^5 [357,358]. A formation of Fe^{II}TPP was observed in a photolysis of Fe^{III}TPP in ethanol [359]. Oxidations of cyclohexane [360] and pinene [362] were discussed in relation to a rebound mechanism [361]. Oxidation processes of alkane initiated by the photochemical homolytic dissociation of the oxygen–metal bond are rather complicated as postulated in Fig. 35 [364]. A similar mechanism was also discussed in the case of Mn porphyrins [365].

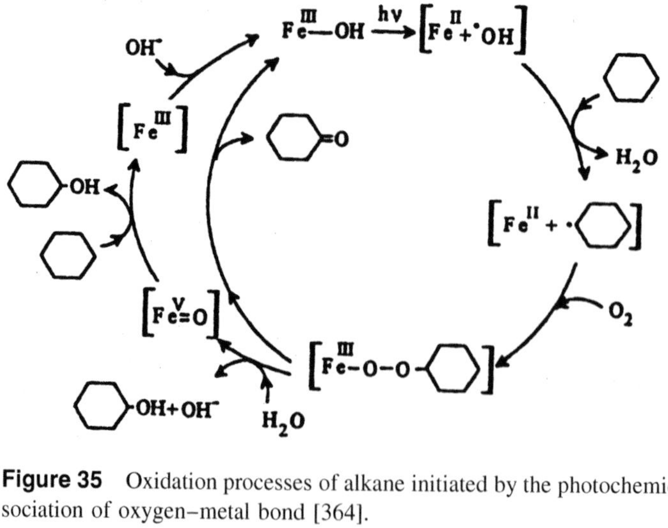

Figure 35 Oxidation processes of alkane initiated by the photochemical homolytic dissociation of oxygen–metal bond [364].

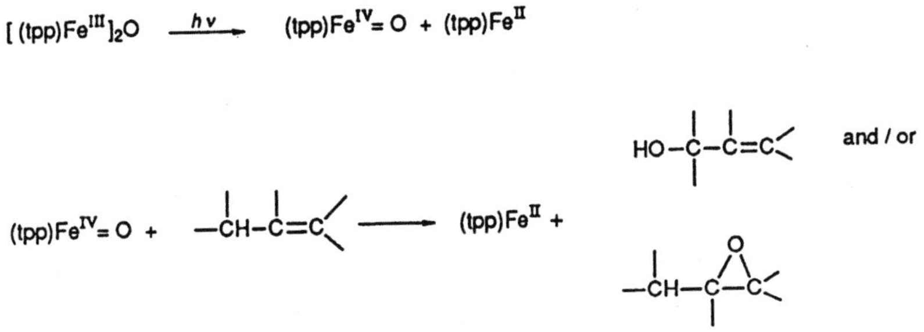

$$[(tpp)Fe^{III}]_2O \xrightarrow{h\nu} (tpp)Fe^{IV}=O + (tpp)Fe^{II}$$

$$(tpp)Fe^{IV}=O + \underset{|}{-}CH\underset{|}{-}C=C\overset{\diagup}{\diagdown} \longrightarrow (tpp)Fe^{II} +$$

$$HO\underset{|}{-}C\underset{|}{-}C=C\overset{\diagup}{\diagdown} \quad \text{and / or}$$

$$\underset{|}{-}CH\underset{|}{-}C\overset{O}{\diagup\diagdown}C\overset{}{\diagdown}$$

Figure 36 Oxygen atom transfer to alkene initiated by a light irradiation to Fe porphyrin μ-oxo dimer [363].

2. Activation of Axial Ligand

Oxygen atom transfer to alkene [363] and phosphine [366] initiated by a light irradiation to Fe porphyrin *m*-oxo dimer was reported (Fig. 36). Epoxidation of alkene initiated by light irradiation to an oxygen complex of Nb or Mo porphyrins was observed to be promoted by molecular oxygen [367–369] (Fig. 37).

Oxidations of alkane and alkene were attempted by a system involving Sn porphyrins as a sensitizer and Fe porphyrins as an oxygen transfer agent, where

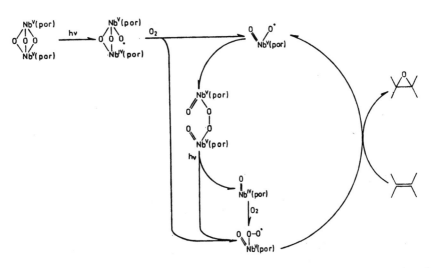

Figure 37 Epoxidation of alkene initiated by light irradiation to oxygen complex of Nb porphyrins [367].

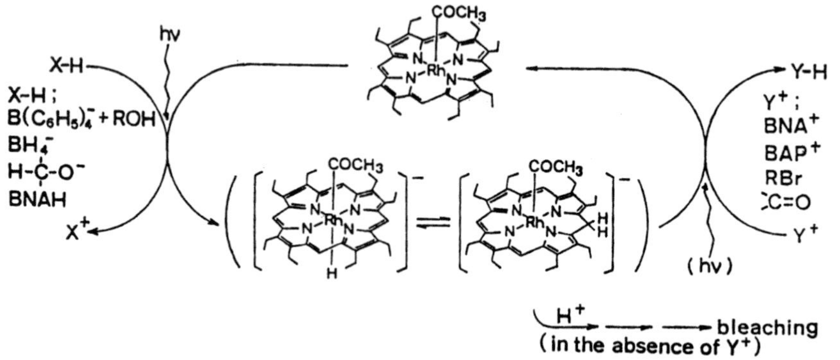

Figure 38   Hydride transfer from hydride donor through Rh porphyrin to substrate [371].

molecular oxygen was indispensable [370]. Axial ligand was revealed to play crucial roles in the photoredox reaction sensitized by SbTPP [415–418,791,792]. Oxygen atom transfer to alkene via a metal–oxo complex generated in a photoredox system with water as both oxygen and electron donor was reported; axial ligand hydroxy group on metalloporphyrins such as Sb porphyrins was reported to be transformed into active oxygen atom in a metal–oxo complex by one-electron transfer and successive deprotonation as described in Section III.C [415–418,791].

Interesting hydride transfer upon light irradiation to Rh^III porphyrin was reported [371]. Hydride was transferred from a hydride donor such as BH^-_4 through Rh porphyrin to substrate such as dodecyl bromide to form dodecane as indicated in Fig. 38.

3.   Other Photochemical Reactions

Singlet oxygen formation is a well known process in porphyrin photochemistry [Eq. (20)]:

$$^3\text{Porphyrin}^* + {}^3\text{O}_2 \rightarrow \text{porphyrin} + {}^1\text{O}_2 \cdot \tag{20}$$

Photochemical oxidations of various substrates such as amines [763], alkenes [769,770,772,775,778–780], furan [764,765], sulfide [766,773], phytol [767], bi-adamantylidene [768], thiophenolate [771], catechol [774], allylsilane [776], and imines [777], etc., through reactions with singlet oxygen sensitized by porphyrins have been extensively studied. Photochemical atrop isomerization of picket-fence-type porphyrin should be noted here as another interesting type of photore-action [793].

IV. SUPRAMOLECULAR PHOTOCHEMISTRY OF PORPHYRINS

A. Introduction

"Supramolecular system" has been defined as a unique molecular assembly exhibiting unique functions [419–423]. Supramolecular system is composed of unit molecules associated with each other through noncovalent interactions such as hydrogen bonding, electrostatic, van der Waals, and other weak intermolecular interactions. The unit molecules have their own molecular functions, while the supramolecular system has unique and additional functions not only limited by a simple combination of functions of the unit molecule but also extended to the concept of a molecular system beyond a molecule. Among the grand targets, an artificial enzymatic system and molecular devices for various objects have attracted special attention. Intensive approaches for constructing supramolecular systems have been focused on synthesis, structural analysis, molecular recognition, energy transfer, electron transfer, and other reactions. Supramolecular photochemistry could be simply defined as the photochemistry of a supramolecular system. Porphyrins and metalloporphyrins have also played crucial roles in various approaches for supramolecular photochemistry. In this section recent advances in the field will be surveyed.

Natural photosynthesis undoubtedly represents an exemplary system for supramolecular photochemistry. In a series of irreversible electron transport processes in bacterial photosynthesis, an electron was ejected from bacteriochlorophyll dimer (special pair) [435–438] and transferred to quinone [439–441] via bacteriopheophytin [442–444]. Ferrocytochrome c supplies an electron to the hole of a special pair [445]. The charge separation and each electron transfer have been supposed to proceed at almost 100% efficiency. Those postulates were actually verified in a series of elegant works on structural analyses of reaction center from *Rhodopseudomonas (Rps.) viridis* and *Rb. sphaeroides* by Deisenhofer et al. [428–430]. In 1984 they found that the special pair and bacteriopheophytin were beautifully aligned and oriented with each other in the system [428]. The intermolecular center-to-center distance within the special pair was revealed to be 7.0 Å and the distances between the two molecular planes were 3.0 Å for *Rps.* and 3.5 Å for *Rb.*, respectively [428–434,446–450] (Fig. 39).

The charge separation within the special pair and successive electron transfers have been studied in detail [424–427,454,455]. As indicated in Fig. 40, the forward electron transfer was revealed to be two to three orders of magnitude faster than the back electron transfer in each process; along with the energy gaps among the porphyrin chromophores, the beautifully aligned structure in Fig. 39 afforded sufficient molecular orientation and intermolecular distance for efficient charge separation and irreversible electron transfer. Marcus theory predicted that the rate constant of electron transfer reached a maximum when the energy gap

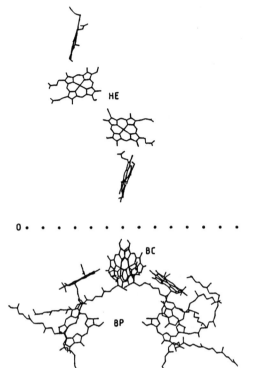

Figure 39   Stereo drawing of the prosthetic groups of the RC showing BChl-b (BC), BPh-b (BP), nonheme iron (Fe), quinone (MQ), and heme groups (HE) in *Rhodopseudomonas viridis* [428].

was equal to the reorganization energy λ (Section III.B). The condition was fulfilled in the natural systems. Protein molecules surrounding the chromophores have attracted attention [455–458,461–467]. A superexchange mechanism via tryptophan moiety of protein has been discussed in terms of electron transfer to the quinone [459].

Another highlight for exemplary supramolecular system would be the light-harvesting system in natural photosynthesis. In green plants approximately 200 molecules of light-harvesting chlorophyll surround one reaction center and the captured light energy is transferred to the reaction center. Recently, the microstructure and light-harvesting mechanism in red bacteria *Rb. acidophilia* have been revealed [451,452] (Fig. 41).

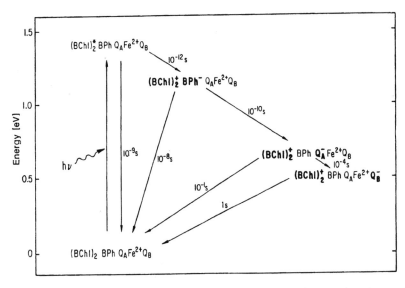

Figure 40   The charge separation within the special pair and successive electron transfers in the reaction center of photosynthetic bacteria. $(BChl)_2$, bacteriochlorophyll dimer; BChl, bacteriochlorophyll monomer; QA, QB, ubiquinones [424].

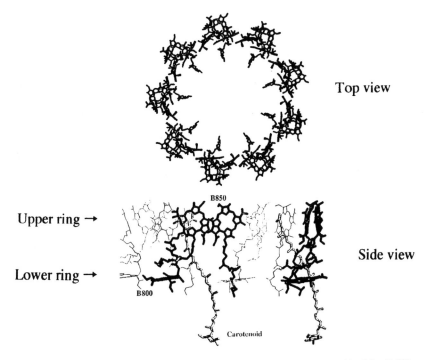

Figure 41   Structure of light-harvesting system 2 (LH2) in *Rps. acidophila* [452].

In the light-harvesting system LH2, two sets of nonameric complex of porphyrins were observed. The upper ring was composed of nine dimers of bacteriochlorophyll (B850); 18 molecules of B850 circled with periodic intermolecular distances of 8.7 and 9.7 Å, respectively. In the lower ring, nine bacteriochlorophyll molecules (B800) were in a coplanar situation and were separated by an equal distance of 21 Å. Nine carotenoide molecules were sandwiched between the nine B800 molecules. Light energy is absorbed by B800 molecules and transferred to B850 within 650 fs. The excited energy circles round the B850 ring and is further transferred in 5–20 ps to the adjacent light-harvesting system LH1, which contains a reaction center where the special pair is finally excited [453].

The amazing structure with precise molecular alignment and reaction mechanism of energy and electron transfer in natural photosynthesis undoubtedly affords one of the most promising examples for photochemically active supramolecular systems. In the following section, recent approaches involving molecular systems linked by covalent bonds and chromophoric assemblies will be surveyed.

B. Chromophoric Assemblies and Supramolecular Systems

Challenging studies on chromophoric assemblies of porphyrins and supramolecular systems have been extensively reported. Most of them have been intended for extension to possible light-harvesting systems, electron transfer systems, and supramolecular devices. Porphyrin aggregates have attracted special attention partly in relation to the special pair in photosynthesis. An exciton coupling model has been accepted as an example of the absorption spectral change on aggregation [668–672]. The exciton band structures in molecular dimers with various geometrical arrangements of transition dipoles are explained in Fig. 42 [668]. A parallel

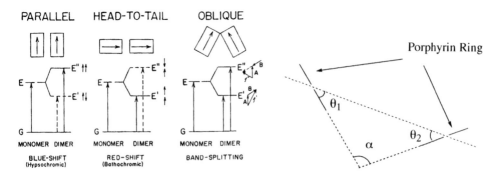

Figure 42   The exciton band structures in molecular dimers with various geometrical arrangements of transition dipoles [668].

orientation induces a blue shift of absorption spectrum, whereas a head-to-tail arrangement leads to a red shift and an oblique structure has a band splitting. The band splitting (ΔE) is generally expressed as Eq. (21).

$$\Delta E = \frac{2|M|^2 (\cos \alpha + 3 \cos \theta_1 \cos \theta_2)}{r^3} \tag{21}$$

where M is the transition dipole moment of the monomer and r, α, θ_1, θ_2 are indicated in Fig. 42. For a linear polymeric aggregates the equation is modified into Eq. (22).

$$\Delta E = \frac{2|M|^2 (1 - 3 \cos^2 \gamma)}{r^3} \tag{22}$$

where γ denotes the tilt angle between the line of centers and molecular long axes.

The validity of the exciton coupling model was beautifully verified by a series of Osuka molecules [673] (Fig. 43). Molecular orientation was also studied in phenyl-bridged porphyrins [674].

Among the various approaches for porphyrin dimers, porphyrins having a crown ether moiety within the molecule have shown interesting dimer formation through complexation with cations such as K^+ and Ca^{2+}. Different types of dimer structure have been proposed (**7, 8**) [675,676].

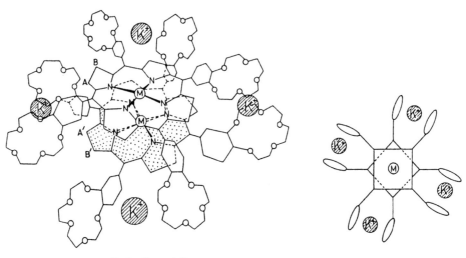

M = 2H , Co , Cu and Zn

7

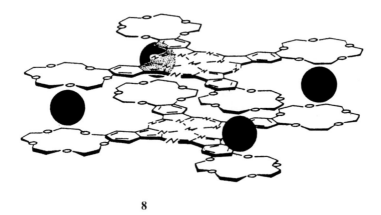

8

The dimer formation of anionic metalloporphyrins [677], cationic porphyrins [678], heterodimer formation [679–682], an aggregate between porphyrin and cytochrome *c* [683], and dimer formation through metal–metal interaction [684] has been reported. The 1:2 complex formation between the metal ions such as Th^{III} and Ce^{IV} and porphyrins afforded an interesting double-decker sandwich complex [685–687] and a triple-decker one [688].

A series of cofacial porphyrins connected with covalent bonds have been reported on dimer [658–662,673,689–692], trimer [693], and oligomer [694] (**9–12**).

M—M = Ru≡Ru

M—M = Mo≡Mo

9

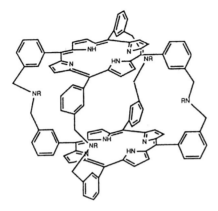

R = $-SO_2$[pyridine]   R = $-SO_2$[N-methylpyridinium]$^+$ CH_3

R = C(O)NH$_2$ R = CN

R = $-SO_2$[benzene]$-CH_3$ R = C(NH)NH$_2$

R = C(NH)OEt

10

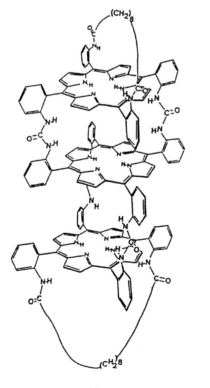

11

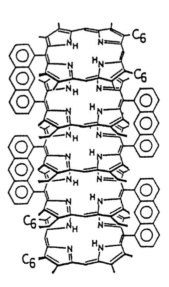

12

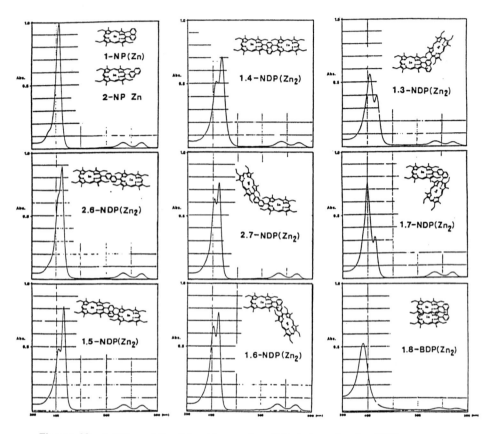

Figure 43   Splitting of Soret band in a series of Osuka's molecules [673].

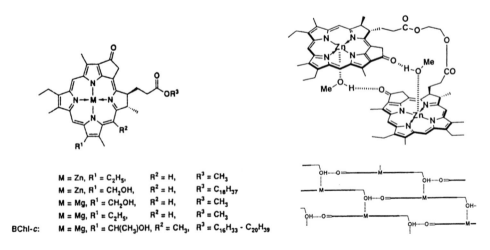

BChl-c:

$M = Zn, R^1 = C_2H_5, \quad R^2 = H, \quad R^3 = CH_3$

$M = Zn, R^1 = CH_2OH, \quad R^2 = H, \quad R^3 = C_{18}H_{37}$

$M = Mg, R^1 = CH_2OH, \quad R^2 = H, \quad R^3 = CH_3$

$M = Mg, R^1 = C_2H_5, \quad R^2 = H, \quad R^3 = CH_3$

$M = Mg, R^1 = CH(CH_3)OH, R^2 = CH_3, \quad R^3 = C_{16}H_{33} - C_{20}H_{39}$

Figure 44   From Ref. [699].

Thorium porphyrins oligomer was observed to have an unique structure where the three subunit centers placed on each vertex of an equilateral triangle and the metallic center of each chromophore being linked via μ-hydroxo bonds [695] (**13**). Though the usual cofacial porphyrins generally exhibit a strong exciton coupling, the thorium porphyrins oligomer has a very weak coupling.

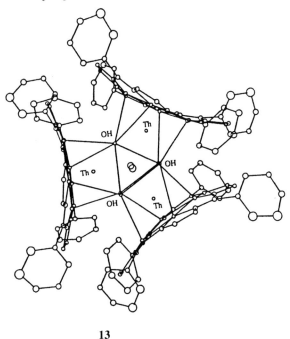

13

A slipped cofacial dimer porphyrin was reported that the substituent coordinated to Zn metal of another porphyrin [696]. A strong coupling was observed in that case (**14**). Metallochlorin was also reported to form multilayered slipped cofacial aggregates [699] (Fig. 44).

14

   The interaction between diol and boronic acid served as a bridging spacer
with fixed angles as shown in Fig. 45 [697].
   Interesting porphyrin assemblies were prepared through sophisticated multiple hydrogen bonding by triaminotriazine [698] (Fig. 46). Axial coordination
interaction has made possible various unique porphyrin assemblies such as wheel-
and-axle [700–702] and cyclic types [703–706] (Figs. 47–50, **15**, **16**).

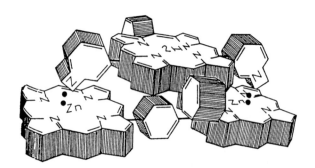

15

16

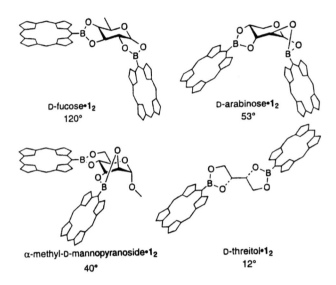

D-fucose•1₂
120°

D-arabinose•1₂
53°

α-methyl-D-mannopyranoside•1₂
40°

D-threitol•1₂
12°

Figure 45   From Ref. [697].

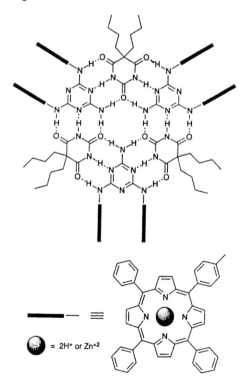

= 2H⁺ or Zn⁺²

Figure 46   From Ref. [698].

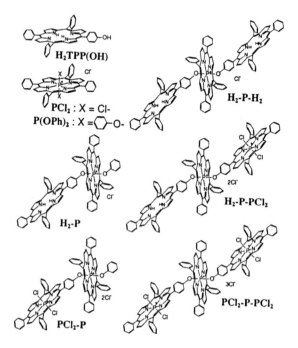

Figure 47 From Ref. [700].

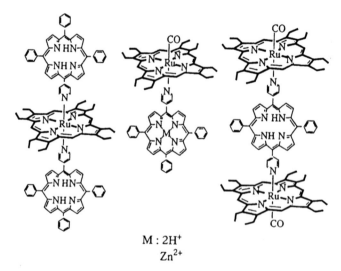

M : 2H⁺
Zn²⁺

Figure 48 From Ref. [702].

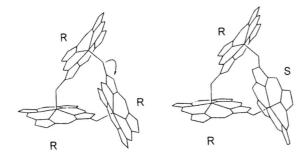

Figure 49 From Ref. [703].

C. Excited Energy Transfer in Chromophoric Assemblies and Supramolecular System

Artificial light-harvesting and energy transfer systems have attracted much attention in relation to natural photosynthesis [634]. Osuka et al. reported an interesting molecular system linked by covalent bonds in relation to a model system of a special pair in photosynthesis [642] (**17**). The singlet excited energy was observed to be transferred by the Forster mechanism through-space from monomer to dimer moiety with a rate constant of 3.4×10^{10} s^{-1}, while the rate constant was decreased to 1.4×10^{10} s^{-1} with the spacer group of 4,4′-biphenylene.

Molecular systems involving carotenoide chromophore have also been studied [643–648]. The singlet excited energy on the carotenoide moiety was observed to be transferred to porphyrin with 30% efficiency competing with the rapid internal conversion within 10 ps [648] (**18**).

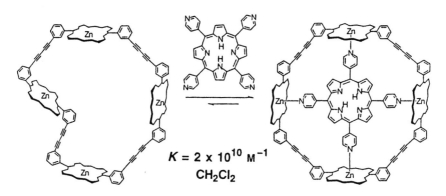

$K = 2 \times 10^{10}$ M^{-1}
CH$_2$Cl$_2$

Figure 50 From Ref. [706].

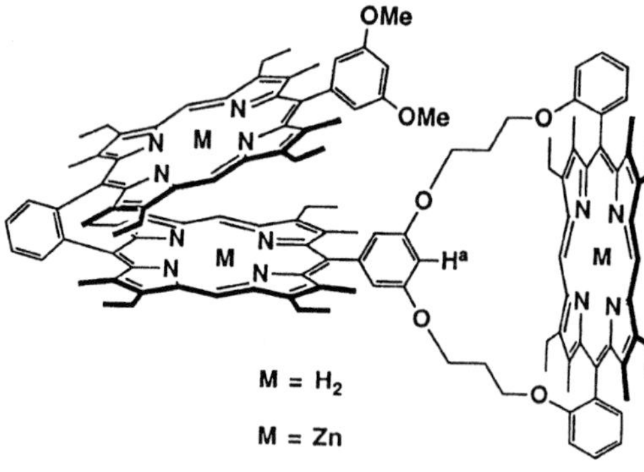
OMe
OMe
N
M
N
N
Hª
O
O
O
O
M = H_2
M = Zn

17

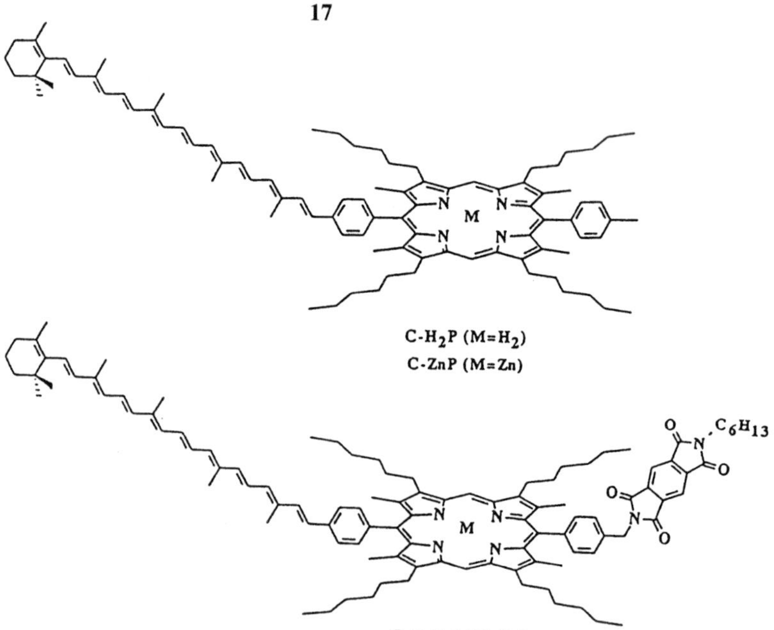
M
C-H_2P (M=H_2)
C-ZnP (M=Zn)
C_6H_{13}
C-H_2P-I (M=H_2)
C-ZnP-I (M=Zn)

18

Systematic studies about distance-dependent energy transfer was reported on linked porphyrin molecules by Osuka et al. [527]. Both Forster and Dexter mechanisms depend on the intermolecular distance and the rate constant k_{ET} is expressed as

$$\ln k_{ET} = A - 6 \ln r \text{ Forster mechanism} \quad (23)$$

$$\ln k_{ET} = A' - r/L \text{ Dexter mechanism} \quad (24)$$

where r and L denote the intermolecular distance and average effective Bohr radius.

In a series of compounds **19**, the distance dependency on the energy transfer was interpreted by the Dexter mechanism [527]. Energy transfer in a supramolecular system linked by hydrogen bonds was studied by Harriman et al [635–637] (**20–22**). Hydrogen bonds were supposed to contribute substantially to through-bond interactions for the triplet energy transfer by the Dexter mechanism.

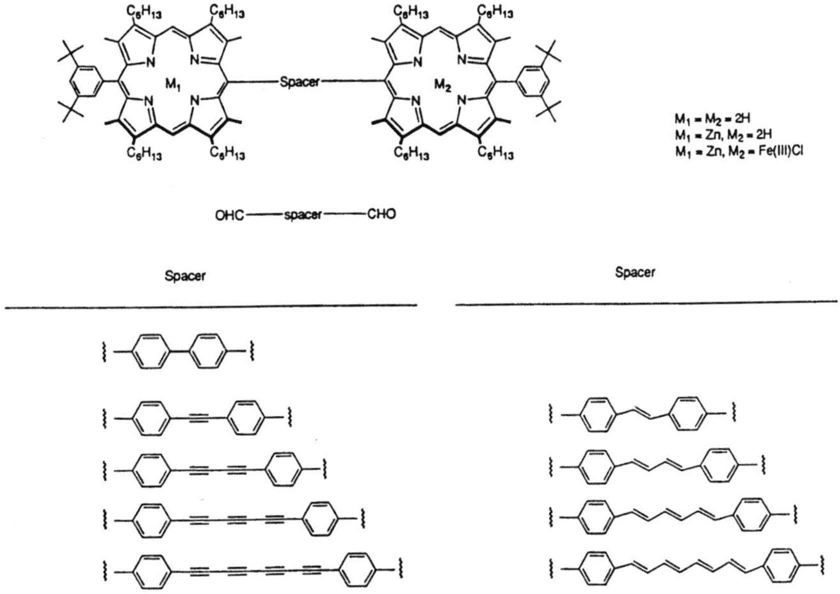

$M_1 = M_2 = 2H$
$M_1 = Zn, M_2 = 2H$
$M_1 = Zn, M_2 = Fe(III)Cl$

OHC────spacer────CHO

Spacer   Spacer

19

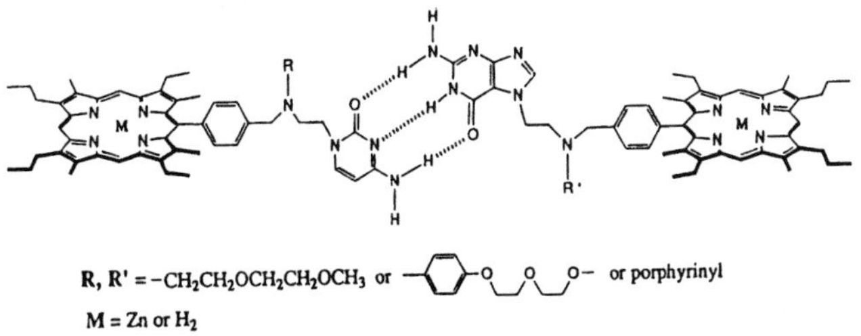

R, R' $= -CH_2CH_2OCH_2CH_2OCH_3$ or ────O───O───O─ or porphyrinyl

M $=$ Zn or H_2

20

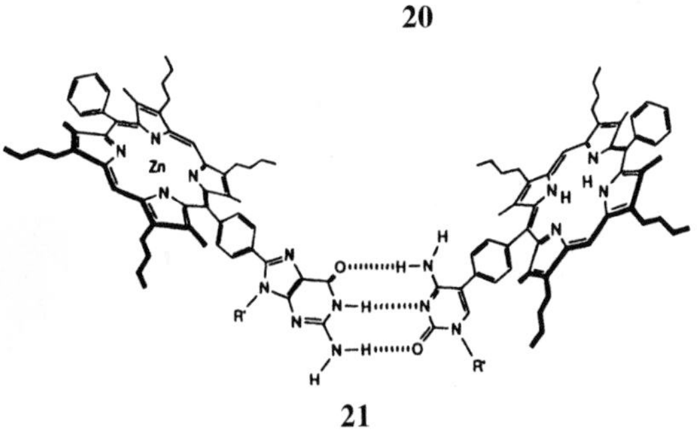

21

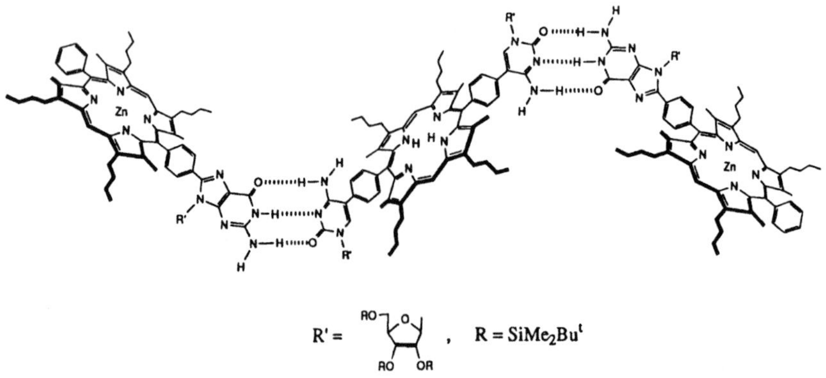

R' = [structure] , R = SiMe₂Buᵗ

22

A light-harvesting array was attempted in a five-porphyrins system linked by a diphenylethyne group [638,639] (**23**). The center-to-center distance was around 20 Å. Excitation of the surrounding four Zn porphyrins induced singlet energy transfer with 90% efficiency to emit fluorescence from the central free-base porphyrin.

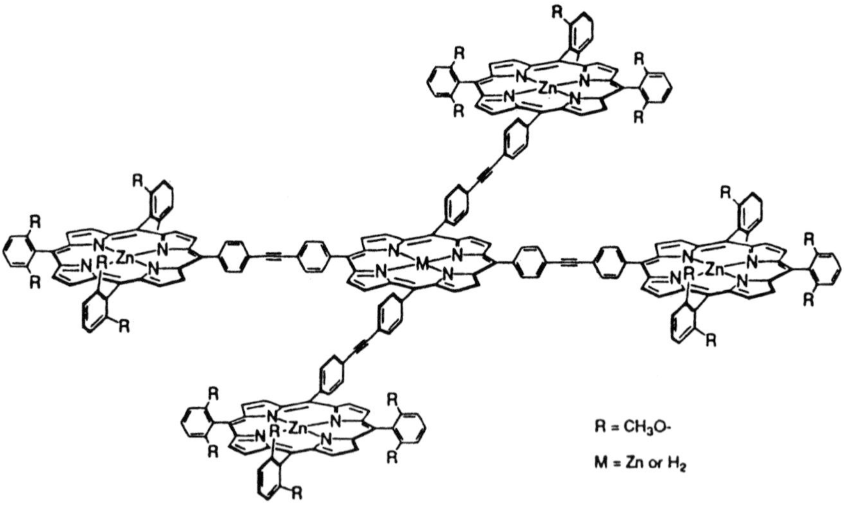

R = CH₃O-

M = Zn or H₂

23

A unique porphyrin trimer having a broad absorption spectrum up to 800 nm was proposed for a light-harvesting system [719].

An interesting non-cofacial salt-bridged system with an association constant of 2600 ± 500 M^{-1} involving pentapyrrolic macrocycles sapphyrin was reported [640]. The singlet energy transfer was well explained by the Forster mechanism (**24**).

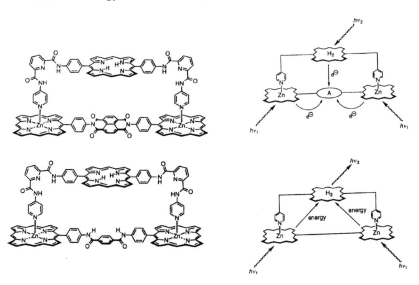

2 (R = butyl)

24

A supramolecular system with coordination interaction in axial position has appeared [641]. The system was reported to have a large association constant of $3 \pm 1 \times 10^8$ M^{-1} and exhibit the singlet energy from Zn porphyrins to the free-base porphyrin (Fig. 51).

Enhanced energy transfer on bilayer membrane was reported [650] (Fig.

Figure 51 A supramolecular system with coordination interaction in the axial position [641].

52). Polymerization of surfactant within a bilayer membrane induced a formation of specific domain that concentrated cyanine and porphyrin molecules to enhance the energy transfer. The energy transfer efficiency increased up to 71% at a 100% polymerization, whereas the efficiency was 9% without polymerization.

A light-harvesting system in solid state has been studied [651] and a molecular design for a novel light-harvesting array has been proposed [652].

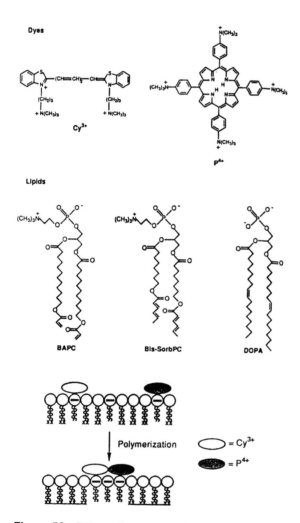

Figure 52   Enhanced energy transfer on a bilayer membrane [650].

D. Electron Transfer in Chromophoric Assemblies and Supramolecular System

1. Linked Molecular System Associated with Covalent Bond

Among intensive studies on linked molecular systems, chromophoric assemblies, and supramolecular systems, most efforts have been focused on electron transfer processes in those systems [460,468,470–478,501]. As described in Section III.B, the "inverted region" predicted by the Marcus theory on electron transfer was found in the linked molecular system having donor and acceptor connected with the appropriate spacer group to verify the theory [479,480,488,490,530,540,733–735]. The inverted region was also observed in various linked molecular systems having porphyrins and metalloporphyrins as shown in Fig. 53 (25, 26) [481–484].

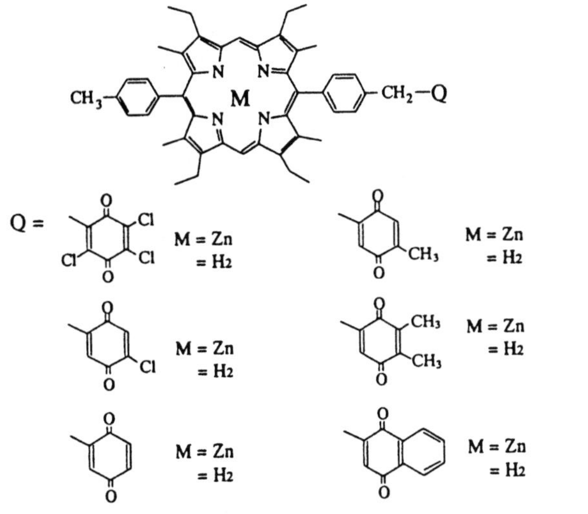

25

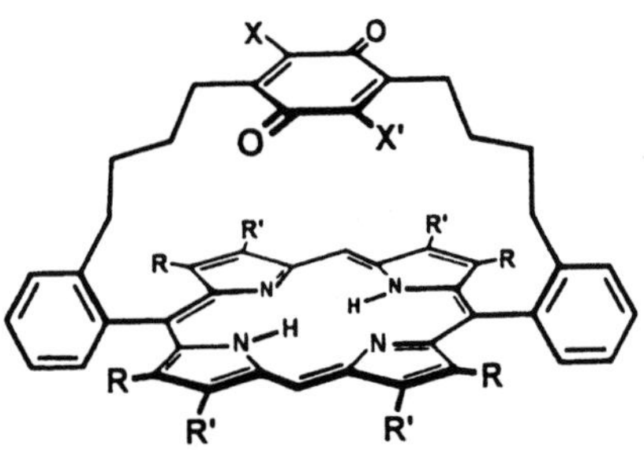

	R,R'	X,X'
1	Me	Me
2	Me	OMe
3	Me	H
4	Me	H,Cl
5	Me,Et	Cl
6	Me	Br
7	Me,Et	CF₃

26

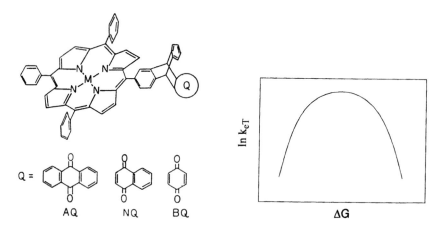

Figure 53   The bell-shaped curve of rate constant of electron transfer observed in linked molecular systems with porphyrin [482].

Attention has been shifted to a more detailed understanding of the electron transfer process. It is accepted that solvent molecules affect the charge separation and charge recombination processes [506–511]. An interesting approach for estimating the solvent effect on the electronic and nuclear terms in the semiclassical Marcus equation [Eq. (14)] was reported [502,503]. Temperature dependence on the rate constant of electron transfer (k_{eT}) could differentiate them as the first term (electronic) and the second term (nuclear) in Eq. (25).

$$\ln(k_{eT}\lambda^{1/2}) = \ln\left(\frac{2\pi}{h}\frac{H^2}{(4\pi k_B T)^{1/2}}\right) - \frac{(\Delta G + \lambda)^2}{4\lambda k_B T} \tag{25}$$

Detailed study on the electron transfer from the excited porphyrin to quinone in the linked molecular systems **27** revealed that the electronic term was the larger in the more polar solvent, while the nuclear term was not affected by solvent polarity. Through-bond interaction and the superexchange mechanism involving interactions with the aromatic moiety of spacer group and solvent were discussed [503].

Another interesting solvent effect was reported on the electron transfer in a cofacial diporphyrin [512–514] (**28**). The rate constant of charge recombination was observed to be well correlated with the relaxational motion of solvent but not with solvent polarity. For a more detailed understanding of electron transfer, a long-range electron transfer would promisingly afford substantial information on whether the two chromophores have 1.) through-space interaction or 2.) super-exchange through-bond interaction, and even on whether the surrounding mole-cules such as solvent, protein, etc., contribute to the interaction or not. Linked molecular systems connected with covalent bonds and supramolecular systems associated with hydrogen bonding, electrostatic interaction, and other intermolec-ular interactions are good candidates for the study on long-range electron transfer. Though challenging studies on linked molecules with flexible spacer groups in the early stage have still left some ambiguity regarding the mechanism, recent series of linked molecules with rigid spacer groups have proven to be substantial contributors of superexchange mechanism in electron transfer as described below.

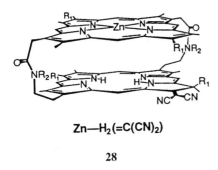

$$Zn—H_2(=C(CN)_2)$$

28

The interaction between the two chromophores in the linked molecular sys-tem has been estimated by the value α in the distance-dependent Eq. (26), which is a modified expression of Eq. (15) [488–490]:

$$k_{eT} = k^0 \exp(-\alpha d) \tag{26}$$

where d denotes edge-to-edge distance between the two chromophores.

The value α has been reported to be dependent on the rigidity of spacers, [500], be within 0.8–2.5 Å^{-1} for aliphatic spacer groups [491–494], and be smaller for conjugated spacers (around 0.4 Å^{-1}) [495–499,528]. The through-space and the through-bond interactions were reported on a series of elegantly designed molecules with rigid aromatic spacer groups by Harriman et al [515–520] (Fig. 54). The electron transfers among Zn and Au porphyrins were observed to be too fast for only the through-space interactions. The authors concluded that the electron transfer from both the excited singlet and triplet states of Zn porphy-rin to Au porphyrin proceeded through a superexchange interaction with the LUMO of the aromatic spacer group, whereas the excited triplet Au porphyrin had a superexchange interaction with the HOMO of the spacer group.

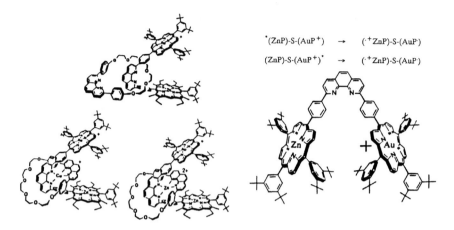

$^\bullet$(ZnP)-S-(AuP$^+$)   →   ($^{\cdot+}$ZnP)-S-(AuP$^\cdot$)

(ZnP)-S-(AuP$^+$)$^\bullet$ →   ($^{\cdot+}$ZnP)-S-(AuP$^\cdot$)

Figure 54 The through-bond interactions in a series of elegantly designed molecules with rigid aromatic spacer groups [518].

Sakata et al. observed the electron transfer from porphyrin to quinone in a series of compounds **29** [521,522,524]. The electronic coupling matrix elements derived from 1.) a direct interaction between the donor and acceptor through-space, 2.) an indirect interaction through solvent molecules, and 3.) an interaction through-bond were discussed.

29 top

R$_1$ = H, R$_2$ = Ph
R$_1$ = R$_2$ = H

29 bottom

An efficient electron transfer through diacetylene as a spacer group with a rate constant of 5.0×10^{10} s^{-1} was reported by the same authors [526] (**30**). The superexchange mechanism was evidently observed in the compounds **31** by Osuka et al. [527]. The decay parameter α in Eq. (26) was observed to be as small as 0.08 Å^{-1} for polyene spacers and 0.1 Å^{-1} for polyyne ones, suggesting substantial contribution of the superexchange mechanism through-bond.

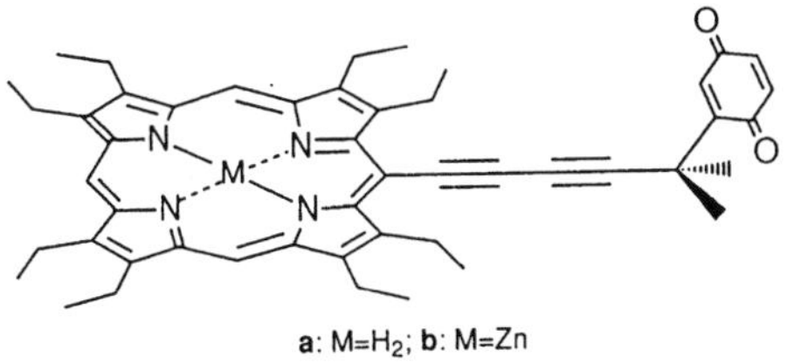

a: M=H$_2$; **b**: M=Zn

30

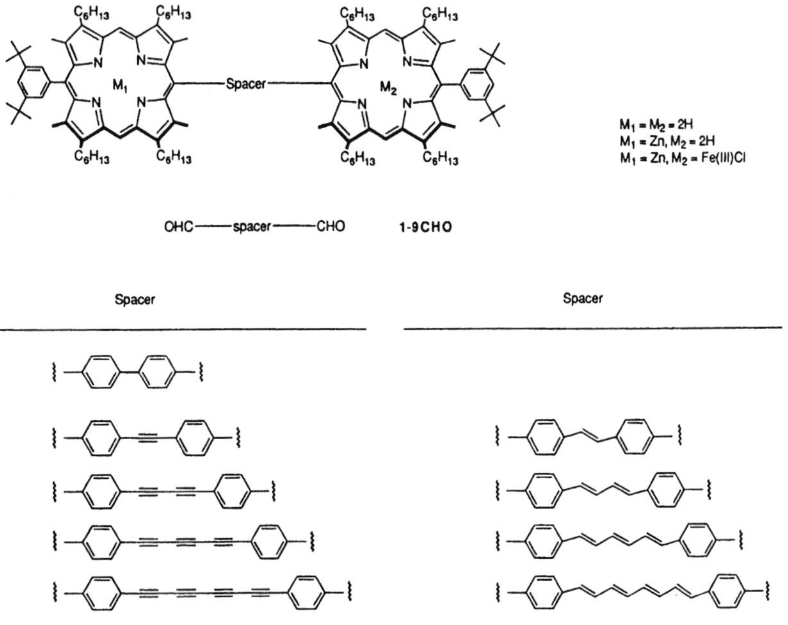

$M_1 = M_2 = 2H$
$M_1 = Zn, M_2 = 2H$
$M_1 = Zn, M_2 = Fe(III)Cl$

OHC———spacer———CHO   **1-9CHO**

Spacer   Spacer

31

Electron transfer in various linked molecular systems has been revealed to be affected by the free energy gap [483–485,540–544], the distance between the two chromophores [486,487,529–533], the relative orientation [522–525,532, 534–539], spacers and surrounding medium such as solvent [502,503,521,524, 526,527,532], respectively. In natural photosynthesis surrounding media, such as protein, has been supposed to contribute to the electron transfer [563–566]. An electron transfer in reconstituted myoglobin was also reported [552].

Attention has further shifted to how an efficient charge separation can be attained as observed in natural photosynthesis. Intensive studies on bichromophoric linked molecular systems have revealed that the lifetime of ion pairs can be lengthened by long spacer groups intervening between them. The forward electron transfer, however, would be reduced at the same time in such bichromophoric systems with long spacer groups. For an irreversible electron transport on the basis of proper design of the energy gap relation among chromophores, polychromophoric linked molecular systems such as triad, tetrad, pentad, etc., would surely be interesting targets.

Osuka et al. reported a long lifetime of ion pair state in a triad system of Zn porphyrin(ZnP)-oxochlorin(H_2C)-pyromellitdiimide(Im) [567,568]. Excitation of ZnP or H_2C led to the ion pair state of $(ZnP)^+$-H_2C-$(Im)^-$ with a long lifetime of 0.24 μs in DMF [567] (Fig. 55). Interesting series of tetrad and pentad molecules have been studied by the same authors [569–572,574] (**32–34**).

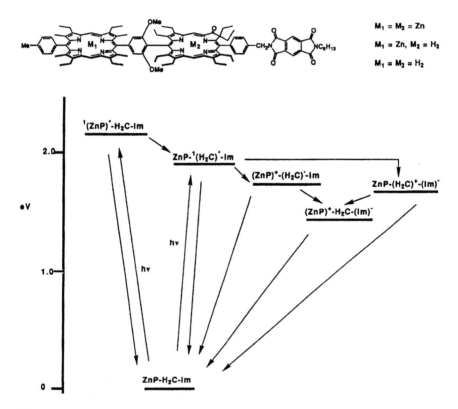

Figure 55 Ion pair state with a long lifetime in a triad system of Zn porphyrin(ZnP)-oxochlorin(H_2C)-pyromellitdiimide(Im) [567].

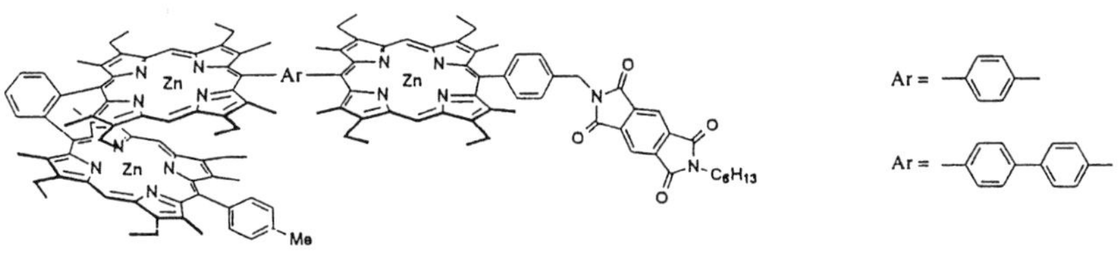

32

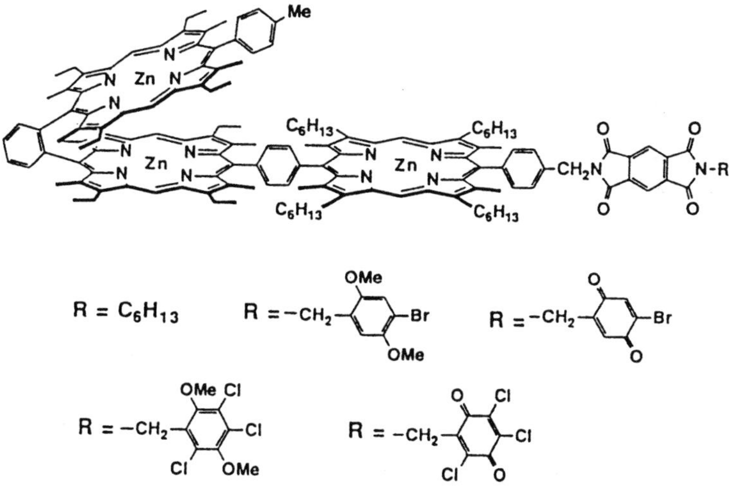

$R = C_6H_{13}$ $R = -CH_2-$ (OMe, OMe) Br $R = -CH_2-$ (quinone) Br

$R = -CH_2-$ (OMe Cl, Cl OMe) $R = -CH_2-$ (quinone Cl, Cl)

33

$M = H_2$

$M = Zn$

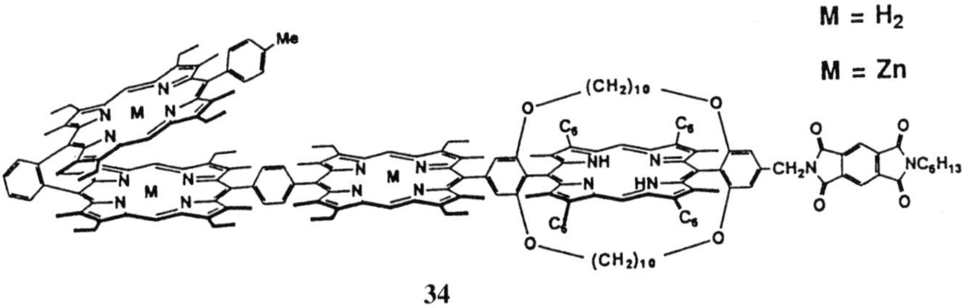

34

The electron transport from dimer porphyrin to the terminal pyromellitdiimide or quinone afforded the ion pair with a lifetime in the order of μs (**32, 33**) [569,570] and long lifetime of 16 μs in THF (**34**) [574]. Obviously, the lifetimes of ion pairs were roughly in the order of diad < triad < tetrad < pentad systems. Other approaches such as proton coupling to the electron transfer [576] as observed in photosynthesis [575], enhancement of electron transfer by pyridine coordination to the central metal [577], and linked molecular systems having oligomeric porphyrins [572–574], carotenoide [543,582–589], aromatic molecules [590–594,753,754], cleft-type [789], MV^{2+} [578–581], and C$_{60}$ [477,595–599, 753,760,762,790] (**35**) have been reported.

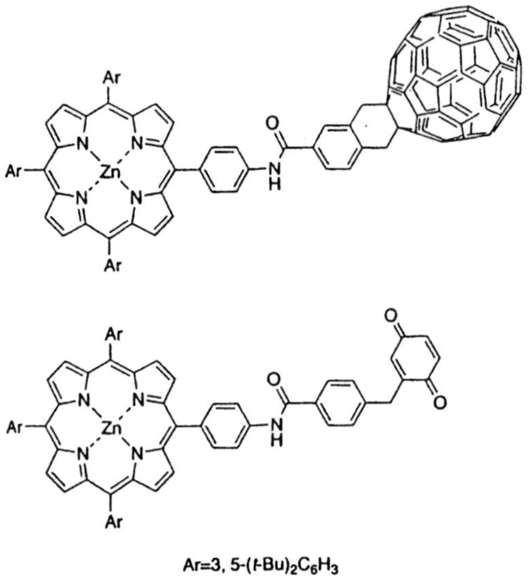

Ar=3, 5-(t-Bu)$_2$C$_6$H$_3$

35

2. Supramolecular System Associated with Non-covalent Interaction

On the basis of intensive studies on electron transfer in linked molecular systems as described above, rather extensive approaches for the supramolecular systems associated with non-covalent interactions have started in this decade. Most of the intermolecular interactions involved in those supramolecular systems are hydrogen bonding as well as electrostatic and coordinating interactions.

Hydrogen Bonding Interaction. A supramolecular system with multiple hydrogen bonds was reported to have a large association constant of 1.0×10^6 M^{-1} (**36**) [604]. Various hydrogen bonding supramolecular systems having large

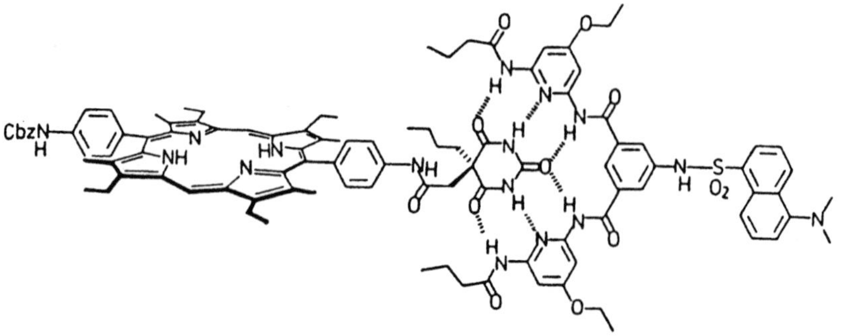

36

association constants (**37, 38**) [602,605], cofacial interaction [600,601], and ca-lixarene moiety for the binding site (**39**) [603] have been reported. Among the hydrogen-bonded supramolecules, an interesting deuterium isotope effect on both the charge separation and recombination between two chromophores was ob-served (**40**) [514].

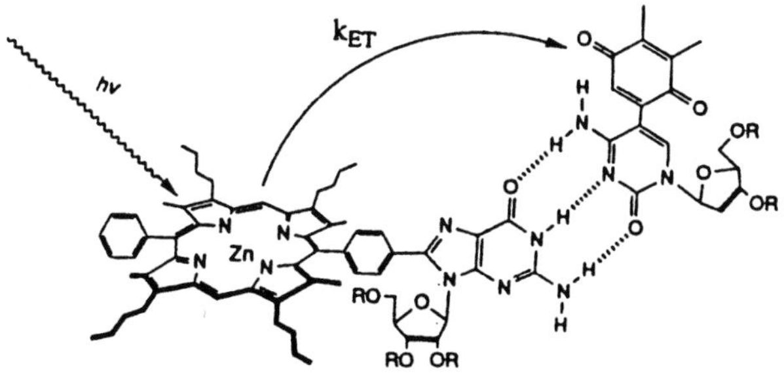

R = SiMe$_2$Bu^t

37

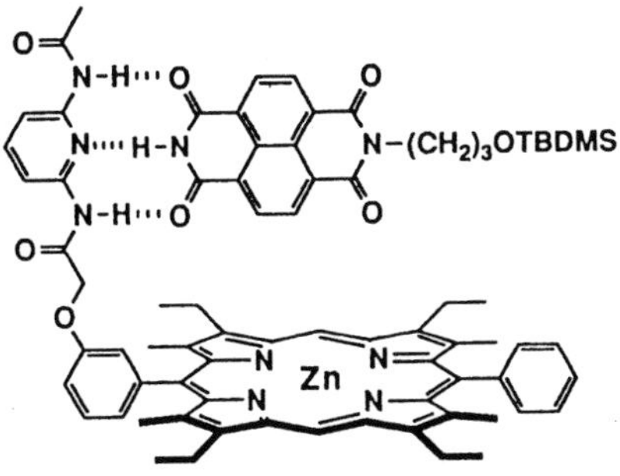

38

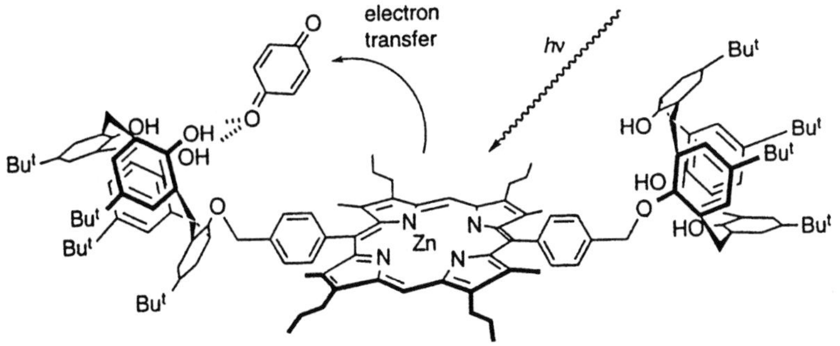

39

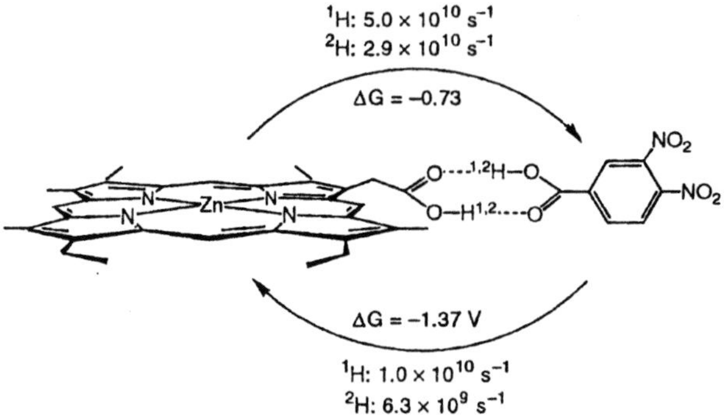
¹H: 5.0 x 10¹⁰ s⁻¹
²H: 2.9 x 10¹⁰ s⁻¹
ΔG = −0.73
NO₂
NO₂
Zn
ΔG = −1.37 V
¹H: 1.0 x 10¹⁰ s⁻¹
²H: 6.3 x 10⁹ s⁻¹

40

Electrostatic Interaction. Electron transfers in supramolecular systems with electrostatic interaction have also been reported. A rapid electron transfer within a cofacial heteroporphyrin dimer was reported [621]. A cofacial ionic interaction between porphyrin and anthraquinone sulfonate was observed to be controlled by sugar hydroxylate [620]. As a model system of cytochrome *c*–peroxidase interaction [608,609], a reconstituted myoglobin with Zn porphyrin was associated with methylviologen through electrostatic interaction and the forward electron transfer was observed to be more favored than the back electron transfer [610] (Fig. 56). In a salt-bridged supramolecules having Zn amidinium porphyrin and aromatic carboxylate as a model of arginine–aspartate interaction [606], the electron transfer was considerably more decreased than expected [607].

Coordination Interaction. Various series of supramolecular systems associated with coordination interaction have been studied [611–617,622] (**41**–**45**).

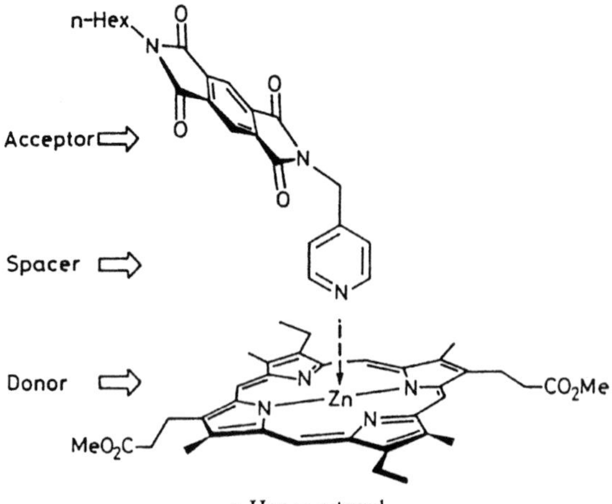
n-Hex
Acceptor
Spacer
Donor
Zn
CO₂Me
MeO₂C

n-Hex = n-hexyl

41

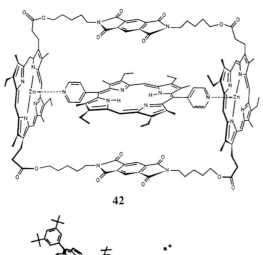

42

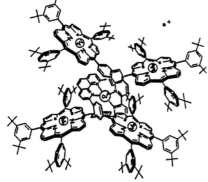

M1 = M2 = Au
M1 = Au: M2 = Zn

43

44

R¹

45

R¹ = ⟨benzene ring⟩—Pr^i

An interesting system with both hydrogen bonding and coordination interactions has been reported [619] (**46**). In bacterial photosynthesis, both histidine
coordination to bacteriochlorophyll and quinone binding through hydrogen bonding with protein have crucial roles [618].

Other supramolecular systems with a moiety of inclusion complex
[623,624] and crown ethers [626] have appeared. An electron transfer from the

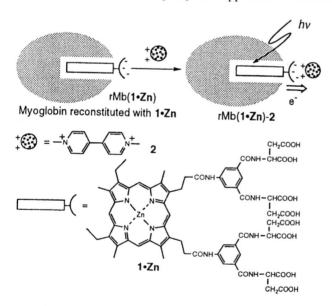

Figure 56 A reconstituted myoglobin with Zn porphyrin associated with methylviologen through electrostatic interaction [610].

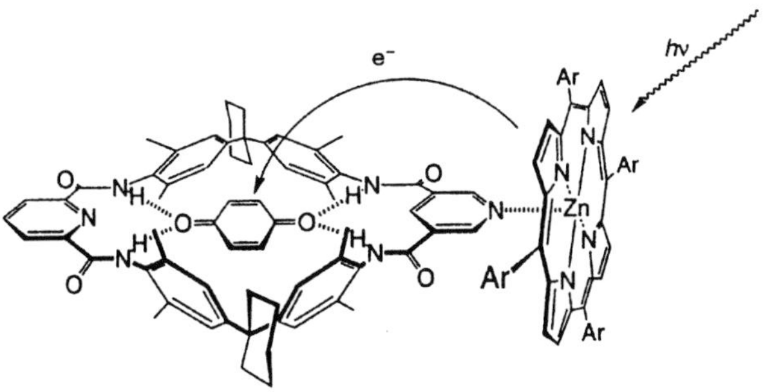

46

excited Zn porphyrin to Ag^+ ion in the crown ether formed Ag^0 to be dissociated from the crown ether moiety. The resultant cation radical of Zn porphyrin was observed to have a long lifetime around 1.6 μs (**47**). The similar complexation and electron transfer to Ag^+ was reported on Zn porphyrin having tetrahydrothiophene pickets on the molecule [625].

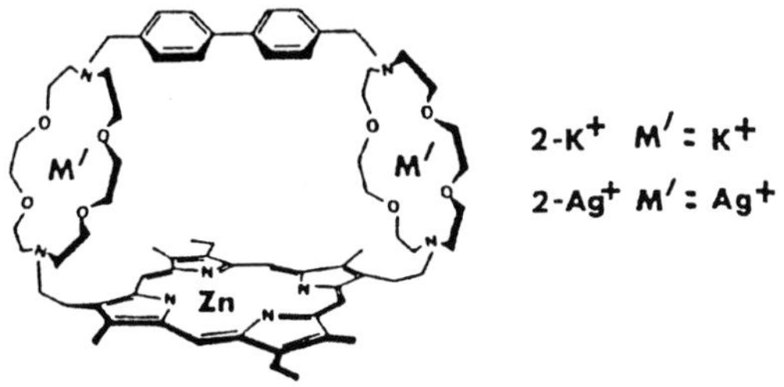

2-K^+ $M' = K^+$
2-Ag^+ $M' = Ag^+$

47

E. Supramolecular Devices

On the basis of extensive challenging approaches for chromophoric assemblies and supramolecular systems, supramolecular devices such as photoresponsive organic conductors and photochemically activated molecular wire have been proposed.

A linearly linked nonamer of porphyrin with 122 Å length has been reported [594,718] (**48**). A synthetic method with the use of the Heck reaction for linear polymer porphyrins has been reported [724].

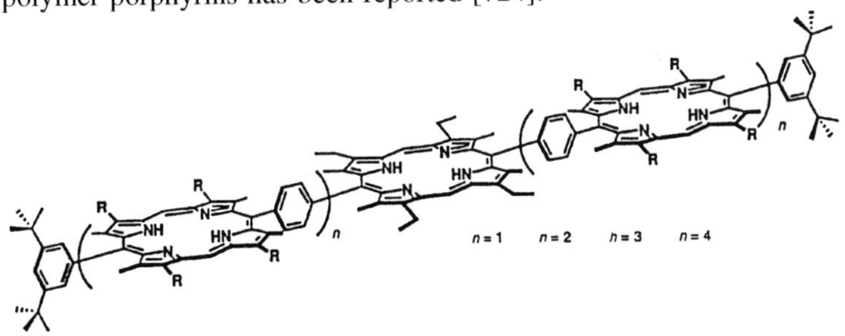

$n = 1$ $n = 2$ $n = 3$ $n = 4$

48

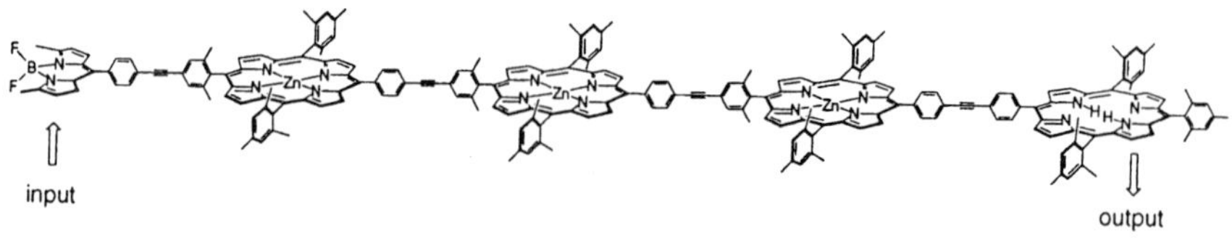

Figure 57   Supramolecular system equipped with input and output device [723].

The linearly linked porphyrins have been extended to a concept of supramolecular systems equipped with an input device and an output device [723] (Fig.
57). Boron-dipyrromethene as an input device mainly absorbed photon and the
excited energy was transferred through Zn porphyrins as transmission elements
to the terminal free-base porphyrin as an output device to fluoresce. The total
molecular length was around 90 Å. The concept has been further proposed to be
extended to systems with redox switched sites [755–758]. Axially linked oligomeric porphyrins known as a "shish-kebab porphyrins" [708–710] have been
proposed to be a "molecular wire" insulated by the porphyrin rings [707] (Fig.
58). The similar axially linked Ru porphyrin oligomer has been reported to be
prepared electrochemically [715]. "Wheel-and-axle"-type oligomeric phosphorous porphyrins have been also pro posed for a conducting molecular wire
[711,712]. The conductivity of electrochemically prepared oligomeric porphyrins
with oligothienyl moiety were observed to be increased upon light irradiation
[713,714] (**49, 50**).

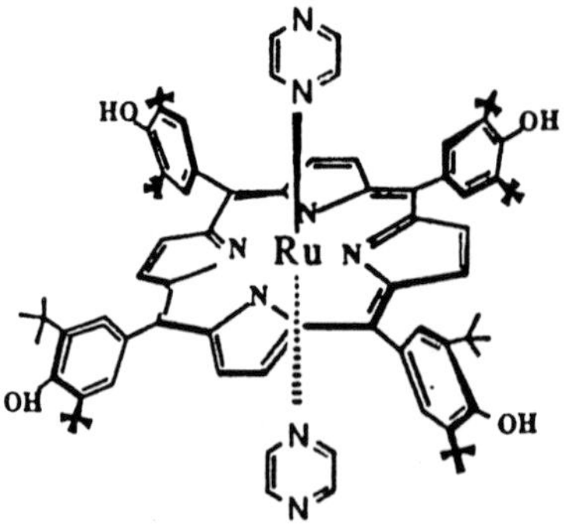

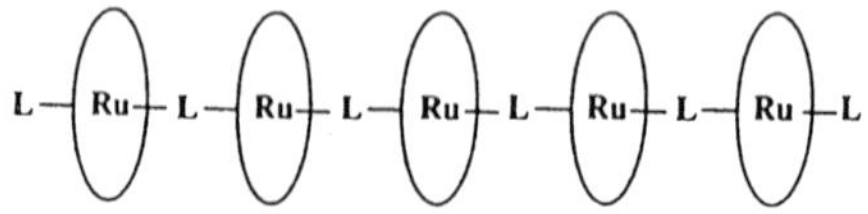

Figure 58   A "shish-kebab porphyrin" [708–710] proposed for a "molecular wire"
insulated by the porphyrin rings [707].

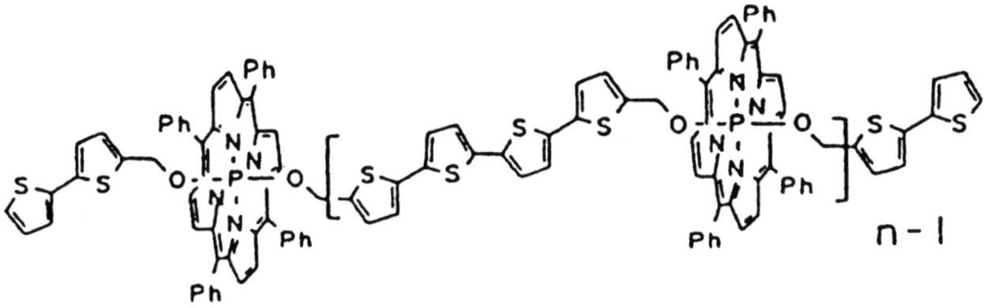

49

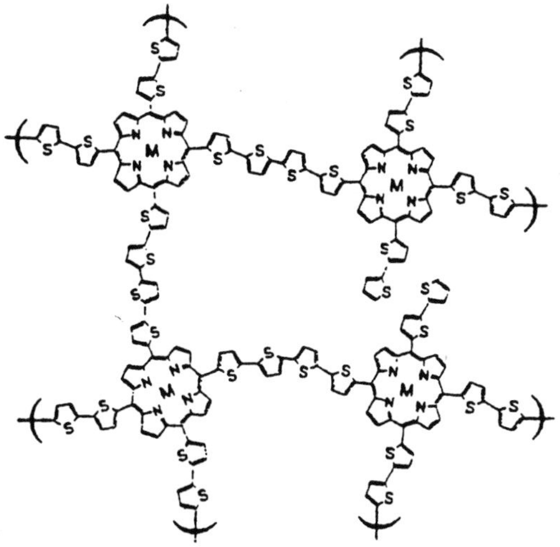

50

F. Other Supramolecular Photochemistry

Photochemical reactions in supramolecular systems have also appeared. Interesting regioselective hydroperoxidation of linoleic acid by singlet oxygen sensitized by cyclodextrin-sandwiched porphyrin has been reported [649]. The hydrophobic cavity of cyclodextrin was supposed to regulate the regioselective attack of singlet oxygen (Fig. 59). Enantio-discriminating oxidation of pinene was also observed.

Asymmetric electron transfer was reported on a hybrid system of RuTCPP(CO)(Py) and bovine serum albumin [632]. Enantioselective photochemical electron transfer to L-Co(acac)$_3$ rather than D-Co(acac)$_3$ was observed.

The photochemistry of supramolecular systems having bilayer membrane [629,630] and molecular assemblies formed by surfactants [631] have also been reported. Electron transfers in zeolite has been reviewed [633]. Photochemical electron transfer in dendrimer has also been reported [759].

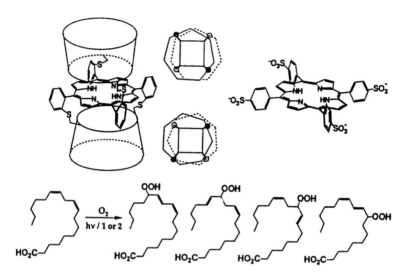

Figure 59   Regioselective hydroperoxidation of linoleic acid by singlet oxygen sensitized by cyclodextrin-sandwiched porphyrin [649].

Diaminoporphyrin was reported to form face-to-face aggregate tubules with an outer diameter around 30 nm and an average length of 800 nm in aqueous solution at pH 5 [716]. Light irradiation to the tubules was observed to induce a formation of porphyrin cation and anion radicals by charge separation (**51**).

51

Octopus porphyrin with eight alkylphosphocholine side chains was observed to form a stable micellar fiber with a thickness of 7 nm in water [717] (Fig. 60). Light irradiation to the molecular assembly in the presence of hydrophobic electron acceptor was observed to induce efficient irreversible electron

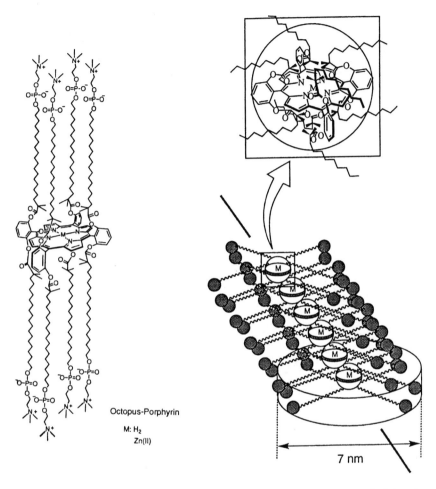

Octopus-Porphyrin

M: H₂
   Zn(II)

Figure 60   A stable micellar fiber formed by octopus porphyrin with eight alkylphospho-choline side chains [717].

transfer from the excited triplet porphyrin moiety. The aggregate was proposed for hydrogen-evolving catalyst.

An interesting electron transfer by a supramolecular system in bilayer membrane has been reported [562,653,663–667,786]. The nonphotochemical electron transfer was observed to be independent on the chain length of spacer alkyl group between the two porphyrin moieties, suggesting a medium-mediated electron transfer via multiple pathways (Fig. 61).

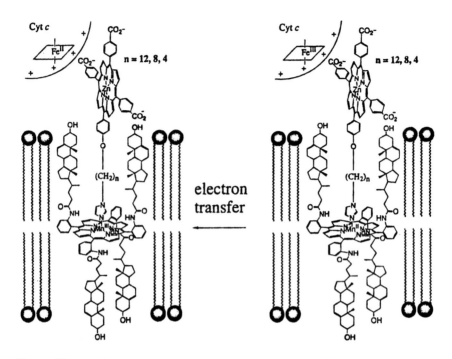

Figure 61 A medium-mediated electron transfer via multiple pathways by a supramolecular system in bilayer membrane [786].

V. CONCLUSION

Molecular photochemistry of porphyrins and metalloporphyrins has exhibited novel aspects even at the current state whereby comprehensive efforts have been accumulated to understand their chemistry in the excited states. As described in the Introduction, the chemistry of porphyrins is inherently an aspect of supramolecular chemistry, since the central metal ion, porphyrin macrocycles, and axial ligands are all associated with coordination interaction. Porphyrin chemistry has multiple aspects of chemistry on 1.) the large π-electron system, 2.) the substituents, 3.) the central metal ion, 4.) the axial ligands, and 5.) the special space for chemical reactions provided by the porphyrin plane and the metal–axial ligand microenvironment. The molecular photochemistry based on their characteristic electronic structures, photophysical properties, and redox properties, which are summarized in Tables 1–3, has been surveyed in this chapter. Comprehensive understanding of porphyrin photochemistry would provide impetus or creation of unit devices in prospective supramolecular systems. Though we are obliged to say that we have not yet had a satisfactory supramolecular system, recent efforts could eventually lead to a real supramolecular system. Molecular photochemistry of porphyrins and metalloporphyrins is undoubtedly opening a main gate leading to a grand field of supramolecular photochemistry.

REFERENCES

1. L. Stryer, *Biochemistry*, W.H. Freeman and Company, New York, 1995.
2. H. Scheer, *Chlorophylls*, CRC Press, Boca Raton, Florida, 1991.
3. M.O. Senge, *Photochem. Photobiol.*, **1993**, *57*, 189.
4. R.E. Dickerson, I. Geis, *Hemoglobin: Structure, Function, Evolution, and Pathology*, Benjamin/Cummings, Menlo Park, California, 1983.
5. R.A. Binstead, M.J. Crossley, N.S. Hush, *Inorg. Chem.*, **1991**, *30*, 1259.
6. O. Ohno, Y. Kaizu, H. Kobayashi, *J. Chem. Phys.*, **1985**, *82*, 1779.
7. M. Gouterman, The Porphyrins, Ed. D. Dolphin, Academic Press, New York, 1978, vol. 3, p. 1.
8. G. Knor, A. Vogler, *Inorg. Chem.*, **1994**, *33*, 314.
9. F. D'Souza, A. Villard, E.V. Caemelbecke, M. Franzen, T. Boschi, P. Tagliatesta, K.M. Kadish, *Inorg. Chem.*, **1993**, *32*, 4042.
10. K.M. Kadish, F. D'Souza, A. Villard, M. Autret, E.V. Caemelbecke, P. Bianco, A. Antonini, P. Tagliatesta, *Inorg. Chem.*, **1994**, *33*, 5169.
11. T. Imamura, T. Jin, T. Suzuki, M. Fujimoto, *Chem. Lett.*, **1985**, 847.
12. R.A. Binstead, M.J. Crossley, N.S. Hush, *Inorg. Chem.*, **1991**, *30*, 1259.
13. A. Harriman, *J. Chem. Soc., Faraday Trans. 2*, **1981**, *77*, 1281.
14. M. Gouterman, *J. Mol. Spectrosc.*, **1961**, *6*, 138.
15. M. Gouterman, G.H. Wagniere, L.C. Snyder, *J. Mol. Spectrosc.*, **1963**, *11*, 108.
16. P.J. Spellane, M. Gouterman, A. Antipas, S. Kim, Y.C. Liu, *Inorg. Chem.*, **1980**, *19*, 386.
17. A. Antipas, M. Gouterman, *J. Am. Chem. Soc.*, **1983**, *105*, 4896.
18. A. Antipas, D. Dolphin, M. Gouterman, E.C. Johnson, *J. Am. Chem. Soc.*, **1978**, *100*, 7705.
19. O. Ohno, Y. Kaizu, H. Kobayashi, *J. Chem. Phys.*, **1985**, *82*, 1779.
20. S.J. Strickler, R.A. Berg, *J. Chem. Phys.*, **1962**, *37*, 814.
21. J.-H. Fuhrhop, in *Porphyrin and Metalloporphyrins*, ed. K. M. Smith, Elsevier, New York, 1975, Chapter 14, pp. 666–698 and references therein.
22. M. Gouterman, L.K. Hanson, G.-E. Khalil, J.W. Buchler, K. Rohbock, D. Dolphin, *J. Am. Chem. Soc.*, **1975**, *97*, 3142.
23. P. Engleman, J. Jortner, *Mol. Phys.*, **1970**, *18*, 145.
24. S.P. McGlynn, T. Azumi, M. Kinoshita, *Molecular Spectroscopy of the Triplet State*, Prentice-Hall, Englewood Cliffs, New Jersey, 1969, p. 16.
25. A. Harriman, *J. Chem. Soc., Faraday Trans. 1*, **1980**, *76*, 1978.
26. S.P. McGlynn, T. Azumi, M. Kinoshita, *Molecular Spectroscopy of the Triplet State*, Prentice Hall, Englewood Cliffs, New Jersey, 1969.
27. A. Harriman, *J. Chem. Soc., Faraday Trans. 2*, **1981**, *77*, 1281.
28. L.M.A. Levine, D. Holten, *J. Phys. Chem.*, **1988**, *92*, 714.

29. C.D. Tait, D. Holten, M.H. Barley, D. Dolphin, B.R. James, *J. Am. Chem. Soc.*, **1985**, *107*, 1930.
30. A. Antipas, J.W. Buchler, M. Gouterman, P.D. Smith, *J. Am. Chem. Soc.*, **1978**, *100*, 3015.
31. A. Antipas, J.W. Buchler, M. Gouterman, P.D. Smith, *J. Am. Chem. Soc.*, **1980**, *102*, 198.
32. T. Kobayashi, K.D. Straub, P.M. Rentzepis, *Photochem. Photobiol.*, **1979**, *29*, 925.
33. V.S. Chirvonyi, B.M. Dzhagarov, A.M. Shul'ga, G.P. Gurinovich, *Dokl. Biophys. (Engl. Transl.)*, **1982**, *259*, 144.
34. D. Kim, C. Kirmaier, D. Holten, *Chem. Phys.*, **1983**, *75*, 305.
35. D. Kim, D. Holten, *Chem. Phys. Lett.*, **1983**, *98*, 584.
36. B.M. Dzhagarow, Yu. V. Timiniskii, V.W. Chirvonyi, G.P. Gurinovich, *Dokl. Biophys. (Engl. Transl)*, **1979**, *247*, 138.
37. D. Tait, D. Holten, M. Gouterman, *Chem. Phys. Lett.*, **1983**, *100*, 268.
38. C.D. Tait, D. Holten, M.H. Barley, D. Dolphin, B.R. James, *J. Am. Chem. Soc.*, **1985**, *107*, 1930.
39. N. Serpone, T.L. Netzel, M. Gouterman, *J. Am. Chem. Soc.*, **1982**, *104*, 246.
40. G. Ponterini, N. Serpone, M.A. Bergkamp, T.L. Netzel, *J. Am. Chem. Soc.*, **1983**, *105*, 4639.
41. D. Kim, D. Holten, M. Gouterman, *J. Am. Chem. Soc.*, **1984**, *106*, 2793.
42. F.D. Straub, P.M. Rentzepis, *Biophys. J.*, **1983**, *41*, 411a.
43. D. Kim, D. Holten, M. Gouterman, J.W. Buchler, *J. Am. Chem. Soc.*, **1984**, *106*, 4015.
44. D. Eastwood, M. Gouterman, *J. Mol. Spectrosc.*, **1970**, *35*, 359.
45. L.K. Hanson, M. Gouterman, J.C. Hanson, *J. Am. Chem. Soc.*, **1973**, *95*, 4822.
46. P. Bhyrappa, V. Krishnan, *Inorg. Chem.*, **1991**, *30*, 239.
47. J.-H. Fuhrhop, K.M. Kadish, D.G. Davis, *J. Am. Chem. Soc.*, **1973**, *95*, 5140.
48. K.M. Kadish, X.Q. Lin, J.Q. Ding, Y.T. Wu, C. Araullo, *Inorg. Chem.*, **1986**, *25*, 3236.
49. K.M. Kadish, F. D'Souza, A. Villard, M. Autret, E.V. Caemelbecke, P. Bianco, A. Antonini, P. Tagliatesta, *Inorg. Chem.*, **1994**, *33*, 5169.
50. M. Autret, Z. Ou, A. Antonini, T. Boschi, P. Tagliatesta, K.M. Kadish, *J. Chem. Soc., Dalton Trans.*, **1996**, 2793.
51. F. D'Souza, A. Villard, E.V. Caemelbecke, M. Franzen, T. Boschi, P. Tagliatesta, K.M. Kadish, *Inorg. Chem.*, **1993**, *32*, 4042.
52. A. Giraudeau, H.J. Callot, J. Jordan, I. Ezhar, M. Gross, *J. Am. Chem. Soc.*, **1979**, *101*, 3875.
53. A. Giraudeau, H.J. Callot, M. Gross, *Inorg. Chem.*, **1979**, *18*, 201.
54. K.M. Barkigia, L. Chantranupong, K.M. Smith, J. Fajer, *J. Am. Chem. Soc.*, **1988**, *110*, 7566.
55. K.M. Barkigia, M.W. Renner, L.R. Furenlid, C.J. Medforth, K.M. Smith, J. Fajer, *J. Am. Chem. Soc.*, **1993**, *115*, 3627.
56. S. Jeon, T.C. Bruice, *Inorg. Chem.*, **1992**, *31*, 4843.
57. S. Jeon, T.C. Bruice, *Inorg. Chem.*, **1991**, *30*, 4311.
58. C.A.-McAdams, K.M. Kadish, *Inorg. Chem.*, **1990**, *29*, 2749.

59. D. Lexa, P. Maillard, M. Momenteau, J.-M. Saveant, *J. Phys. Chem.*, **1987**, *91*, 1951.
60. A.B.P. Lever, *Inorg. Chem.*, **1990**, *29*, 1271.
61. K.M. Kadish, L.R. Shiue, *Inorg. Chem.*, **1982**, *21*, 3623.
62. G.R. Seely, D. Gust, T.A. Moore, A.L. Moore, *J. Phys. Chem.*, **1994**, *98*, 10659.
63. M. Autret, Z. Ou, A. Antonini, T. Boschi, P. Tagliatesta, K.M. Kadish, *J. Chem. Soc., Dalton Trans.*, **1996**, 2793.
64. K. Kalyanasundaram, *J. Chem. Soc., Faraday Trans. 2*, **1983**, *79*, 1365.
65. S. Gaspard, T.-H T.-Thi, *J. Chem. Soc., Perkin Trans. II*, **1989**, 383.
66. S. Yamauchi, Y. Matsukawa, Y. Ohba, M. Iwaizumi, *Inorg. Chem.*, **1996**, *35*, 2910.
67. J.-H. Fuhrhop, D. Mauzerall, *J. Am. Chem. Soc.*, **1968**, *90*, 3875.
68. A. Harriman, A.D. Osborne, *J. Chem. Soc., Faraday Trans. 1*, **1983**, *79*, 765.
69. K.M. Smith, Ed., *Porphyrins and Metalloporphyrins*, Elsevier, New York, 1975, and references therein.
70. T. Barbour, W.J. Belcher, P.J. Brothers, C.E.F. Rickard, D.C. Ware, *Inorg. Chem.*, **1992**, *31*, 746.
71. P. Sayer, M. Gouterman, C.R. Connell, *Acc. Chem. Res.*, **1982**, *15*, 73.
72. P. Sayer, M. Gouterman, C.R. Connell, *J. Am. Chem. Soc.*, **1977**, *99*, 1082.
73. M. Gouterman, L.K. Hanson, G.-E. Khalil, J.W. Buchler, K. Rohbock, D. Dolphin, *J. Am. Chem. Soc.*, **1975**, *97*, 3142.
74. M. Hoshino, K. Yamamoto, J.P. Lillis, T. Chijimatsu, J. Uzawa, *Inorg. Chem.*, **1993**, *32*, 5002.
75. J.-C. Marchon, J.-C. Latour, A. Grand, M. Belakhovsky, M. Loos, *Inorg. Chem.*, **1990**, *29*, 57.
76. M. Hoshino, K. Yamamoto, J.P. Lillis, T. Chijimatsu, J. Izawa, *Coord. Chem. Rev.*, **1994**, *132*, 235.
77. P. Bergamini, S. Sostero, O. Traverso, P. Deplano, L.J. Wilson, *J. Chem. Soc., Dalton Trans.*, **1986**, 2311.
78. L.M. Berreau, J.A. Hays, V.G. Young, Jr., L.K. Woo, *Inorg. Chem.*, **1994**, *33*, 105.
79. C.E. Schulz, H. Song, Y.J. Lee, J.U. Mondal, K. Mohanrao, C.A. Reed, F.A. Walker, W.R. Scheidt, *J. Am. Chem. Soc.*, **1994**, *116*, 7196.
80. M. Lin, M. Lee, K.T. Yue, L.G. Marzilli, *Inorg. Chem.*, **1993**, *32*, 3217.
81. K.M. Kadish, S. Sazou, C. Araullo, Y.M. Liu, A. Saoiabi, M. Ferhat, R. Guilard, *Inorg. Chem.*, **1988**, *27*, 2313.
82. S.E. Creager, R.W. Murray, *Inorg. Chem.*, **1985**, *24*, 3824.
83. T. Imamura, T. Jin, T. Suzuki, M. Fujimoto, *Chem. Lett.*, **1985**, *847*.
84. D. Gao, D. Liu, R.-Q. Yu, G.-D. Zheng, *Fresenius J. Anal. Chem.*, **1995**, *351*, 484.
85. A. Harriman, G. Porter, *J. Chem. Soc., Faraday Trans. 2*, **1979**, *75*, 1543.
86. D. Lexa, M. Momenteau, J. Mispelter, J.-M. Saveant, *Inorg. Chem.*, **1989**, *28*, 30.
87. F. D'Souza, A. Villard, E.V. Caemelbecke, M. Franzen, T. Boschi, P. Tagliatesta, K.M. Kadish, *Inorg. Chem.*, **1993**, *32*, 4042.
88. S.G. DiMagno, A.K. Wertsching, C.R. Ross, II, *J. Am. Chem. Soc.*, **1995**, *117*, 8279.
89. A. Antipas, M. Gouterman, *J. Am. Chem. Soc.*, **1983**, *105*, 4896 and references therein.
90. D. Chang, T. Malinski, A. Ulman, K.M. Kadish, *Inorg. Chem.*, **1984**, *23*, 817.

91. A.M. Stolzenberg, L.J. Schussel, *Inorg. Chem.*, **1991**, *30*, 3205.
92. K.M. Kadish, L.R. Shiue, *Inorg. Chem.*, **1982**, *21*, 3623.
93. K.M. Kadish, B.B.-Cocolios, A. Coutsolelos, P. Mitaine, R. Guilard, *Inorg. Chem.*, **1985**, *24*, 4521.
94. K.M. Kadish, Q.Y. Xu, J.-M. Barbe, J.E. Anderson, E. Wang, R. Guilard, *Inorg. Chem.*, **1988**, *27*, 691.
95. L.A. Martarano, C.-P. Wong, W.D. Horrocks, Jr., A.M.P. Goncalves, *J. Org. Chem.*, **1976**, *80*, 2389.
96. M. Gouterman, C.D. Schumaker, T.S. Srivastava, T. Yonetani, *Chem. Phys. Lett.*, **1976**, *40*, 456.
97. J.W. Buchler, B. Scharbert, *J. Am. Chem. Soc.*, **1988**, *110*, 4272.
98. J.W. Buchler, G. Heinz, *Chem. Ber.*, **1996**, *129*, 201.
99. J.W. Buchler, A.D. Cian, J. Fischer, P. Hammerschmitt, R. Weiss, *Chem. Ber.*, **1991**, *124*, 1051. and references there in.
100. J.E. Anderson, Y.H. Liu, R. Guilard, J.M. Barbe, K.M. Kadish, *Inorg. Chem.*, **1986**, *25*, 3786.
101. J.E. Anderson, Y.H. Liu, R. Guilard, J.M. Barbe, K.M. Kadish, *Inorg. Chem.*, **1986**, *25*, 2250.
102. R. Guilard, P. Richard, M. EL Borai, E. Laviron, *J. Chem. Soc., Chem. Comm.*, **1980**, 516.
103. M. Osada, J. Tachibana, T. Imamura, Y. Sasaki, *Chem. Lett.*, **1996**, 713.
104. M. Hoshino, Y. Iimura, *J. Phys. Chem.*, **1992**, *96*, 179.
105. T. Malinski, P.M. Hanley, K.M. Kadish, *Inorg. Chem.*, **1986**, *25*, 3229.
106. M. Tsutsui, C.P. Hrung, D. Ostfeld, T.S. Srivastava, D.L. Cullen, E.F. Meyer, Jr., *J. Am. Chem. Soc.*, **1975**, *97*, 3952.
107. D.P. Rillema, J.K. Nagle, L.F. Barringer, T.M. Meyer, *J. Am. Chem. Soc.*, **1981**, *103*, 56.
108. W.-H. Leung, *Polyhedron*, **1993**, *12*, 2331.
109. K.J. Berry, B. Moubaraki, K.S. Murray, P.J. Nichols, L.D. Schulz, B.O. West, *Inorg. Chem.*, **1995**, *34*, 4123.
110. K. Kalyanasundaram, *Chem. Phys. Lett.*, **1984**, *104*, 357.
111. A.G. Coutsolelos, D. Lux, *Polyhedron*, **1996**, *15*, 705.
112. R.C. Young, T.M. Meyer, D.G. Whitten, *J. Am. Chem. Soc.*, **1976**, *98*, 286.
113. K.M. Kadish, X.Q. Lin, J.Q. Ding, Y.T. Wu, C. Araullo, *Inorg. Chem.*, **1986**, *25*, 3236.
114. J.-H. Fuhrhop, K.M. Kadish, D.G. Davis, *J. Am. Chem. Soc.*, **1973**, *95*, 5140.
115. A. Shamim, P. Hambright, *Inorg. Chem.*, **1980**, *19*, 564.
116. S. Takagi, Y. Kato, H. Furuta, S. Onaka, *J. Organometal. Chem.*, **1992**, *429*, 287.
117. A. Tabard, R. Guilard, K.M. Kadish, *Inorg. Chem.*, **1986**, *25*, 4277.
118. K.M. Kadish, Q.Y.Y. Xu, G.B. Maiya, *J. Chem. Soc., Dalton Trans.*, **1989**, 1531.
119. L.M. Berreau, L.K. Woo, *J. Am. Chem. Soc.*, **1995**, *117*, 1314.
120. K.M. Kadish, M. Autret, Z. Ou, K. Akiba, S. Masumoto, R. Wada, Y. Yamamoto, *Inorg. Chem.*, **1996**, *35*, 5564.
121. G. Knor, A. Vogler, *Inorg. Chem.*, **1994**, *33*, 314.
122. K. Kalyanasundaram, J.A. Shelnutt, M. Gratzel, *Inorg. Chem.*, **1988**, *27*, 2820.
123. Y. Kaizu, M. Asano, H. Kobayashi, *J. Phys. Chem.*, **1986**, *90*, 3906.

124. R.J. Donohoe, J.K. Duchowski, D.F. Bocian, *J. Am. Chem. Soc.*, **1988**, *110*, 6119.
125. E.B. Fleischer, R.D. Chapman, M. Krishnamurthy, *Inorg. Chem.*, **1979**, *18*, 2156.
126. S. Kato, M. Tsutsui, D.L. Cullen, E.F. Meyer, Jr., *J. Am. Chem. Soc.*, **1977**, *99*, 620.
127. W.-H. Leung, T.S.M. Hun, K.-Y. Wong, W.-T. Womg, *J. Chem. Soc., Dalton Trans.*, **1994**, 2713.
128. C.-M. Che, W.-H. Leung, W.-C. Chung, *Inorg. Chem.*, **1990**, *29*, 1841.
129. C.M. Che, W.-C. Chung, T.-F. Lai, *Inorg. Chem.*, **1988**, *27*, 2801.
130. S. Mosseri, P. Neta, P. Hambright, D.Y. Sabry, A. Harriman, *J. Chem. Soc., Dalton Trans.*, **1988**, 2705.
131. J.P. Collman, L.L. Chung, D.A. Tyvoll, *Inorg. Chem.*, **1995**, *34*, 1311.
132. K.M. Kadish, Y.J. Deng, J.D. Korp, *Inorg. Chem.*, **1990**, *29*, 1036.
133. T. Shimidzu, H. Segawa, T. Iyoda, K. Honda, *J. Chem. Soc., Faraday Trans. 2*, **1987**, *83*, 2191.
134. L.R. Robinson, P. Hambright, *Inorg. Chem.*, **1992**, *31*, 652.
135. K.M. Smith, J.-J. Lai, *Tetrahedron Lett.*, **1980**, *21*, 433.
136. G.-P. Chacko, P. Hambright, *Inorg. Chem.*, **1994**, *33*, 5595.
137. A. Dormond, B. Belkalem, P. Charpin, M. Lance, D. Vigner, G. Folcher, G. Guilard, *Inorg. Chem.*, **1986**, *25*, 4785.
138. K.M. Kadish, G. Moninot, Y. Hu, D. Dubois, A. Ibnlfassi, J.-M. Barbe, R. Guilard, *J. Am. Chem. Soc.*, **1993**, *115*, 8153.
139. K. Kalyanasundaram, *J. Chem. Soc., Faraday Trans. 2*, **1983**, *79*, 1365.
140. S. Gaspard, T.-H. T.-Thi, *J. Chem. Soc. Perkin Trans. II*, **1989**, 383.
141. G.R. Loppnow, D. Melamed, A.R. Leheny, A.D. Hamilton, T.G. Spiro, *J. Phys. Chem.*, **1993**, *97*, 8969.
142. L.A. Martarano, C.-P. Wong, W.D. Horrocks, Jr., A.M.P. Goncalves, *J. Org. Chem.*, **1976**, *80*, 2389.
143. M. Gouterman, C.D. Schumaker, T.S. Srivastava, T. Yonetani, *Chem. Phys. Lett.*, **1976**, *40*, 456.
144. K. Kalyanasundaram, J.A. Shelnutt, M. Gratzel, *Inorg. Chem.*, **1988**, *27*, 2820.
145. G. Ponterini, N. Serpone, M.A. Bergkamp, T.L. Netzel, *J. Am. Chem. Soc.*, **1983**, *105*, 4639.
146. O. Bilsel, J. Rodriguez, D. Holten, *J. Am. Chem. Soc.*, **1990**, *112*, 4075.
147. A. Harriman, *J. Chem. Soc., Faraday Trans. 1*, **1980**, *76*, 1978.
148. G.S. Neta, P. Neta, P. Hambright, A.N. Thompson, Jr., A. Harriman, *J. Phys. Chem.*, **1989**, *93*, 6181.
149. F.R. Hopf, D.G. Whitten, in *Porphyrins and Metalloporphyrins*, Ed. K.M. Smith, Elsevier, New York, 1975, Chapter 16, pp. 666–698 and references therein.
150. S.C. Jeoung, D. Kim, S.J. Hahn, S.Y. Ryu, M. Yoon, *J. Phys. Chem. A.*, **1998**, *102*, 315.
151. A. Harriman, *J. Chem. Soc., Faraday Trans. 1*, **1981**, *77*, 369.
152. S.G. Kruglik, P.A. Apanasevich, V.S. Chirvony, V.V. Kvach, V.A. Orlovch, *J. Phys. Chem.*, **1995**, *99*, 2978 and references therein.
153. A. Harriman, G. Porter, P. Walters, *J. Chem. Soc., Faraday Trans. 1*, **1983**, *79*, 1335.
154. M. Gouterman, D.B. Howell, *J. Chem. Phys.*, **1974**, *61*, 3491.

155. S. Gentemann, J. Albaneze, R.G. Ferrer, S. Knapp, J.A. Potenza, H.J. Schugar, D. Holten, *J. Am. Chem. Soc.*, **1994**, *116*, 281.
156. Z.A.-Gamra, A. Harriman, *J. Chem. Soc., Faraday Trans.* 2, **1986**, *82*, 2337.
157. R. Bonnett, D.J. McGarvey, A. Harriman, E.J. Land, T.G. Truscott, U-J. Winfield, *Photochem. Photobiol.*, **1988**, *48*, 271.
158. O. Ohno, Y. Kaizu, H. Kobayashi, *J. Chem. Phys.*, **1985**, *82*, 1779.
159. A. Harriman, *J. Chem. Soc. Faraday Trans.* 2, **1981**, *77*, 1281.
160. X. Yan, D. Holten, *J. Phys. Chem.*, **1988**, *92*, 409.
161. A. Harriman, *J. Chem. Soc. Faraday Trans.* 1, **1982**, *78*, 2727.
162. R. Grigg, W.D.J.A. Norbert, *J. Chem. Soc. Chem. Commun.*, **1992**, 1298.
163. K. Susumu, H. Segawa, T. Shimidzu, *Chem. Lett.*, **1995**, 929.
164. A. Harriman, *J. Chem. Soc., Faraday Trans.* 1, **1980**, *76*, 1978.
165. A. Harriman, G. Porter, A. Wilowska, *J. Chem. Soc., Faraday Trans.* 2, **1983**, *79*, 807.
166. A. Harriman, G. Porter, M.-C. Richoux, *J. Chem. Soc., Faraday Trans.* 2, **1981**, *77*, 833.
167. A. Antipas, D. Dolphin, M. Gouterman, E.C. Johnson, *J. Am. Chem. Soc.*, **1978**, *100*, 7705.
168. H.N. Fonda, J.V. Gilbert, R.A. Cormier, J.R. Sprague, K. Kamioka, J.S. Connolly, *J. Phys. Chem.*, **1993**, *97*, 7024.
169. G. Ponterini, N. Serpone, M.A. Bergkamp, T.L. Netzel, *J. Am. Chem. Soc.*, **1983**, *105*, 4639 and references therein.
170. H.-J. Kim, K. Kim, *J. Phys. Chem.*, **1992**, *96*, 8374.
171. V.S. Chirvonyi, B.M. Dzhagarov, Yu. V. Timibskii, G.P. Gurinovich, *Chem. Phys. Lett.*, **1980**, *70*, 79.
172. O. Bilsel, J. Rodriguez, S.N. Milam, P.A. Gorlin, G.S. Girolami, K.S. Suslick, D. Holten, *J. Am. Chem. Soc.*, **1992**, *114*, 6528.
173. M.P. Irvine, R.J. Harrison, M.A. Strahand, G.S. Beddard, *Ber. Bunsenges. Phys. Chem.*, **1985**, *89*, 226.
174. O. Bilsel, J. Rodriguez, D. Holten, *J. Phys. Chem.*, **1990**, *94*, 3508.
175. D. Kim, D. Holten, M. Gouterman, J.W. Buchler, *J. Am. Chem. Soc.*, **1984**, *106*, 4015.
176. C. Drew Tait, D. Holten, M.H. Barley, D. Dolphin, B.R. James, *J. Am. Chem. Soc.*, **1985**, *107*, 1930.
177. L.M.A. Levine, D. Holten, *J. Phys. Chem.*, **1988**, *82*, 714.
178. M. Hoshino, M. Imamura, S. Watanabe, T. Hama, *J. Phys. Chem.*, **1984**, *88*, 45.
179. N. Serpone, H. Ledon, T.L. Netzel, *Inorg. Chem.*, **1984**, *23*, 454.
180. X. Yan, D. Holten, *J. Phys. Chem.*, **1988**, *92*, 5982.
181. L.M.A. Levine, D. Holten, *J. Phys. Chem.*, **1998**, *92*, 714.
182. H.N. Fonda, J.V. Gilbert, R.A. Cormier, J.R. Sprague, K. Kamioka, J.S. Connolly, *J. Phys. Chem.*, **1993**, *97*, 7024.
183. M. Tezuka, Y. Ohkatsu, T. Osa, *Bull. Chem. Soc. Jpn.*, **1976**, *49*, 1435.
184. M. Autret, Z. Ou, A. Antonini, T. Boschi, P. Tagliatesta, K.M. Kadish, *J. Chem. Soc., Dalton Trans.*, **1996**, 2793 and references therein.
185. J.A. Ferguson, T.J. Meyer, D.J. Whitten, *Inorg. Chem.*, **1972**, *11*, 2767.
186. J.-H. Fuhrhop, K.M. Kadish, D.G. Davis, *J. Am. Chem. Soc.*, **1973**, *95*, 5140.

187. M.C. Richoux, P. Neta, A. Harriman, S. Baral, P. Hambright, *J. Phys. Chem.*, **1986**, *90*, 2462.

188. K. Kalyanasundaram, *J. Chem. Soc., Faraday Trans. 2*, **1983**, *79*, 1365.

189. J. Takeda, M. Sato, *Chem. Lett.*, **1995**, 939.

190. T.A. Rao, B.G. Maiya, *J. Chem. Soc. Chem. Commun.*, **1995**, 939.

191. Y.H. Liu, M.-F. Benassy, S. Chojnacki, F. D'Souza, T. Barbour, W.J. Belcher, P.J. Brothers, K.M. Kadish, *Inorg. Chem.*, **1994**, *33*, 4480.

192. J.-C. Marchon, J.-C. Latour, A. Grand, M. Belakhovsky, M. Loos, *Inorg. Chem.*, **1990**, *29*, 57.

193. R. Guilard, J.-M. Latour, C. Lecomte, J.-M. Marchon, J. Protas, D. Ripoll, *Inorg. Chem.*, **1978**, *17*, 1228.

194. K.M. Kadish, S. Sazou, C. Araullo, Y.M. Liu, A. Saoiabi, M. Ferhat, R. Guilard, *Inorg. Chem.*, **1988**, *27*, 2313.

195. S. Jeon, T.C. Bruice, *Inorg. Chem.*, **1991**, *30*, 4311.

196. S. Jeon, T.C. Bruice, *Inorg. Chem.*, **1992**, *31*, 4843.

197. D.M. Guldi, P. Hambright, D. Lexa, P. Neta, J.-M. Saveant, *J. Phys. Chem.*, **1992**, *96*, 4459.

198. S.L. Kelly, K.M. Kadish, *Inorg. Chem.*, **1984**, *23*, 679.

199. H. Sugimoto, N. Ueda, M. Mori, *Bull. Chem. Soc. Jpn.*, **1981**, *54*, 3425.

200. M. Tezuka, Y. Ohkatsu, T. Osa, *Bull. Chem. Soc. Jpn.*, **1976**, *49*, 1435.

201. F. D'Souza, A. Villard, E.V. Caemelbecke, M. Franzen, T. Boschi, P. Tagliatesta, K.M. Kadish, *Inorg. Chem.*, **1993**, *32*, 4042.

202. R.A. Binstead, M.J. Crossley, N.S. Hush, *Inorg. Chem.*, **1991**, *30*, 1259.

203. A.M. Stolzenberg, L.J. Schussel, *Inorg. Chem.*, **1991**, *30*, 3205.

204. A. Harriman, G. Porter, P. Walters, *J. Chem. Soc., Faraday Trans. 1*, **1983**, *79*, 1335.

205. K.M. Kadish, B.B.-Cocolios, A. Coutsolelos, P. Mitaine, R. Guilard, *Inorg. Chem.*, **1985**, *24*, 4521.

206. K.M. Kadish, Q.Y. Xu, J.-M. Barbe, J.E. Anderson, E. Wang, R. Guilard, *Inorg. Chem.*, **1988**, *27*, 691.

207. J.W. Buchler, B. Scharbert, *J. Am. Chem. Soc.*, **1988**, *110*, 4272.

208. G.S. Girolami, C.L. Hein, K.S. Suslick, *Angew. Chem. Int. Ed. Engl.*, **1996**, *35*, 1223.

209. J.W. Buchler, A.D. Cian, J. Fischer, P. Hammerschmitt, R. Weiss, *Chem. Ber.*, **1991**, *124*, 1051 and references therein.

210. J.E. Anderson, Y.H. Liu, R. Guilard, J.M. Barbe, K.M. Kadish, *Inorg. Chem.*, **1986**, *25*, 3786.

211. J.E. Anderson, Y.H. Liu, R. Guilard, J.M. Barbe, K.M. Kadish, *Inorg. Chem.*, **1986**, *25*, 2250.

212. T. Malinski, P.M. Hanley, K.M. Kadish, *Inorg. Chem.*, **1986**, *25*, 3229.

213. X.H. Mu, K.M. Kadish, *Langmuir*, **1990**, *6*, 51.

214. D.P. Rillema, J.K. Nagle, L.F. Barringer, T.M. Meyer, *J. Am. Chem. Soc.*, **1981**, *103*, 56.

215. X.H. Mu, K.M. Kadish, *Langmuir*, **1990**, *6*, 51.

216. E.R. Birnbaum, W.P. Schaefer, J.A. Labinger, J.E. Bercaw, H.B. Gray, *Inorg. Chem.*, **1995**, *34*, 1751.

217. K.M. Kadish, C.-L. Yao, J.E. Anderson, P. Cocolios, *Inorg. Chem.*, **1985**, *24*, 4515.
218. K.M. Kadish, Y. Hu, T. Boschi, P. Tagliatesta, *Inorg. Chem.*, **1993**, *32*, 2996.
219. K.M. Kadish, X.Q. Lin, J.Q. Ding, Y.T. Wu, C. Araullo, *Inorg. Chem.*, **1986**, *25*, 3236.
220. A. Giraudeau, A. Louati, H.J. Callot, M. Gross, *Inorg. Chem.*, **1981**, *20*, 769.
221. K.M. Kadish, L.R. Shiue, *Inorg. Chem.*, **1982**, *21*, 3623.
222. J.H. Fuhrhop, in *Porphyrin and Metalloporphyrins*, Ed. K.M. Smith, Elsevier, New York, Chap. 14, 1975, pp. 593–622.
223. A. Stanienda, Z. *Phys. Chem.*, **1964**, *229*, 259.
224. J.-L. Cornillon, J.E. Anderson, K.M. Kadish, *Inorg. Chem.*, **1986**, *25*, 2611.
225. K.M. Kadish, J.L. Cornillon, P. Cocolios, A. Tabard, R. Guilard, *Inorg. Chem.*, **1985**, *24*, 3645.
226. K.M. Kadish, Q.Y.Y. Xu, G.B. Maiya, *J. Chem. Soc., Dalton Trans.*, **1989**, 1531.
227. K.M. Kadish, D. Dubois, J.-M. Barbe, R. Guilard, *Inorg. Chem.*, **1991**, *30*, 4498.
228. K.M. Kadish, M. Autret, Z. Ou, K. Akiba, S. Masumoto, R. Wada, Y. Yamamoto, *Inorg. Chem.*, **1996**, *35*, 5564.
229. K. Kalyanasundaram, J.A. Shelnutt, M. Gratzel, *Inorg. Chem.*, **1988**, *27*, 2820.
230. R.J. Donohoe, J.K. Duchowski, D.F. Bocian, *J. Am. Chem. Soc.*, **1988**, *110*, 6119.
231. W.-H. Leung, T.S.M. Hun, K.-Y. Wong, W.-T. Womg, *J. Chem. Soc., Dalton Trans.*, **1994**, 2713.
232. S. Mosseri, P. Neta, P. Hambright, D.Y. Sabry, A. Harriman, *J. Chem. Soc., Dalton Trans.*, **1988**, 2705.
233. K.M. Kadish, Y.J. Deng, J.D. Korp, *Inorg. Chem.*, **1990**, *29*, 1036.
234. A. Antipas, D. Dolphin, M. Gouterman, E.C. Johnson, *J. Am. Chem. Soc.*, **1978**, *100*, 7705.
235. T. Shimidzu, H. Segawa, T. Iyoda, K. Honda, *J. Chem. Soc., Faraday Trans. 2*, **1987**, *83*, 2191.
236. A. Giraudeau, A. Louati, H.J. Callot, M. Gross, *Inorg. Chem.*, **1981**, *20*, 769.
237. R.H. Felton, H. Linschitz, *J. Am. Chem. Soc.*, **1966**, *88*, 1113.
238. K.M. Kadish, G. Moninot, Y. Hu, D. Dubois, A. Ibnlfassi, J.-M. Barbe, R. Guilard, *J. Am. Chem. Soc.*, **1993**, *115*, 8153.
239. J.M. Zaleski, C. Turro, R.D. Mussell, D.G. Nocera, *Coord. Chem. Rev.*, **1994**, *132*, 249.
240. (a) R.A. Marcus, *J. Chem. Phys.*, **1956**, *24*, 966. (b) R.A. Marcus, *J. Chem. Phys.*, **1956**, *24*, 979.
241. R.A. Marcus, *J. Chem. Phys.*, **1965**, *43*, 2654.
242. R.A. Marcus, *Pure and Appl. Chem.*, **1997**, *69*, 13.
243. (a) G.J. Kavarnos, N.J. Turro, *Chem. Rev.*, **1986**, *86*, 401. (b) H.D. Roth, *Top. Curr. Chem.*, **1990**, *156*, 1. (c) G.J. Kavarnos, *Fundamentals of Photoinduced Electron Transfer*, Wiley-VCH, New York, 1993.
244. D. Rehm, A. Weller, *Isr. J. Chem.*, **1970**, *8*, 259.
245. J.-y. Liu, J.R. Bolton, *J. Phys. Chem.*, **1992**, *96*, 1718.
246. R.A. Marcus, P. Siders, *J. Phys. Chem.*, **1982**, *86*, 622.
247. S.S. Isied, A. Vaddilian, J.F. Wishart, C. Creutz, H.A. Schwarz, N. Sutin, *J. Am. Chem. Soc.*, **1988**, *110*, 635.
248. M.J. Weaver, *Chem. Rev.*, **1992**, *92*, 463.

249. M.J. Weaver, G.E. McManis III, *Acc. Chem. Res.*, **1990**, *23*, 294.
250. M. Maroncelli, J. MaCinnis, G.R. Fleming, *Science*, **1989**, *243*, 1674.
251. P.F. Barbara, G.C. Walker, T.P. Smith, *Science*, **1992**, *256*, 975.
252. R.I. Cukier, D.G. Nocera, *J. Chem. Phys.*, **1992**, *97*, 7371.
253. R.A. Marcus, N. Sutin, *Biochim. Biophys. Acta*, **1985**, *811*, 265.
254. S.S. Isied, A. Vassilian, J.F. Wishart, *J. Am. Chem. Soc.*, **1988**, *110*, 635.
255. A. Haim, *Pure Appl. Chem.*, **1983**, *55*, 89.
256. S.S. Isied, A. Vassilian, R.H. Magnuson, H.A. Schwarz, *J. Am. Chem. Soc.*, **1986**, *107*, 7432.
257. H. Oevering, M.N. PaddonRow, M. Heppener, A.M. Oliver, E. Cotsaris, J.W. Verhoeven, N.S. Hush, *J. Am. Chem. Soc.*, **1987**, *109*, 3258.
258. M.R. Wasielewski, M.P. Niemczyk, W.A. Svec, E.B. Pewitt, *J. Am. Chem. Soc.*, **1985**, *107*, 5562.
259. B.S. Brunschwig, P.J. DeLaive, A.M. Goldberg, H.B. Gray, S.L. Mayo, N. Sutin, *Inorg. Chem.*, **1985**, *24*, 3473.
260. R. Bechtold, C. Kuehm, C. Lepre, S.S. Isied, *Nature*, **1986**, *322*, 286.
261. P.S. Ho, C. Sutoris, N. Liang, E. Margoliash, B.M. Hoffman, *J. Am. Chem. Soc.*, **1985**, *107*, 1070.
262. K.T. Conklin, G. McLendon, *Inorg. Chem.*, **1986**, *25*, 4084.
263. L.J. Schaffer, H. Taube, *J. Phys. Chem.*, **1986**, *90*, 3669.
264. A.M. Brun, A. Harriman, S.M. Hubig, *J. Phys. Chem.*, **1992**, *96*, 254.
265. I.R. Gould, J.E. Moser, D. Ege, S. Farid, *J. Am. Chem. Soc.*, **1988**, *110*, 1991.
266. S. Ojima, H. Miyasaka, N. Mataga, *J. Phys. Chem.*, **1990**, *94*, 7534.
267. H. Seki, *J. Chem. Soc., Faraday Trans.*, **1992**, *88*, 35.
268. A. Harriman, G. Porter, A. Wilowska, *J. Chem. Soc., Faraday Trans. 2*, **1983**, *79*, 807.
269. A.M. Brun, A. Harriman, S.M. Hubig, *J. Phys. Chem.*, **1992**, *96*, 254.
270. Y. Degani, I. Willner, *J. Phys. Chem.*, **1985**, *89*, 5685.
271. I. Willner, Y. Degani, *J. Chem. Soc., Chem. Commun.*, **1982**, 1249.
272. R.H. Schmell, D.G. Whitten, *J. Phys. Chem.*, **1981**, *85*, 3473.
273. M.P. Pileni, M. Gratzel, *J. Phys. Chem.*, **1980**, *84*, 1822.
274. N.J. Turro, M. Gratzel, A.M. Braun, *Angew. Chem. Int. Ed. Engl.*, **1980**, *19*, 675.
275. P.A. Brugger, P.P. Infelta, A.M. Braun, M. Gratzel, *J. Am. Chem. Soc.*, **1981**, *103*, 320.
276. T. Matsuo, K. Takuma, Y. Tsusui, T. Nishijima, *Coord. Chem. Rev.*, **1980**, *10*, 195.
277. I. Willner, J.W. Otvos, M. Calvin, *J. Am. Chem. Soc.*, **1981**, *103*, 3203.
278. I. Willner, J.M. Yang, J.W. Otvos, M. Calvin, *J. Phys. Chem.*, **1981**, *85*, 3277.
279. Y. Degani, I. Willner, *J. Am. Chem. Soc.*, **1983**, *105*, 6228.
280. T. Nagamura, N. Takeyama, K. Tanaka, T. Matsuo, *J. Phys. Chem.*, **1986**, *90*, 2247.
281. J.A. Barltrop, J.D. Coyle, *Excited States In Organic Chemistry*, John Wiley, New York, 1975.
282. N.J. Turro, *Modern Molecular Photochemistry*, University Science Book, Mill Valley, California, 1991.
283. T. Kakitani, Hikari, Bussitsu, *Seimei to Hannou*, Maruzen, 1998.
284. M. Kasha, *Radiat. Res.*, **1963**, *20*, 55.

285. D.L. Dexter, *Chem. Phys.*, **1953**, *21*, 836.
286. T. Forster, *Disc. Faraday Soc.*, **1959**, *27*, 7.
287. J. Martensson, *Chem. Phys. Lett.*, **1994**, *229*, 449.
288. A. Osuka, K. Maruyama, I. Yamazaki, N. Tamai, *Chem. Phys. Lett.*, **1990**, *165*, 392.
289. A. Osuka, K. Maruyama, *J. Am. Chem. Soc.*, **1988**, *110*, 4454.
290. T. Nagata, A. Osuka, K. Maruyama, *J. Am. Chem. Soc.*, **1990**, *112*, 3054.
291. A. Jablonski, *Z. Physik.*, **1935**, *96*, 236.
292. N.J. Turro, *Modern Molecular Photochemistry*, 300.
293. M. Kasha, M. Ashraf El-Bayoumi, W. Rhodes, *J. Chim. Phys.*, **1961**, *58*, 916.
294. J.P. Collman, R.R. Gagne, C.A. Reed, T.R. Halbert, G. Lang, W.T. Robinson, *J. Am. Chem. Soc.*, **1975**, *97*, 1427.
295. A. Vogler, H. Kunkely, *Ber. Bunsen-Ges. Phys. Chem.*, **1976**, *80*, 425.
296. D.K. White, J.B. Cannon, T.G. Traylor, *J. Am. Chem. Soc.*, **1979**, *101*, 443.
297. E.A. Morlino, L.A. Walker II, R.J. Sension, M.A.J. Rodgers, *J. Am. Chem. Soc.*, **1995**, *117*, 4429.
298. P.A. Cornelius, R.M. Hochstrasser, A.W. Steele, *J. Mol. Biol.*, **1983**, *163*, 119.
299. J.W. Petrich, J.-C. Lambry, K. Kuczera, M. Karplus, C. Poyart, J.-L. Martin, *Biochemistry*, **1991**, *30*, 3975.
300. T.G. Traylor, D. Magde, D.J. Taube, K.A. Jongeward, *J. Am. Chem. Soc.*, **1987**, *109*, 5864.
301. T.G. Traylor, D. Magde, D.J. Taube, K.A. Jongeward, D. Bandyopadhyay, J. Luo, K.N. Walda, *J. Am. Chem. Soc.*, **1992**, *114*, 417.
302. T. Mincey, T.G. Traylor, *J. Am. Chem. Soc.*, **1979**, *101*, 765.
303. T.G. Traylor, D. Magde, J. Marsters, K. Jongeward, G.-Z. Wu, K. Walda, *J. Am. Chem. Soc.*, **1993**, *115*, 4808.
304. T.G. Traylor, D. Magde, J. Luo, K.N. Walda, D. Bandyopadhyay, G.-Z. Wu, V.S. Sharma, *J. Am. Chem. Soc.*, **1992**, *114*, 9011.
305. W. Doster, S.F. Bowne, H. Frauenfelder, L. Reinisch, E. Scyamsunder, *J. Mol. Biol.*, **1987**, *194*, 299.
306. A. Ansari, E.E. DiIorio, D.D. Dlott, H. Frauenfelder, I.E.T. Iben, P. Langer, H. Roder, T.B. Sauke, E. Scyamsunder, *Biochemistry*, **1986**, *25*, 3139.
307. J.W. Petrich, J.L. Martin, *Chem. Phys.*, **1989**, *131*, 31.
308. T.G. Traylor, *Pure Appl. Chem.*, **1991**, *63*, 265.
309. R.H. Austin, K. Beeson, L. Eisenstein, H. Frauenfelder, I.C. Gunsalus, V.P. Marshall, *Phys. Rev. Lett.*, **1974**, *32*, 403.
310. R.H. Austin, K. Beeson, L. Eisenstein, H. Frauenfelder, I.C. Gunsalus, *Biochemistry*, **1975**, *14*, 5355.
311. T. Bucher, J. Kaspers, *Biochim. Biophys. Acta.*, **1947**, *1*, 21.
312. M. Hoshino, K. Ueda, M. Takahashi, M. Yamaji, Y. Hama, Y. Miyazaki, *J. Phys. Chem.*, **1992**, *96*, 8863.
313. D. Beece, L. Eisenstein, H. Frauenfelder, D. Good, M.C. Marden, L. Reinisch, A.H. Reynolds, L.B. Sorensen, K.T. Yue, *Biochemistry*, **1980**, *19*, 5147.
314. C. Tetreau, M. Momenteau, D. Lavalette, *Inorg. Chem.*, **1990**, *29*, 1727.
315. T.G. Traylor, D. Campbell, S. Tsuchiya, M. Mitchell, D.V. Stynes, *J. Am. Chem. Soc.*, **1980**, *102*, 5939.

316. T. Hashimoto, R.L. Dyer, M.J. Crossly, J.E. Baldwin, F. Basolo, *J. Am. Chem. Soc.*, **1982**, *104*, 2101.

317. M. Hoshino, Y. Kashiwagi, *J. Phys. Chem.*, **1990**, *94*, 673.

318. G.B. Jameson, J.A. Ibers, in *Bioinorganic Chemistry*, I. Bertini, H.B. Gray, J. Valentine, Eds., University Science Book, Mill Valley, California, 1991.

319. T.G. Grogan, N. Bag, T.G. Traylor, D. Magde, *J. Phys. Chem.*, **1994**, *98*, 13791.

320. M. Hoshino, K. Yamamoto, J.P. Lillis, T. Chijimatsu, J. Uzawa, *Coord. Chem. Rev.*, **1994**, *132*, 235.

321. M. Hoshino, K. Yamamoto, J.P. Lillis, T. Chijimatsu, J. Uzawa, *Inorg. Chem.*, **1993**, *32*, 5002.

322. H. Ledon, M. Bonnet, *J. Chem. Soc., Chem. Commun.*, **1979**, 702.

323. H.J. Ledon, M. Bonnet, *J. Am. Chem. Soc.*, **1981**, *103*, 6209.

324. J. Zakrzewski, C. Giannotti, *J. Chem. Soc., Chem. Commun.*, **1990**, 743.

325. S. Yamauchi, Y. Sizuki, T. Ueda, K. Akiyama, Y. Ohba, M. Iwaizumi, *Chem. Phys. Lett.*, **1995**, *232*, 121.

326. H. Segawa, A. Nakamoto, T. Shimidzu, *J. Chem. Soc., Chem. Commun.*, **1992**, 1066.

327. H. Segawa, K. Kunimoto, A. Nakamoto, T. Shimidzu, *J. Chem. Soc., Perkin Trans. 1*, **1992**, 939.

328. H. Murayama, Shiragami, H. Inoue, the 8th Symposium on Photochemistry of Coordination Compounds, **1995**, 59.

329. T. Imamura, Y. Yamamoto, T. Suzuki, M. Fujimoto, *Chem. Lett.*, **1987**, 2185.

330. D.G. Whitten, P.D. Wildes, C.A. DeRosier, *J. Am. Chem. Soc.*, **1972**, *94*, 7811.
D.G. Whitten, P.D. Woldes, I.G. Lopp, *J. Am. Chem. Soc.*, **1969**, *91*, 3393.

331. Y. Iseki, E. Watanabe, A. Mori, S. Inoue, *J. Am. Chem. Soc.*, **1993**, *115*, 7313.

332. Y. Iseki, S. Inoue, *J. Chem. Soc., Chem. Commun.*, **1994**, *2577*.

333. D.N. Henderickson, M.G. Kinnard, K.S. Suslick, *J. Am. Chem. Soc.*, **1987**, *109*, 1243.

334. M. Hoshino, K. Ueda, M. Takahashi, M. Yamaji, Y. Hama, *J. Chem. Soc., Faraday Trans.*, **1992**, *88*, 405.

335. C. Bartocci, F. Scandola, A. Ferri, V. Carassiti, *J. Am. Chem. Soc.*, **1980**, *102*, 7067.

336. C. Bartocci, R. Amadelli, A. Maldotti, V. Carassiti, *Polyhedron*, **1986**, *5*, 1297.

337. D.N. Hendrickson, M.G. Kinnaird, K.S. Suslick, *J. Am. Chem. Soc.*, **1987**, *109*, 1243.

338. M. Hoshino, T. Hirai, *J. Phys. Chem.*, **1987**, *91*, 4510.

339. M. Hoshino, M. Yamaji, Y. Hama, *Chem. Phys. Lett.*, **1986**, *125*, 369.

340. M. Hoshino, K. Yasufuku, H. Seki, H. Yamazaki, *J. Phys. Chem.*, **1985**, *89*, 3080.

341. J.F. Endicott, T.L. Netzel, *J. Am. Chem. Soc.*, **1979**, *101*, 4000.

342. R.T. Traylor, L. Smucker, M.L. Hanna, J. Gill, *Arch. Biochem. Biophys.*, **1973**, *156*, 521.

343. H. Murayama, S. Inoue, Y. Ohkatsu, *Chem. Lett.*, **1983**, 381.

344. X.-X. Zhang, B.B. Wayland, *J. Am. Chem. Soc.*, **1994**, *116*, 7897.

345. B.B. Wayland, S. Ba, A.E. Sherry, *J. Am. Chem. Soc.*, **1991**, *113*, 5305.

346. B.B. Wayland, S. Ba, A.E. Sherry, *Inorg. Chem.*, **1992**, *31*, 148.

347. L.I. Simandi, E.B.-Zahonyi, Z. Szeverenyi, S. Nemeth, *J. Chem. Soc., Dalton Trans.*, **1980**, 276.

348. J. Halpern, *Inorg. Chim. Acta*, **1982**, *62*, 31.
349. H. Murayama, S. Inoue, *Chem. Lett.*, **1985**, 1377.
350. Y. Watanabe, T. Aida, S. Inoue, *Macromolecules*, **1990**, *23*, 2612.
351. S. Inoue, N. Takeda, *Bell. Chem. Soc. Jpn.*, **1977**, *50*, 984.
352. Y. Watanabe, T. Aida, S. Inoue, *Macromolecules*, **1991**, *24*, 3970.
353. Y. Hirai, T. Aida, S. Inoue, *J. Am. Chem. Soc.*, **1989**, *111*, 3062.
354. M. Komatsu, T. Aida, S. Inoue, *J. Am. Chem. Soc.*, **1991**, *113*, 8492.
355. Y. Ito, K. Kunimoto, S. Miyachi, T. Kako, *Tetrahedron Lett.*, **1991**, *32*, 4007.
356. Y. Ito, J. Chem. Soc., *Chem. Commun.*, **1991**, 622.
357. B.C. Gilbert, J.R. Lindsey Smith, P. MacFaul, P. Traylor, *J. Chem. Soc., Perkin Trans. 2*, **1996**, 511.
358. B.C. Gilbert, J.R. Lindsey Smith, P. MacFaul, P. Traylor, *J. Chem. Soc., Perkin Trans. 2*, **1996**, 519.
359. C. Bartocci, A. Maldotti, G. Varani, P. Battioni, V. Carassiti, D. Mansuy, *Inorg. Chem.*, **1991**, *30*, 1255.
360. A. Maldotti, C. Bartocci, R. Amadelli, E. Polo, P. Battioni, D. Mansuy, *J. Chem. Soc. Chem. Commun.*, **1991**, 1487.
361. J.T. Groves, D.V. Subramanian, *J. Am. Chem. Soc.*, **1984**, *106*, 2177.
362. L. Weber, R. Hommel, J. Behling, G. Haufe, H. Hennig, *J. Am. Chem. Soc.*, **1994**, *116*, 2400.
363. L. Weber, G. Haufe, D. Rehorek, H. Henning, *J. Chem. Soc., Chem. Commun.*, **1991**, 502.
364. A. Maldotti, C. Bartocci, G. Varani, A. Molinari, *Inorg. Chem.*, **1996**, *35*, 1126.
365. H. Hennig, J. Behling, R. Meusinger, L. Weber, *Chem. Ber.*, **1995**, *128*, 229.
366. P. Bergamini, S. Sostero, O. Traverso, P. Deplano, L.J. Wilson, *J. Chem. Soc., Dalton Trans.*, **1986**, 2311.
367. Y. Matsuda, S. Sakamoto, H. Koshima, Y. Murakami, *J. Am. Chem. Soc.*, **1985**, *107*, 6415.
368. Y. Matsuda, H. Koshima, K. Nakamura, Y. Murakami, *Chem. Lett.*, **1988**, 625.
369. L. Weber, G. Haufe, D. Rehorek, H. Hennig, *J. Mol. Cat.*, **1990**, *60*, 267.
370. J.A. Shelnutt, D.E. Trudell, *Tetrahedron Lett.*, **1989**, 5231.
371. Y. Aoyama, K. Midorikawa, H. Toi, H. Ogoshi, *Chem. Lett.*, **1987**, 1651.
372. J. Zakrzewski, C. Giannotti, *J. Chem. Soc., Chem. Commun.*, **1992**, 662.
373. (a)A. Fujishima, K. Honda, *Nature*, **1972**, *238*, 37. (b)A.J. Bard, M.A. Fox., *Acc. Chem. Res.*, **1995**, *28*, 141.
374. J.M. Lehn, J.P. Sauvage, *Nuov. J. Chim.*, **1977**, *1*, 449.
375. K. Kalyanasundaram, J. Kiwi, M. Gratzel, *Helv. Chim. Acta*, **1978**, *61*, 2720.
376. A. Harriman, G. Porter, M.-C. Richoux, *J. Chem. Soc., Faraday Trans. 2*, **1981**, *77*, 833.
377. I. Okura, N.K.-Thuan, *J. Chem. Soc., Chem. Commun.*, **1980**, 84.
378. J.-H. Fuhrhop, W. Kruger, H.H. David, *Liebigs Ann. Chem.*, **1983**, 204.
379. J. Darwent, P. Douglas, A. Harriman, G. Porter, M.C. Richoux, *Coord. Chem. Rev.*, **1982**, *44*, 83.
380. K. Kalyanasundaram, M. Gratzel, *Helv. Chim. Acta*, **1980**, *63*, 478.
381. P. Cuendet, M. Gratzel, *Photochem. Photobiol.*, **1984**, *39*, 609.
382. J. Handman, A. Harriman, G. Porter, *Nature*, **1984**, *307*, 534.

383. I. Okura, S. Aono, T. Kita, *Chem. Lett.*, **1984**, 57.
384. S. Aono, T. Kita, I. Okura, A. Yamada, *Photochem. Photobiol.*, **1986**, *43*, 1.
385. L. Persaud, A.J. Bard, A. Campion, M.A. Fox, T.E. Mallouk, S.E. Webber, J.M. White, *J. Am. Chem. Soc.*, **1987**, *109*, 7309.
386. K.B. Yoon, *Chem. Rev.*, **1993**, *93*, 321.
387. D.S. Robins, P.K. Dutta, *Langmuir*, **1996**, *12*, 402.
388. S. Aono, N. Kaji, I. Okura, *J. Chem. Soc., Chem. Commun.*, **1986**, 170.
389. I. Okura, S. Aono, M. Hoshino, A. Yamada, *Inorg. Chim. Acta*, **1984**, *86*, L55.
390. I. Okura, Y. Kinumi, *Bull. Chem. Soc. Jpn.*, **1990**, *63*, 2922.
391. H. Hosono, T. Tani, I. Uemura, *Chem. Commun.*, **1996**, 1893.
392. T. Kamachi, T. Hiraishi, I. Okura, *Chem. Lett.*, **1995**, 33.
393. S. Tazuke, N. Kitamura, *Nature*, **1978**, *275*, 301.
394. B.P. Sullivan, T.J. Meyer, *J. Chem. Soc., Chem. Commun.*, **1984**, 1244.
395. F.R. Remke, D.L. DeLaet, J. Gao, C.P. Kubiak, *J. Am. Chem. Soc.*, **1988**, *110*, 6904.
396. Y. Ito, Y. Uozo, T. Matsumura, *J. Chem. Soc., Chem. Commun.*, **1988**, 562.
397. J.-M. Lehn, R. Ziessel, *J. Organometal. Chem.*, **1990**, *382*, 157.
398. M. Ishida, T. Terada, K. Tanaka, T. Tanaka, *Inorg. Chem.*, **1990**, *29*, 905.
399. T.J. Meyer, *J. Chem. Soc., Chem. Commun.*, **1985**, 1416.
400. S. Matsuoka, K. Yamamoto, T. Ogawa, M. Kusaba, N. Nakashima, E. Fujita, S. Yanagida, *J. Am. Chem. Soc.*, **1993**, *115*, 601.
401. J. Hewecker, J.-M. Lehn, R. Ziessel, *J. Chem. Soc., Chem. Commun.*, **1983**, 536.
402. I. Willner, D. Mandler, *J. Am. Chem. Soc.*, **1989**, *111*, 1330.
403. J. Costamagna, G. Ferraudi, J. Canales, J. Vargas, *Coord. Chem. Rev.*, **1996**, *148*, 221.
404. N. Takeda, S. Inoue, *Bull. Chem. Soc. Jpn.*, **1977**, *50*, 984.
405. Y. Hirai, T. Aida, S. Inoue, *J. Am. Chem. Soc.*, **1989**, *111*, 3062.
406. M. Komatsu, T. Aida, S. Inoue, *J. Am. Chem. Soc.*, **1991**, *113*, 8492.
407. W. Ruttinger, G.C. Dismukes, *Chem. Rev.*, **1997**, *97*, 1 and references therein.
408. P.A. Christensen, A. Harriman, G. Porter, *J. Chem. Soc., Faraday Trans. 2*, **1984**, *80*, 1451.
409. A. Harriman, G.S. Nahor, S. Mosseri, P. Neta, *J. Chem. Soc., Faraday Trans. 1*, **1988**, *84*, 2821.
410. G.S. Nahor, S. Mosseri, P. Neta, A. Harriman, *J. Phys. Chem.*, **1988**, *92*, 4499.
411. P.A. Christensen, A. Harriman, G. Porter, P. Neta, *J. Chem. Soc., Faraday Trans. 2*, **1984**, *80*, 1451.
412. G.S. Nahor, P. Neta, P. Hambright, A.N. Thompson, Jr., A. Harriman, *J. Phys. Chem.*, **1989**, *93*, 6181.
413. S. Takagi, T. Okamoto, T. Shiragami, H. Inoue, *Chem. Lett.*, **1993**, 793.
414. L. Eberson, *Acta Chem. Scand. B*, **1984**, *38*, 439.
415. H. Inoue, M. Sumitani, A. Sekita, M. Hida, *J. Chem. Soc., Chem. Commun.*, **1987**, 1681.
416. H. Inoue, T. Okamoto, Y. Kameo, M. Sumitani, A. Fujiwara, D. Ishibashi, M. Hida, *J. Chem. Soc., Perkin Trans. 1*, **1994**, 105.
417. S. Takagi, M. Suzuki, T. Shiragami, H. Inoue, *J. Am. Chem. Soc.*, **1997**, *119*, 8712.
418. T. Shiragami, K. Kubomura, D. Ishibashi, H. Inoue, *J. Am. Chem. Soc.*, **1996**, *118*, 6311.

419. J.-M. Lehn, *Science*, **1985**, *227*, 849.
420. J.-M. Lehn, *Angew. Chem. Int. Ed. Engl.*, **1990**, *29*, 1304.
421. V. Balzani, *Tetrahedron*, **1992**, *48*, 10443.
422. J.-M. Lehn, *Angew. Chem. Int. Ed. Engl.*, **1988**, *27*, 89.
423. J.-M. Lehn, *Supramolecular Chemistry: Concepts and Perspectives*, VCH, New York, 1995.
424. G. Feher, J.P. Allen, M.Y. Okamura, D.C. Rees, *Nature*, **1989**, *339*, 111.
425. J. Diesenhofer, H. Michel, *Angew. Chem., Int. Ed. Engl.*, **1989**, *28*, 829.
426. R. Huber, *Angew. Chem., Int. Ed. Engl.*, **1989**, *28*, 848.
427. G. McDermott, S.M. Prince, A.A. Freer, A.M. Hawthornthwaite-Lawless, M.Z. Papiz, R.J. Cogdell, N.W. Isaacs, *Nature*, **1995**, *374*, 517.
428. J. Deisenhofer, O. Epp, K. Miki, R. Huber, H. Michel, *J. Mol. Biol.*, **1984**, *180*, 385.
429. J. Deisenhofer, O. Epp, K. Miki, R. Huber, H. Michel, *Nature*, **1985**, *318*, 618.
430. H. Michel, O. Epp, J. Deisenhofer, *EMBO J.*, **1986**, *5*, 2445.
431. J.P. Allen, G. Feher, T.O. Yeates, H. Komiya, D.C. Rees, *Proc. Natl. Acad. Sci. U.S.A.*, **1987**, *84*, 5730.
432. J.P. Allen, G. Feher, T.O. Yeates, H. Komiya, D.C. Rees, *Proc. Natl. Acad. Sci. U.S.A.*, **1987**, *84*, 6162.
433. T.O. Yeates, et. al., *Proc. Natl. Acad. Sci. U.S.A.*, **1988**, *85*, 7993.
434. J.P. Allen, G. Feher, T.O. Yeates, H. Komiya, D.C. Rees, *Proc. Natl. Acad. Sci. U.S.A.*, **1988**, *85*, 8487.
435. J.D. McElroy, G. Feher, D.C. Mauzerall, *Biochim. Biophys. Acta*, **1969**, *172*, 180.
436. J.R. Norris, R.A. Uphaus, H.L. Crespi, J.J. Katz, *Proc. Natl. Acad. Sci. U.S.A.*, **1971**, *68*, 625.
437. G. Feher, A.J. Hoff, R.A. Isaacson, L.C. Ackerson, *Ann. N.Y. Acad. Sci.*, **1975**, *244*, 239.
438. J.R. Norris, H. Scheer, J.J. Katz, *Ann. N.Y. Acad. Sci.*, **1975**, *244*, 260.
439. P.A. Loach, R.L. Hall, *Proc. Natl. Acad. Sci. U.S.A.*, **1972**, *69*, 786.
440. G. Feher, M.Y. Okamura, J.D. McElroy, *Biochim. Biophys. Acta*, **1972**, *267*, 222.
441. M.Y. Okamura, R.A. Isaacson, G. Feher, *Proc. Natl. Acad. Sci. U.S.A.*, **1975**, *72*, 3491.
442. J. Fajer, D.C. Brune, M.S. Davis, A. Forman, L.D. Spaulding, *Proc. Natl. Acad. Sci. U.S.A.*, **1975**, *72*, 4956.
443. V.A. Shuvalov, V.V. Klimov, *Biochim. Biophys. Acta*, **1976**, *440*, 587.
444. D.M. Tiede, R.C. Prince, G.H. Reed, P.L. Dutton, *FEBS Lett.*, **1976**, *65*, 301.
445. C. Kirmaier, D. Holten, *Photosynthesis Res.*, **1987**, *13*, 225.
446. H. Michel, *L. Mol. Biol.*, **1982**, *158*, 567.
447. T.O. Yeates, H. Komiya, D.C. Rees, J.P. Allen, G. Feher, *Proc. Natl. Acad. Sci. U.S.A.*, **1987**, *84*, 6438.
448. H. Komiya, T.O. Yeates, D.C. Rees, J.P. Allen, G. Feher, *Proc. Natl. Acad. Sci. U.S.A.*, **1988**, *85*, 9012.
449. C.-H. Chang, D.M. Tiede, J. Tang, J.R. Norris, M. Schiffer, *FEBS Lett.*, **1986**, *205*, 82.
450. J. Deisenhofer, H. Michel, *Angew. Chem. Int. Ed. Engl.*, **1989**, *28*, 829.

451. Th. Forster, *Z. Electrochem.*, **1949**, *53*, 93.
452. R. van Grondelle, R. Monshouwer, L. Valkunas, *Pure Appl. Chem.*, **1997**, *69*, 1211.
453. R. van Grondelle, J.P. Dekker, T. Gillbro, V. Sundstrom, *Biochim. Biophys. Acta*, **1994**, *1187*, 1.
454. C.C. Moser, J.M. Keske, K. Warncke, R.S. Farid, P.L. Dutton, *Nature*, **1992**, *355*, 796.
455. D.N. Beratan, J.N. Betts, J.N. Onuchic, *Science*, **1991**, *252*, 1285.
456. D.N. Beratan, J.N. Onuchic, J.R. Winkler, H.B. Gray, *Science*, **1992**, *258*, 1740.
457. J.R. Winkler, H.B. Gray, *Chem. Rev.*, **1992**, *92*, 369.
458. G. McLendon, R. Hake, *Chem. Rev.*, **1992**, *92*, 481.
459. M. Plato, M.E. Michel-Beyerle, M. Bixon, J. Jortner, *FEBS Lett.*, **1989**, *249*, 70.
460. M.R. Wasielewski, *Chem. Rev.*, **1992**, *92*, 435.
461. C.C. Moser, J.M. Keske, K. Warncke, R.S. Farid, P.L. Dutton, *Nature*, **1992**, *355*, 796.
462. D.N. Beratan, J.N. Onuchic, J. Winkler, H. Gray, *Science*, **1992**, *355*, 796.
463. H. Pelletier, J. Kraut, *Science*, **1992**, *258*, 1748.
464. C.C. Moser, J.M. Keske, K. Warncke, R.S. Farid, P.L. Dutton, *Nature*, **1992**, *355*, 796.
465. D.N. Beratan, J.N. Betts, J.N. Onuchic, *Science*, **1991**, *252*, 1285.
466. D.N. Beratan, J.N. Onuchic, J.R. Winkler, H.B. Gray, *Science*, **1992**, *258*, 1740.
467. J.R. Winkler, H.B. Gray, *Chem. Rev.*, **1992**, *92*, 369.
468. M.R. Wasielewski, M.P. Niemczyk, W.A. Svec, *J. Am. Chem. Soc.*, **1985**, *107*, 1080.
469. M.R. Wasielewski, W.A. Svec, *J. Am. Chem. Soc.*, **1984**, *106*, 5043.
470. J.S. Lindsey, J.K. DAaney, D.C. Mauzerall, et. al., *J. Am. Chem. Soc.*, **1988**, *110*, 3610.
471. M.R. Wasielewski, *Chem. Rev.*, **1992**, *92*, 435.
472. H. Kurreck, M. Huber, *Angew. Chem. Int. Ed. Engl.*, **1995**, *34*, 849.
473. J.L. Sessler, *Isr. J. Chem.*, **1992**, *32*, 449.
474. V.V. Borovkov, R.P. Evstigneeva, L.N. Strekova, E.I. Filippovich, *Russ. Chem. Rev.*, **1989**, *58*, 602.
475. A. Osuka, N. Mataga, T. Okada, *Pure Appl. Chem.*, **1997**, *69*, 797.
476. N. Mataga, *Pure Appl. Chem.*, **1997**, *69*, 729.
477. Y. Sakata, H. Imahori, H. Tsue, S. Higashida, T. Akiyama, E. Yoshizawa, M. Aoki, K. Yamada, K. Hagiwara, S. Taniguchi, T. Okada, *Pure Appl. Chem.*, **1997**, *69*, 1951.
478. A.L. Moore, T.A. Moore, D. Gust, J.J. Silber, L. Sereno, F. Fungo, L. Otero, G. Steinberg-Yfrach, P.A. Liddell, S.-C. Hung, H. Imahori, S. Cardoso, D. Tatman, A.N. Macpherson, *Pure Appl. Chem.*, **1997**, *69*, 2111.
479. J.R. Miller, *J. Am. Chem. Soc.*, **1984**, *106*, 3047.
480. G.L. Closs, J.R. Miller, *Science*, **1988**, *240*, 440.
481. R.A. Marcus, N. Sutin, *Biochim. Biophys. Acta*, **1985**, *811*, 265.
482. M.R. Wasielewski, M.P. Niemczyk, W.A. Svec, E.B. Pewitt, *J. Am. Chem. Soc.*, **1985**, *107*, 1080.

483. T. Asahi, M. Ohkohchi, R. Matsusaka, N. Mataga, R.P. Zhang, A. Osuka, K. Maruyama, *J. Am. Chem. Soc.*, **1993**, *115*, 5665.

484. G.L. Gaines, III, M.P. O'Neil, W.A. Svec, M.P. Niemczyk, M.R. Wasielewski, *J. Am. Chem. Soc.*, **1991**, *113*, 719.

485. F. Pollinger, C. Musewald, H. Heitele, M.E. Michel-Beyerle, C. Anders, M. Futscher, G. Voit, H.A. Staab, *Ber. Bunsenges. Phys. Chem.*, **1996**, *100*, 2076.

486. H.A. Staab, A. Feurer, C. Krieger, A.S. Kumar, *Liebigs Ann./Recueil*, **1997**, 2321.

487. H.A. Staab, R. Hauck, B. Popp, *Eur. J. Org. Chem.*, **1998**, 631.

488. J.R. Miller, J.V. Beitz, R.K. Huddleston, *J. Chem. Phys.*, **1984**, *106*, 5057.

489. M.R. Wasielewski, in *Photoinduced Electron Transfer*, M.A. Fox, M. Chanon, Eds., Elsevier, New York, 1988, part A, pp. 161–206.

490. G.L. Closs, J.R. Miller, *Science*, **1988**, *240*, 440.

491. J.J. Hopfield, *Proc. Natl. Acad. Sci. U.S.A.*, **1974**, *71*, 4135.

492. J. Jortner, *J. Chem. Phys.*, **1976**, *64*, 4860.

493. J.M. Warman, K.J. Smit, S.A. Jonker, J.W. Verhoeven, H. Oevering, J. Kroon, M.N. Paddon-Row, A.M. Oliver, *Chem. Phys.*, **1993**, *170*, 359.

494. B.A. Leland, A.D. Jaran, P.M. Felker, J.J. Hopfield, A.H. Zewail, P.B. Dervan, *J. Phys. Chem.*, **1985**, *89*, 5571.

495. P. Finckh, H. Heitele, M.M. Beyerle, *J. Phys. Chem.*, **1988**, *92*, 6584.

496. A. Helms, D. Heiler, G. McLendon, *J. Am. Chem. Soc.*, **1992**, *114*, 6227.

497. A. Osuka, K. Maruyama, N. Mataga, T. Asahi, I. Yamazaki, N. Tamai, *J. Am. Chem. Soc.*, **1990**, *112*, 4958.

498. A. Helms, D. Heiler, G. McLendon, *J. Am. Chem. Soc.*, **1991**, *113*, 4325.

499. Y. Won, R.A. Friesner, *Biochim. Biophys. Acta*, **1988**, *935*, 9.

500. S. Knapp, T.G.M. Dhar, J. Albaneze, S. Gentemann, J.A. Potenza, D. Holten, H.J. Schugar, *J. Am. Chem. Soc.*, **1991**, *113*, 4010.

501. M. Migita, T. Okada, N. Mataga, S. Nishitani, N. Kurata, Y. Sakata, S. Misumi, *Chem. Phys. Lett.*, **1981**, *84*, 263.

502. J.A. Schmidt, J.-Y. Liu, J.R. Bolton, M.D. Archer, V.P.Y. Gadzekpo, *J. Chem. Soc., Faraday Trans. 1*, **1989**, *85*, 1027.

503. J.-Y. Liu, J.A. Schmidt, J.R. Bolton, *J. Phys. Chem.*, **1991**, *95*, 6924.

504. J.-Y. Liu, J.R. Bolton, *J. Phys. Chem.*, **1992**, *96*, 1718.

505. H. Tamiaki, K. Nomura, K. Maruyama, *Bull. Chem. Soc. Jpn.*, **1994**, *67*, 1863.

506. G.L. Gaines, M.P. O'Neil, W.A. Svec, M.P. Niemczyk, M.R. Wasielewski, *J. Am. Chem. Soc.*, **1991**, *113*, 719.

507. G. Tapolsky, R. Duesing, T.J. Meyer, *J. Phys. Chem.*, **1991**, *95*, 1105.

508. J. Liu, J.A. Schmidt, J.R. Bolton, *J. Phys. Chem.*, **1991**, *95*, 6924.

509. A.R. Katritzky, D.W. Zhu, K.S. Schanze, *J. Phys. Chem.*, **1991**, *95*, 5737.

510. M.R. Wasielewski, D.G. Johnson, M.P. Niemczyk, G.L. Gaines, M.P. O'Neil, W.A. Svec, *J. Am. Chem. Soc.*, **1990**, *112*, 6482.

511. S.S. Isied, A. Vassilian, J.F. Wishart, C. Creutz, H.A. Schwarz, N. Sutin, *J. Am. Chem. Soc.*, **1988**, *110*, 635.

512. J.M. Zaleski, C.K. Chang, G.E. Leroi, R.I. Cukier, D.G. Nocera, *J. Am. Chem. Soc.*, **1992**, *114*, 3564.

513. C. Turro, C.K. Chang, G.E. Leroi, R.I. Cukier, D.G. Nocera, *J. Am. Chem. Soc.*, **1992**, *114*, 4013.

514. J.M. Zaleski, C. Turro, R.D. Mussell, D.G. Nocera, *Coord. Chem. Rev.*, **1994**, *132*, 249.

515. A. Harriman, V. Heitz, J.-P. Sauvage, *J. Phys. Chem.*, **1993**, *97*, 5940.

516. A.M. Brun, A. Harriman, V. Heitz, J.-P. Sauvage, *J. Am. Chem. Soc.*, **1991**, *113*, 8657.

517. S. Larsson, *J. Am. Chem. Soc.*, **1981**, *103*, 4034.

518. A. Harriman, V. Heitz, J.-C. Chambron, J.-P. Sauvage, *Coord. Chem. Rev.*, **1994**, *132*, 229.

519. J.-C. Chambron, S.C.-Noblat, A. Harriman, V. Heitz, J.P. Sauvage, *Pure Appl. Chem.*, **1993**, *65*, 2343.

520. J.-C. Chambron, A. Harriman, V. Heitz, J.-P. Sauvage, *J. Am. Chem. Soc.*, **1993**, *115*, 7419.

521. S. Higashida, H. Tsue, K. Sugiura, T. Kaneda, Y. Sakata, Y. Tanaka, S. Taniguchi, T. Okada, *Bull. Chem. Soc. Jpn.*, **1996**, *69*, 1329.

522. Y. Sakata, H. Tsue, Y. Goto, S. Misumi, T. Asahi, S. Nishikawa, T. Okada, N. Mataga, *Chem. Lett.*, **1991**, 1307.

523. A.M. Oliver, D.C. Craig, M.N. Paddon-Row, J. Kroon, J.W. Verhoeven, *Chem. Phys. Lett.*, **1988**, *150*, 366.

524. Y. Sakata, H. Tsue, M.P. O'Neil, G.P. Wiederrecht, M.R. Wasielewski, *J. Am. Chem. Soc.*, **1994**, *116*, 6904.

525. V.V. Borovkov, A. Ishida, S. Takamuku, Y. Sakata, *Chem. Lett.*, **1993**, 145.

526. H. Imahori, Y. Tanaka, T. Okada, Y. Sakata, *Chem. Lett.*, **1993**, 1215.

527. A. Osuka, N. Tanabe, S. Kawabata, I. Yamazaki, Y. Nishimura, *J. Org. Chem.*, **1995**, *60*, 7177.

528. A. Osuka, K. Maruyama, N. Mataga, T. Asahi, I. Yamazaki, N. Tamai, *J. Am. Chem. Soc.*, **1990**, *112*, 4958.

529. M.N. Psddon-Low, J.W. Verhoeven, *N. J. Chem.*, **1991**, *15*, 107.

530. G.L. Closs, M.D. Johnson, J.R. Miller, N.S. Green, *J. Phys. Chem.*, **1989**, *93*, 1173.

531. B.A. Leland, A.D. Joran, P.M. Felker, J.J. Hopfield, A.H. Zewail, P.B. Dervan, *J. Phys. Chem.*, **1985**, *89*, 5571.

532. A. Osuka, F. Kobayashi, K. Maruyama, N. Mataga, T. Asahi, T. Okada, I. Yamazaki, Y. Nishimura, *Chem. Phys. Lett.*, **1993**, *201*, 223.

533. A. Osuka, K. Maruyama, S. Hirayama, *Tetrahedron*, **1989**, *45*, 4815.

534. Y. Sakata, T. Tsue, M.P. O'Neil, G.P. Wiederrecht, M.R. Wasielewski, *J. Am. Chem. Soc.*, **1994**, *116*, 6904.

535. J.L. Sessler, M.R. Johnson, T.-Y. Lin, S.E. Creager, *J. Am. Chem. Soc.*, **1988**, *110*, 3659.

536. A. Osuka, K. Maruyama, N. Mataga, T. Asahi, I. Yamazaki, N. Tamai, *J. Am. Chem. Soc.*, **1990**, *112*, 4958.

537. A. Helms, D. Heiler, G. McLendon, *J. Am. Chem. Soc.*, **1991**, *113*, 4325.

538. H. Tsue, S. Nakashima, Y. Goto, H. Tatemitsu, S. Misumi, R.J. Abraham, T. Asahi, Y. Tanaka, T. Okada, *Bull. Chem. Soc. Jpn.*, **1994**, *67*, 3067.

539. A. Osuka, T. Nagata, F. Kobayashi, R.P. Zhang, K. Maruyama, N. Mataga, T. Asahi, T. Ohno, K. Nozaki, *Chem. Phys. Lett.*, **1992**, *199*, 302.

540. G.L. Closs, J.R. Miller, *Science*, **1988**, *240*, 440.

541. M.R. Wasielewski, M.P. Niemczyk, W.A. Suec, E.B. Pewitt, *J. Am. Chem. Soc.*, **1985**, *107*, 1080.

542. T. Asahi, M. Ohkohchi, R. Matsusaka, N. Mataga, R. Zhang, A. Osuka, K. Maruyama, *J. Am. Chem. Soc.*, **1993**, *115*, 5665.

543. J.M. DeGraziano, P.A. Liddell, L. Leggett, A.L. Moore, T.A. Moore, D. Gust, *J. Phys. Chem.*, **1994**, *98*, 1758.

544. H. Heitele, F. Pollinger, T. Haberle, M.E. Michel-Beyerle, H.A. Staab, *J. Phys. Chem.*, **1994**, *98*, 7402.

545. S.S. Ishied, M.Y. Ogawa, J.F. Wishart, *Chem. Rev.*, **1992**, *92*, 381.

546. J.R. Winkler, H.B. Gray, *Chem. Rev.*, **1992**, *92*, 369.

547. S.E. Peterson-Kennedy, J.L. McGourty, P.S. Ho, C.J. Sutoris, N. Liang, H. Zemel, N.V. Blough, E. Mrgoliash, B.M. Hoffman, *Coord. Chem. Rev.*, **1985**, *64*, 125.

548. G. McLendon, R. Hake, *Chem. Rev.*, **1992**, *92*, 481.

549. K.S. Schanze, L.A. Cabana, *J. Phys. Chem.*, **1990**, *94*, 2740.

550. V.V. Borovkov, A. Ishida, S. Takamuku, Y. Sakata, *Chem. Lett.*, **1993**, 737.

551. H. Tamiaki, K. Maruyama, *Chem. Lett.*, **1993**, 1499.

552. T. Hayashi, T. Takimura, T. Ohara, Y. Hitomi, H. Ogoshi, *J. Chem. Soc., Chem. Commun.*, **1995**, 2503.

553. A.N. Macpherson, P.A. Liddell, S. Lin, L. Noss, G.R. Seely, J.M. DeGraziano, A.L. Moore, T.A. Moore, D. Gust, *J. Am. Chem. Soc.*, **1995**, *117*, 7202.

554. H. Heitele, F. Pollinger, K. Kremer, M.E. Michel-Beyerle, M. Futscher, G. Voit, J. Weiser, H.A. Staab, *Chem. Phys. Lett.*, **1992**, *188*, 270.

555. D.D. Fraser, J.R. Bolton, *J. Phys. Chem.*, **1994**, *98*, 1626.

556. D.G. Johnson, M.P. Niemczyk, D.W. Minsek, G.P. Wiederrecht, W.A. Svec, G.L. Gaines, III, M.R. Wasielewski, *J. Am. Chem. Soc.*, **1993**, *115*, 5692.

557. J.-y. Liu, J.R. Bolton, *J. Phys. Chem.*, **1992**, *96*, 1718.

558. J.-y. Liu, J.A. Schmidt, J.R. Bolton, *J. Phys. Chem.*, **1991**, *95*, 6924.

559. J.A. Schmidt, J.-y. Liu, J.R. Bolton, M.D. Archer, V.P.Y. Gadzekpo, *J. Chem. Soc., Faraday Trans. 1*, **1989**, *85*, 1027.

560. M.D. Archer, V.P.Y. Gadzekpo, J.R. Bolton, J.A. Schmidt, A.C. Weedon, *J. Chem. Soc., Faraday Trans. 2*, **1986**, *82*, 2305.

561. J.A. Schmidt, A. Siemiarczuk, A.C. Weedon, J.R. Bolton, *J. Am. Chem. Soc.*, **1985**, *107*, 6112.

562. M. Kaneko, *Proc. Ind. Acad. Sci.*, **1992**, *104*, 723.

563. C.C. Moser, J.M. Keske, K. Warncke, R.S. Farid, P.L. Dutton, *The Photosynthetic Reaction Center*, Vol 2, Academic Press, San Diego, CA, 1993, p. 1.

564. W.W. Parson, Z.T. Chu, A. Warshel, *Biochim. Biophys. Acta*, **1990**, *1017*, 251.

565. D.N. Beratan, J.N. Betts, J.N. Onuchic, *Science*, **1991**, *252*, 1285.

566. A. Kuki, P.G. Wolynes, *Science*, **1987**, *236*, 1647.

567. A. Osuka, S. Marumo, K. Maruyama, N. Mataga, Y. Tanaka, S. Taniguchi, T. Okada, I. Yamazaki, Y. Nishimura, *Bull. Chem. Soc. Jpn.*, **1995**, *68*, 262.

568. A. Osuka, R.-P. Zhang, K. Maruyama, T. Ohno, K. Nozaki, *Bull. Chem. Soc. Jpn.*, **1993**, *66*, 3773.

569. A. Osuka, S. Nakajima, T. Okada, S. Taniguchi, K. Nozaki, T. Ohno, I. Yamazaki, Y. Nishimura, N. Mataga, *Angew. Chem. Int. Ed. Engl.*, **1996**, *35*, 92.

570. A. Osuka, S. Nakajima, K. Maruyama, N. Mataga, T. Asahi, I. Yamazaki, Y. Nishi-mura, T. Ohno, K. Nozaki, *J. Am. Chem. Soc.,* **1993**, *115,* 4577.

571. A. Osuka, H. Yamada, K. Maruyama, T. Ohno, K. Nozaki, T. Okada, Y. Tanaka, N. Mataga, *Chem. Lett.,* **1995,** 591.

572. A. Osuka, S. Marumo, S. Taniguchi, T. Okada, N. Mataga, *Chem. Phys. Lett.,* **1994,** *230,* 144.

573. J.L. Sessler, V.L. Capuano, A. Harriman, *J. Am. Chem. Soc.,* **1993,** *115,* 4618.

574. A. Osuka, T. Okada, S. Taniguchi, K. Nozaki, T. Ohno, N. Mataga, *Tetrahedron Lett.,* **1995,** 5781.

575. M.Y. Okamura, G. Feher, *Annu. Rev. Biochem.,* **1992,** *61,* 861.

576. S.-C. Hung, A.N. Macpherson, S. Lin, P.A. Liddell, G.R. Seely, A.L. Moore, T.A. Moore, D. Gust, *J. Am. Chem. Soc.,* **1995,** *117,* 1657.

577. D. Gust, T.A. Moore, A.L. Moore, H.K. Kang, J.M. DeGraziano, P.A. Liddell, G.R. Seely, *J. Phys. Chem.,* **1993,** 97, 13637.

578. J. Hirota, I. Okura, *J. Phys. Chem.,* **1993,** *97,* 6867.

579. K. Tsukahara, N. Sawai, K. Koji, T. Nakazawa, *Chem. Phys. Lett.,* **1995,** *246,* 331.

580. V. Ya. Shafirovivh, E.E. Batova, P.P. Levin, *J. Phys. Chem.,* **1993,** *97,* 4877.

581. J.D. Batteas, A. Harriman, Y. Kanda, N. Mataga, A.K. Nowak, *J. Am. Chem. Soc.,* **1990,** *112,* 126.

582. H.A. Frank, V. Chynwat, Ruel Z.B. Desamero, R. Farhoosh, J. Erickson, J. Bautista, *Pure Appl. Chem.,* **1997,** *69,* 2117.

583. A.J. Young, D. Phillip, A.V. Ruban, P. Horton, H.A. Frank, *Pure Appl. Chem.,* **1997,** *69,* 2125.

584. G. Dirks, A.L. Moore, T.A. Moore, D. Gust, *Photochem. Photobiol.,* **1980,** *32,* 277.

585. A.L. Moore, G. Dirks, D. Gust, T.A. Moore, *Photochem. Photobiol.,* **1980,** *32,* 691.

586. A. Osuka, H. Yamada, K. Maruyama, N. Mataga, T. Asahi, M. Ohkohchi, T. Okada, I. Yamazaki, Y. Nishimura, *J. Am. Chem. Soc.,* **1993,** *115,* 9439.

587. D. Gust, T.A. Moore, A.L. Moore, A.N. Macpherson, A. Lopez, J.M. DeGraziano, I. Gouni, E. Bittersmann, G.R. Seely, F. Gao, R.A. Nieman, X.C. Ma, L.J. Deman-che, S.-C. Hung, D.K. Luttrull, S.-J. Lee, P.K. Kerrigan, *J. Am. Chem. Soc.,* **1993,** *115,* 11141.

588. D. Gust, T.A. Moore, A.L. Moore, F. Gao, D. Luttrull, J.M. DeGraziano, X.C. Ma, L.R. Makings, S.-J. Lee, T.T. Trier, E. Bittersmann, G.R. Seely, S. Woodward, R.V. Bensasson, M. Rougee, F.C. De Schryver, M. Van der Auweraer, *J. Am. Chem. Soc.,* **1991,** *113,* 3638.

589. D. Gust, T.A. Moore, A.L. Moore, S.-J. Lee, E. Bittersmann, D.K. Luttrull, A.A. Rehms, J.M. DeGraziano, X.C.Ma, F. Gao, R.E. Belford, T.T. Trier, *Science,* **1990,** *248,* 199.

590. F. D'Souza, V. Krishnan, *J. Chem. Soc., Dalton Trans.,* **1992,** 2873.

591. F. D'Souza, V. Krishnan, *Photochem. Photobiol.,* **1992,** *56,* 145.

592. R. Giasson, E.J. Lee, X. Zhao, M.S. Wrighton, *J. Phys. Chem.,* **1993,** *97,* 2596.

593. G.R. Loppnow, D. Melamed, A.D. Hamilton, T.G. Spiro, *J. Phys. Chem.,* **1993,** *97,* 8957.

594. G.R. Loppnow, D. Melamed, A.R. Leheny, A.D. Hamilton, T.G. Spiro, *J. Phys. Chem.*, **1993**, *97*, 8969.

595. R.M. Williams, J.M. Zwier, J.W. Verhoeven, *J. Am. Chem. Soc.*, **1995**, *117*, 4093.

596. T.G. Lissen, K. Durr, M. Hanack, A. Hirsch, *J. Chem. Soc. Chem. Commun.*, **1995**, 103.

597. H. Imahori, K. Hagiwara, T. Akiyama, M. Aoki, S. Taniguchi, T. Okada, M. Shirakawa, Y. Sakata, *Chem. Phys. Lett.*, **1996**, *263*, 545.

598. M.G. Ranasinghe, A.M. Oliver, D.F. Rothenfluh, A. Salek, M.N. Paddon-Row, *Tetrahedron Lett.*, **1996**, *37*, 4797.

599. P.A. Liddell, J.P. Sumida, A.N. Macpherson, L. Noss, G.R. Seely, K.N. Clark, A.L. Moore, T.A. Moore, D. Gust, *Photochem. Photobiol.*, **1994**, *60*, 537.

600. T. Hayashi, T. Miyahara, S. Kumazaki, H. Ogoshi, K. Yoshihara, *Angew. Chem. Int. Ed. Engl.*, **1996**, *35*, 1964.

601. T. Hayashi, T. Miyahara, Y. Aoyama, M. Nonoguchi, H. Ogoshi, *Chem. Lett.*, **1994**, 1749.

602. J.L. Sessler, B. Wang, A. Harriman, *J. Am. Chem. Soc.*, **1993**, *115*, 10418.

603. T. Arimura, C.T. Brown, S.L. Springs, J.L. Sessler, *Chem. Commun.*, **1996**, 2293.

604. P. Tecilla, R.P. Dixon, G. Slobodkin, D.S. Alavi, D.H. Waldeck, A.D. Hamilton, *J. Am. Chem. Soc.*, **1990**, *112*, 9408.

605. A. Osuka, H. Shiratori, R. Yoneshima,T. Okada, S. Taniguchi, N. Mataga, *Chem. Lett.*, **1995**, 913.

606. J.M. Berg, *Acc. Chem. Res.*, **1995**, *28*, 14.

607. Y. Deng, J.A. Roberts, S.-M. Peng, C.K. Chang, D.G. Nocera, *Angew. Chem. Int. Ed. Engl.*, **1997**, *36*, 2124.

608. T. Takano, R.E. Dickerson, *Proc. Natl. Acad. Sci. U.S.A.*, **1980**, *77*, 6371.

609. T.L. Poulos, J. Kraut, *J. Biol. Chem.*, **1980**, *259*, 10322.

610. T. Hayashi, T. Takimura, H. Ogoshi, *J. Am. Chem. Soc.*, **1995**, *117*, 11606.

611. C.A. Hunter, J.K.M. Sanders, G.S. Beddard, S. Evans, *J. Chem. Soc., Chem. Commun.*, **1989**, 1765.

612. H.L. Anderson, C.A. Hunter, J.K.M. Sanders, *J. Chem. Soc., Chem. Commun.*, **1989**, 226.

613. H. Imahori, E. Yoshizawa, K. Yamada, K. Hagiwara, T. Okada, Y. Sakata, *J. Chem. Soc., Chem. Commun.*, **1995**, 1133.

614. A.M. Brun, S.J. Atherton, A. Harriman, V. Heitz, J.-P. Sauvage, *J. Am. Chem. Soc.*, **1992**, *114*, 4632.

615. A. Harriman, F. Odobel, J.-P. Sauvage, *J. Am. Chem. Soc.*, **1995**, *117*, 9461.

616. J.-P. Collin, A. Harriman, V. Heitz, F. Odobel, J.-P. Sauvage, *J. Am. Chem. Soc.*, **1994**, *116*, 5679.

617. J.-P. Collin, A. Harriman, V. Heitz, F. Odobel, J.-P. Sauvage, *Coord. Chem. Rev.*, **1996**, *148*, 63.

618. J.P. Allen, G. Feher, T.O. Yeates, H. Komiya, D.C. Rees, *Proc. Natl. Acad. Sci. U.S.A.*, **1988**, *85*, 8487.

619. C.A. Hunter, R.J. Scannon, *Chem. Commun.*, **1996**, 1361.

620. S. Arimori, H. Murakami, M. Takeuchi, S. Shinkai, *J. Chem. Soc., Chem. Commun.*, **1995**, 961.

621. H. Segawa, C. Takehara, K. Honda, T. Shimidzu, T. Asahi, N. Mataga, *J. Phys. Chem.*, **1992**, *96*, 503.

622. U. Rempel, B. von Maltzan, C. von Borczyskowski, *Pure Appl. Chem.*, **1993**, *65*, 1681.

623. G. Blondeel, A. Harriman, G. Porter, A. Wilowska, *J. Chem. Soc., Faraday Trans. 2*, **1984**, *80*, 867.

624. M.C. Gonzalez, A.R. McIntosh, J.R. Bolton, A.C. Weedon, *J. Chem. Soc., Chem. Commun.*, **1984**, 1138.

625. F. D'Souza, V. Krishnan, *J. Chem. Soc., Dalton Trans.*, **1992**, 2873.

626. M. GubAmann, A. Harriman, J.-M. Lehn, J.L. Sessler, *J. Phys. Chem.*, **1990**, *94*, 308.

627. M. GubAmann, A. Harriman, J.-M. Lehn, J.L. Sessler, *J. Chem. Soc., Chem. Commun.*, **1988**, 77.

628. M. Gubelmann, A. Harriman, J.-M. Lehn, J.L. Sessler, *J. Phys. Chem.*, **1990**, *94*, 308.

629. P.J. Clapp, B. Armitage, P. Roosa, D.F. O'Brien, *J. Am. Chem. Soc.*, **1994**, *116*, 9166.

630. K.C. Hwang, D. Mauzerall, *J. Am. Chem. Soc.*, **1992**, *114*, 9705.

631. V.Y. Shafirovich, E.E. Batova, *Chem. Phys. Lett.*, **1990**, *172*, 10.

632. T. Hamada, H. Ishida, M. Kuwada, K. Ohkubo, *Chem. Lett.*, **1992**, 1283.

633. K.B. Yoon, *Chem. Rev.*, **1993**, *93*, 321.

634. H. Scheer, S. Siegried, Ed. *Photosynthetic Light-Harvesting Systems*, W.de. Gruyter, Berlin, 1988.

635. A. Harriman, D.J. Magda, J.L. Sessler, *J. Chem. Soc., Chem. Commun.*, **1991**, 345.

636. A. Harriman, D.J. Magda, J.L. Sessler, *J. Phys. Chem.*, **1991**, *95*, 1530.

637. J.L. Sessler, B. Wang, A. Harriman, *J. Am. Chem. Soc.*, **1995**, *117*, 704.

638. S. Prathapan, T.E. Johnson, J.S. Jindsey, *J. Am. Chem. Soc.*, **1993**, *115*, 7519.

639. J. Seth, V. Planiappan, T.E. Johnson, S. Prathapan, J.S. Lindsey, D.F. Bocian, *J. Am. Chem. Soc.*, **1994**, *116*, 10578.

640. V. Kral, S.L. Springs, J.L. Sessler, *J. Am. Chem. Soc.*, **1995**, *117*, 8881.

641. C.A. Hunter, R.K. Hyde, *Angrew. Chem. Int. Ed. Engl.*, **1996**, *35*, 1936.

642. S. Osuka, F. Kobayashi, S. Nakajima, K. Maruyama, I. Yamazaki, Y. Nishimura, *Chem. Lett.*, **1993**, 161.

643. D. Gust, T.A. Moore, A.L. Moore, A.A. Krasnovsky, Jr., P.A. Liddell, D. Nicodem, J.M. DeGraziano, P. Kerrigan, L.R. Makings, P.J. Pessiki, *J. Am. Chem. Soc.*, **1993**, *115*, 5684.

644. G. Dirks, A.C. Moore, T.A. Moore, D. Gust, *Photochem. Photobiol.*, **1980**, *32*, 277.

645. F.V.-Maeder, P. Claverie, *J. Am. Chem. Soc.*, **1987**, *109*, 24.

646. D. Gust, T.A. Moore, R.V. Bensasson, P. Mathis, E.J. Land, C. Chachaty, A.L. Moore, P.A. Liddell, G.A. Nemeth, *J. Am. Chem. Soc.*, **1985**, *32*, 691.

647. D. Gust, T.A. Moore, P.A. Liddell, G.A. Nemeth, L.R. Makings, A.L. Moore, D. Barrett, P.J. Pessiki, R.V. Bensasson, M. Rougee, C. Chachaty, F.C. de Schryver, M. van der Auweraer, A.R. Holzwarth, J.S. Connolly, *J. Am. Chem. Soc.*, **1987**, *109*, 846.

648. A. Osuka, H. Yamada, K. Maruyama, N. Mataga, T. Asahi, I. Yamazaki, Y. mura, *Chem. Phys. Lett.*, **1991**, *181*, 419.

649. Y. Kuroda, T. Sera, H. Ogoshi, *J. Am. Chem. Soc.*, **1991**, *113*, 2793.

650. B. Armitage, P.A. Klekotka, E. Oblinger, D.F. O'Brien, *J. Am. Chem. Soc* **3**, *115*, 7920.

651. R.W. Wagner, J.S. Lindsey, H. Turowska-Tyrk, W.R. Scheidt, *Tetrahedr* **94**. *50*, 11097.

652. P.G. Van Patten, A.P. Shreve, J.S. Lindsey, R.J. Donohoe, *J. Phys. Chem* **998**, *102*, 4209.

653. E. Tsuchida, M. Kaneko, H. Nishide, M. Hoshino, *J. Phys. Chem.*, **198(** 2283.

654. D. Rehm, A. Weller, *Ber. Bunsenges. Phys. Chem.*, **1969**, *73*, 834.

655. D. Gust, T.A. Moore, *Science*, **1989**, *244*, 35.

656. M.R. Wasoelewski, *Photochem. Photobiol.*, **1988**, *47*, 923.

657. N. Mataga in *Photochemical Energy Conversion*, J.R. Norris, D. el. Eds. Elsevier, New York, 1989, p. 32.

658. C.K. Chang, M.-S. Kuo, C.-B. Wang, *J. Heterocycl. Chem.*, **1977**,)43.

659. C.K. Chang, *J. Heterocycl. Chem.*, **1977**, *14*, 1285.

660. I. Fujita, T.L. Netzel, C.K. Chang, C.-B. Wang, *Proc. Natl. Acad. ! JSA*, **1982**, *79*, 413.

661. A.D. Hamilton, J.-M. Lehn, J.L. Sessler, *J. Chem. Soc., Chem. Com* ., **1984**, 311.

662. A.D. Hamilton, J.-M. Lehn, J.L. Sessler, *J. Am. Chem. Soc.*, **198** *08*, 5158.

663. M. Nango, A. Mizusawa, T. Miyake, J. Yoshinaga, *J. Am. Chem* ., **1990**, *112*, 1640.

664. I. Tabushi, T. Nishiya, M. Shimomura, T. Kunitake, H. Inokuc' *.* Yagi, *J. Am. Chem. Soc.*, **1984**, *106*, 219.

665. M. Nango, H. Kryu, P.A. Loach, *J. Chem. Soc., Chem. Comn* , **1988**, 697.

666. J.A. Runquist, P.A. Loach, *Biochem. Biophys. Acta*, **1981**, *6?* /31.

667. M. Nango, T. Dannhauser, D. Huang, K. Spears, L. Morrison. A. Loach, *Macromolecules*, **1984**, *17*, 1898.

668. M. Kasha, *Radiat. Res.*, **1963**, *20*, 55.

669. M. Kasha, M. Ashraf El-Bayoumi, W. Rhodes, *J. Chim. Pl* ., **1961**, *58*, 916.

670. E.S. Emerson, M.A. Conlin, A.E. Rosenoff, K.S. Norland, ? R driguez, D. Chin, G.R. Bird, *J. Phys. Chem.*, **1967**, *71*, 2396.

671. R.L. Fulton, M. Gouterman, *J. Chem. Phys.*, **1961**, *35*, 1(.

672. R.L. Fulton, M. Gouterman, *J. Chem. Phys.*, **1964**, *41*, 2).

673. A. Osuka, K. Maruyama, *J. Am. Chem. Soc.*, **1988**, *110*. 4!4.

674. J.L. Sessler, M.R. Johnson, S.E. Creager, J.C. Fettinger. A. Ibers, *J. Am. Chem. Soc.*, **1990**, *112*, 9310.

675. V. Thanabal, V. Krishnan, *J. Am. Chem. Soc.*, **1982**, *1* , 3643.

676. N. Kobayashi, A.B.P. Lever, *J. Am. Chem. Soc.*, **1987** *1*)9, 7433.

677. M. Krishnamurthy, J.R. Sutter, P. Hambright, *J. Ch* n. *Soc., Chem. Commun.*, **1975**, 13.

678. K. Kano, T. Nakajima, M. Takei, S. Hashimoto, *Bul* *Chem. Soc. Jpn.*, **1987**, *60*, 1281.

679. H. Segawa, C. Takehara, K. Honda, T. Shimidzu, *!.* Asahi, N. Mataga, *J. Phys. Chem.*, **1992**, *96*, 503.

680. F.J. Vergeldt, R.B.M. Koehorst, T.J. Schaafsma, J.-C. Lambry, J.-L. Martin, D.G. Johnson, M.R. Wasielewski, *Chem. Phys. Lett.*, **1991**, *182*, 107.

681. T.H. Tran-Thi, J.F. Lipskier, D. Houde, C. Pepin, E. Keszei, J.P. Jay-Gerin, *J. Chem. Soc., Faraday Trans.*, **1992**, *88*, 2129.

682. L. Banci, *Inorg. Chem.*, **1985**, *24*, 782.

683. J.S. Zhou, E.S.V. Granada, N.B. Leontis, M.A.J. Rodgers, *J. Am. Chem. Soc.*, **1990**, *112*, 5074.

684. A.G. Coutsolelos, D. Lux, E. Mikros, *Polyhedron*, **1996**, *15*, 705.

685. O. Bilsel, J. Rodriguez, D. Holten, G.S. Girolami, S.N. Milam, K.S. Suslick, *J. Am. Chem. Soc.*, **1990**, *112*, 4075.

686. J.W. Buchler, B. Scharbert, *J. Am. Chem. Soc.*, **1988**, *110*, 4272.

687. J.-H. Perng, J.K. Duchoeski, D.F. Bocian, *J. Phys. Chem.*, **1991**, *95*, 1319.

688. D. Chabach, A. De Cian, J. Fischer, R. Weiss, M.E.M. Bibout, *Angew. Chem. Int. Ed. Engl.*, **1996**, *35*, 898.

689. R. Karaman, A. Blasko, O. Almarsson, R. Arasasingham, T.C. Bruice, *J. Am. Chem. Soc.*, **1992**, *114*, 4889.

690. R. Karaman, T.C. Bruice, *J. Org. Chem.*, **1991**, *56*, 3470.

691. J.P. Collman, K. Kim, J.M. Garner, *J. Chem. Soc., Chem. Commun.*, **1986**, 1711.

692. J.P. Collman, Y. Ha, R. Guilard, M.-A. Lopez, *Inorg. Chem.*, **1993**, *32*, 1788.

693. T.H. Tran-Thi, J.F. Lipskier, P. Maillard, M. Momenteau, J.-M. Lopez-Castillo, J.-P. Jay-Gerin, *J. Phys. Chem.*, **1992**, *96*, 1073.

694. T. Nagata, A. Osuka, K. Maruyama, *J. Am. Chem. Soc.*, **1990**, *112*, 3054.

695. T.H. Tran-Thi, A. Dormond, R. Guilard, *J. Phys. Chem.*, **1992**, *96*, 3139.

696. Y. Kobuke, H. Miyaji, *J. Am. Chem. Soc.*, **1994**, *116*, 4111.

697. M. Takeuchi, Y. Chin, T. Imada, S. Shinkai, *Chem. Commun.*, **1996**, 1867.

698. C.M. Drain, K.C. Russell, J.-M. Lehn, *Chem. Commun.*, **1996**, 337.

699. P. Hildebrandt, H. Tamiaki, A.R. Holzwarth, K. Schaffner, *J. Phys. Chem.*, **1994**, *98*, 2192.

700. K. Susumu, H. Segawa, T. Shimidzu, *Chem. Lett.*, **1995**, 929.

701. T.A. Rao, B.G. Maiya, *J. Chem. Soc., Chem. Commun.*, **1995**, 939.

702. A. Kimura, K. Funatsu, T. Imamura, H. Kido, Y. Sasaki, *Chem. Lett.*, **1995**, 207.

703. J. Wojaczynski, L. Latos-Grazynski, *Inorg. Chem.*, **1995**, *34*, 1044.

704. A.V. Chernook, A.M. Shulga, E.I. Zenkevich, U. Rempel, C. von Borczyskowski, *Ber. Bunsenges. Phys. Chem.*, **1996**, *100*, 2065.

705. K. Funatsu, A. Kimura, T. Imamura, Y. Sasaki, *Chem. Lett.*, **1995**, 765.

706. S. Anderson, H.L. Anderson, A. Bashall, M. McPartlin, J.K.M. Sanders, *Angew. Chem. Int. Ed. Engl.*, **1995**, *34*, 1096.

707. M. Ikonen, D. Guez, V. Marvaud, D. Markovitsi, *Chem. Phys. Lett.*, **1994**, *231*, 93.

708. M. Hanack, A. Gul, L.R. Subramanian, *Inorg. Chem.*, **1992**, *31*, 1542.

709. J.P. Collman, J.T. McDevitt, C.R. Leidner, G.T. Yee, J.B. Torrance, W.A. Little, *J. Am. Chem. Soc.*, **1987**, *109*, 4606.

710. V. Marvaud, J.P. Launay, *Inorg. Chem.*, **1993**, *32*, 1376.

711. H. Segawa, K. Kunimoto, K. Susumu, M. Taniguchi, T. Shimidzu, *J. Am. Chem. Soc.*, **1994**, *116*, 11193.

712. T. Shimidzu, H. Segawa, *Thin Solid Films*, **1996**, *273*, 14.

713. T. Shimidzu, *Synthetic Metals*, **1996**, *81*, 235.
714. H. Segawa, F.-P. Wu, N. Nakayama, H. Maruyama, S. Sagisaka, N. Higuchi, M. Fujitsuka, T. Shimidzu, *Synthetic Metals*, **1995**, *71*, 2151.
715. M. Saito, A. Endo, K. Shimizu, G.P. Sato, *Chem. Lett.*, **1995**, 1079.
716. J.-H. Fuhrhop, U. Bindig, U. Siggel, *J. Am. Chem. Soc.*, **1993**, *115*, 11036.
717. T. Komatsu, K. Yamada, E. Tsuchida, U. Siggel, C. Bottcher, J.-H. Fuhrhop, *Langmuir*, **1996**, *12*, 6242.
718. A. Osuka, N. Tanabe, S. Nakajima, K. Maruyama, *J. Chem. Soc., Perkin Trans. 2*, **1995**, 199.
719. V.S.-Y. Lin, S.G. DiMagno, M.J. Therien, *Science*, **1994**, *264*, 1105.
720. Z. Cai, C.R. Martin, *J. Am. Chem. Soc.*, **1989**, *111*, 4138.
721. A. Slama-Schwok, M. Blanchard-Desce, J.-M. Lehn, *J. Phys. Chem.*, **1992**, *96*, 10559.
722. C.-G. Wu, T. Bein, *Science*, **1994**, *264*, 1757.
723. R.W. Wagner, J.S. Lindsey, *J. Am. Chem. Soc.*, **1994**, *116*, 9759.
724. Z. Bao, Y. Chen, L. Yu, *Macromolecules*, **1994**, *27*, 4629.
725. J. Zakrzewski, C. Giannotti, *Coord. Chem. Rev.*, **1995**, *140*, 169.
726. D.P. Rillema, J.K. Nagle, L.F. Barringer, Jr., T.J. Meyer, *J. Am. Chem. Soc.*, **1981**, *103*, 56.
727. A.A. Lamola, N.J. Turro, Ed. *Energy Transfer and Organic Photochemistry*, Interscience, New York, 1969.
728. R.A. Marcus, *J. Chem. Soc.*, **1956**, *24*, 966.
729. R.A. Marcus, *Disc. Faraday Soc.*, **1960**, *29*, 21.
730. R.A. Marcus, *J. Phys. Chem.*, **1963**, *67*, 853.
731. R.A. Marcus, P. Siders, *J. Phys. Chem.*, **1982**, *86*, 622.
732. D. Rehm, A. Weller, *Israel J. Chem.*, **1970**, *8*, 259.
733. L.T. Calcaterra, G.L. Closs, J.R. Miller, *J. Am. Chem. Soc.*, **1983**, *105*, 670.
734. J.R. Miller, L.T. Calcaterra, G.L. Closs, *J. Am. Chem. Soc.*, **1984**, *106*, 3047.
735. G.L. Closs, L.T. Calcaterra, N.J. Green, K.W. Penfield, J.R. Miller, *J. Phys. Chem.*, **1986**, *90*, 3673.
736. K. Kalyanasundaram, M. Gratzel, *Helv. Chim. Acta.*, **1980**, *63*, 478.
737. A. Harriman, G. Porter, M.C. Richoux, *J. Chem. Soc., Faraday Trans. II*, **1981**, *77*, 833.
738. K. Kalyanasundaram, *J. Chem. Soc., Faraday Trans. II*, **1983**, *79*, 1365.
739. S.C. Shim, H. Lee, *J. Bull. Korean Chem. Soc.*, **1988**, *9*, 68.
740. S.C. Shim, H. Lee, *J. Bull. Korean Chem. Soc.*, **1988**, *9*, 112.
741. K. Kalyanasundaram, M. Gratzel, *Photosensitization and Photocatalysis Using Inorganic and Organometallic Compounds*, Kluwer, Dordrecht, 1993, and references therein.
742. J.S. Connoly, *Photochemical Conversion and Storage of Solar Energy*, Academic Press, London, 1981.
743. K. Kalyanasundaram, *Photoelectrochemistry, Photocatalysis and Photoreactors; Fundamentals and Developments*, M. Schiarello, Ed., Reidel, Dordrecht, 1985.
744. J.-M. Lehm, J.-P. Sauvage, R. Ziessel, *Nouv. J. Chim.*, **1979**, *3*, 423.
745. W. Erbs, J. Kiwi, M. Gratzel, *Chem. Phys. Lett.*, **1984**, *110*, 648.
746. W. Erbs, J. Desilvestro, E. Borgarello, M. Gratzel, *J. Phys. Chem.*, **1984**, *88*, 4001.

747. G.S. Nahor, S. Mosseri, P. Neta, A. Harriman, *J. Phys. Chem.*, **1988**, *92*, 4499.
748. G.S. Nahor, P. Neta, P. Hambright, A.N. Thompson Jr., A. Harriman, *J. Phys. Chem.*, **1989**, *93*, 6181.
749. M. Kaneko, G.-J. Yao, A. Kira, *J. Chem. Soc., Chem. Commun.*, **1989**, 1338.
750. A. Harriman, *J. Photochem. Photobiol. A*, **1990**, *51*, 41.
751. T.J. Meyer, *Acc. Chem. Res.*, **1989**, *22*, 163.
752. D. Geselowitz, T.J. Meyer, *Inorg. Chem.*, **1990**, *29*, 3894.
753. P.A. Liddell, D. Kuciauskas, J.P. Sumida, B. Nash, D. Nguyen, A.L. Moore, T.A. Moore, D. Gust, *J. Am. Chem. Soc.*, **1997**, *119*, 1400.
754. M.H. Wall, Jr., P. Basu, T. Buranda, B.S. Wicks, E.W. Findsem, M. Ondrias, J.H. Enemark, M.L. Kirk, *Inorg. Chem.*, **1997**, *36*, 5676.
755. R.W. Wagner, J.S. Lindsey, J. Seth, V. Palaniappan, D.F. Bocian, *J. Am. Chem. Soc.*, **1996**, *118*, 3996.
756. R.W. Wagner, T.E. Johnson, J.S. Lindsey, *J. Am. Chem. Soc.*, **1996**, *118*, 11166.
757. N. Nishino, R.W. Wagner, J.S. Lindsey, *J. Org. Chem.*, **1996**, *61*, 7534.
758. J.-P. Strachan, S. Gentemann, J. Seth, W.A. Kalsbeck, J.S. Lindsey, D. Holten, D.F. Bocian, *J. Am. Chem. Soc.*, **1997**, *119*, 11191.
759. R. Sadamoto, N. Tomioka, T. Aida, *J. Am. Chem. Soc.*, **1996**, *118*, 3978.
760. H. Imahori, K. Hagiwara, M. Aoki, T. Akiyama, S. Taniguchi, T. Okada, M. Shirakawa, Y. Sakata, *J. Am. Chem. Soc.*, **1996**, *118*, 11771.
761. D. Kuciauskas, S. Lin, G.R. Seely, A.L. Moore, T.A. Moore, D. Gust, *J. Phys. Chem.*, **1996**, *100*, 15926.
762. H. Imahori, K. Yamada, M. Hasegawa, S. Taniguchi, T. Okada, Y. Sakata, *Angew. Chem. Int. Ed. Engl.*, **1997**, *36*, 2626.
763. C.M. Haugen, W.R. Bergmark, D.G. Whitten, *J. Am. Chem. Soc.*, **1992**, *114*, 10293.
764. E. Reddi, G. Valduga, M.A.J. Rodgers, G. Jori, *Photochem. Photobiol.*, **1991**, *54*, 633.
765. W. Adam, O. Albrecht, E. Feineis, I. Reuther, C.R. Saha-Moller, P. Seufert-Baumbach, D. Wild, *Liebigs Ann. Chem.*, **1991**, 33.
766. T. Akasaka, M. Haranaka, W. Ando, *J. Am. Chem. Soc.*, **1991**, *113*, 9898.
767. E.J.-F. Rontani, *Tetrahedron Lett.*, **1991**, *32*, 6551.
768. C.W. Jefford, A.F. Boschung, *Tetrahedron Lett.*, **1976**, 4771.
769. W. Adam, M. Balci, *J. Am. Chem. Soc.*, **1979**, *101*, 7537.
770. D.M. Floyd, C.M. Cimarusti, *Tetrahedron Lett.*, **1979**, 4129.
771. S.J. Jongsma, J. Cornelisse, *Tetrahedron Lett.*, **1981**, 2919.
772. B. Ohtani, M. Nishida, S. Nishimoto, T. Kagiya, *Photochem. Photobiol.*, **1986**, *44*, 725.
773. T. Takata, K. Ishibashi, W. Ando, *Tetrahedron Lett.*, **1985**, 4609.
774. F. Jensen, C.S. Foote, *Photochem. Photobiol.*, **1987**, *46*, 325.
775. E.M.K. Mansour, P. Maillard, P. Krausz, S. Gaspard, C. Giannotti, *J. Mol. Catal.*, **1987**, *41*, 361.
776. N. Shimizu, F. Shibata, S. Imazu, Y. Tsuno, *Chem. Lett.*, **1987**, 1071.
777. C. Castro, M. Dixon, I. Erden, P. Ergonenc, J.R. Keeffe, A. Sukhovitsky, *J. Org. Chem.*, **1989**, *54*, 3732.
778. M. Sakuragi, K. Ichimura, H. Sakuragi, *Bull. Chem. Soc. Jpn.*, **1992**, *65*, 1944.
779. H. Quast, T. Dietz, A. Witzel, *Liebigs Ann.*, **1995**, 1495.

780. A.P. Schaap, A.L. Thayer, E.C. Blossey, D.C. Neckers, *J. Am. Chem. Soc.*, **1975**, *97*, 3741.

781. S. Takagi, T. Okamoto, T. Shiragami, H. Inoue, *J. Org. Chem.*, **1994**, *59*, 7373.

782. T. Okamoto, S. Takagi, T. Shiragami, H. Inoue, *Chem. Lett.*, **1993**, 687.

783. L. Rothberg, T.M. Jedju, R.H. Austin, *Biophys. J.*, **1990**, *57*, 369.

784. K. Mizuno, N. Ichinose, Y. Otsuji, *J. Org. Chem.*, **1992**, *57*, 1855.

785. J. Deisenhofer, O. Epp, M. Miki, R. Huber, H. Michel, *J. Mol. Biol.*, **1984**, *180*, 385.

786. J. Lahiri, G.D. Fate, S.B. Ungashe, J.T. Groves, *J. Am. Chem. Soc.*, **1996**, *118*, 2347.

787. B.C. Gilbert, J.R. Lindsay Smith, A.F. Parsons, P.K. Setchell, *J. Chem. Soc., Perkin Trans. 2*, **1997**, 1065.

788. J. Grodkowski, D. Behar, P. Neta, P. Hambright, *J. Phys. Chem. A*, **1997**, *101*, 248.

789. J.N.H. Reek, A.E. Rowan, R. de Gelder, P.T. Beurskens, M.J. Crossley, S. De Feyter, F. de Schryver, R.J.M. Nolte, *Angew. Chem. Int. Ed. Engl.*, **1997**, *36*, 361.

790. Y. Sakata, H. Imahori, H. Tsue, S. Higashida, T. Akiyama, E. Yoshizawa, M. Aoki, K. Yamada, K. Hagiwara, S. Taniguchi, T. Okada, *Pure Appl. Chem.*, **1997**, *69*, 1951.

791. H. Inoue, T. Okamoto, M. Hida, *J. Photochem. Photobiol. A: Chem.*, **1992**, *65*, 221.

792. K. Kubomura, H. Inoue, *Res. Chem. Intermed.*, **1995**, *21*, 923.

793. R.A. Freitag, D. Barber, H. Inoue, D.G. Whitten, *Am. Chem. Soc. Symp. Ser.*, **1986**, *321*, 280.

Index

T